Papermaking Science and Technology

a series of 19 books covering the latest technology and future trends

Book 2

Forest Resources and Sustainable Management

Series editors
Johan Gullichsen, Helsinki University of Techology
Hannu Paulapuro, Helsinki University of Techology

Book editor
Seppo Kellomäki, University of Joensuu

Series reviewer
Brian Attwood, St. Anne's Paper and Paperboard Co. Ltd

Book reviewer
Desmond Smith, Acrowood Corporation

Published in cooperation with the Finnish Paper Engineers' Association and TAPPI

Cover photo by Ilkka Konttinen

ISBN 952-5216-00-4 (the series)
ISBN 952-5216-02-0 (book 2)

Published by Fapet Oy
(Fapet Oy, PO BOX 146, FIN-00171 HELSINKI, FINLAND)

Copyright © 1998 by Fapet Oy. All rights reserved.

Printed by Gummerus Oy, Jyväskylä, Finland 1998

 Printed on LumiSilk 115 g/m^2, Enso Fine Papers Oy, Oulu Mills

Foreword

Johan Gullichsen and Hannu Paulapuro

PAPERMAKING SCIENCE AND TECHNOLOGY

Papermaking is a vast, multidisciplinary technology that has expanded tremendously in recent years. Significant advances have been made in all areas of papermaking, including raw materials, production technology, process control and end products. The complexity of the processes, the scale of operation and production speeds leave little room for error or malfunction. Modern papermaking would not be possible without a proper command of a great variety of technologies, in particular advanced process control and diagnostic methods. Not only has the technology progressed and new technology emerged, but our understanding of the fundamentals of unit processes, raw materials and product properties has also deepened considerably. The variations in the industry's heterogeneous raw materials, and the sophistication of pulping and papermaking processes require a profound understanding of the mechanisms involved. Paper and board products are complex in structure and contain many different components. The requirements placed on the way these products perform are wide, varied and often conflicting. Those involved in product development will continue to need a profound understanding of the chemistry and physics of both raw materials and product structures.

Paper has played a vital role in the cultural development of mankind. It still has a key role in communication and is needed in many other areas of our society. There is no doubt that it will continue to have an important place in the future. Paper must, however, maintain its competitiveness through continuous product development in order to meet the ever-increasing demands on its performance. It must also be produced economically by environment-friendly processes with the minimum use of resources. To meet these challenges, everyone working in this field must seek solutions by applying the basic sciences of engineering and economics in an integrated, multidisciplinary way.

The Finnish Paper Engineers' Association has previously published textbooks and handbooks on pulping and papermaking. The last edition appeared in the early 80's. There is now a clear need for a new series of books. It was felt that the new series should provide more comprehensive coverage of all aspects of papermaking science and technology. Also, that it should meet the need for an academic-level textbook and at the same time serve as a handbook for production and management people working in this field. The result is this series of 19 volumes, which is also available as a CD-ROM.

When the decision was made to publish the series in English, it was natural to seek the assistance of an international organization in this field. TAPPI was the obvious partner as it is very active in publishing books and other educational material on pulping and papermaking. TAPPI immediately understood the significance of the suggested new series, and readily agreed to assist. As most of the contributors to the series are Finnish, TAPPI provided North American reviewers for each volume in the series. Mr. Brian Attwood was appointed overall reviewer for the series as a whole. His input is gratefully acknowledged. We thank TAPPI and its representatives for their valuable contribution throughout the project. Thanks are also due to all TAPPI-appointed reviewers, whose work has been invaluable in finalizing the text and in maintaining a high standard throughout the series.

A project like this could never have succeeded without contributors of the very highest standard. Their motivation, enthusiasm and the ability to produce the necessary material in a reasonable time has made our work both easy and enjoyable. We have also learnt a lot in our "own field" by reading the excellent manuscripts for these books.

We also wish to thank FAPET (Finnish American Paper Engineers' Textbook), which is handling the entire project. We are especially obliged to Ms. Mari Barck, the project coordinator. Her devotion, patience and hard work have been instrumental in getting the project completed on schedule.

Finally, we wish to thank the following companies for their financial support:

A. Ahlstrom Corporation

Enso Oyj

Kemira Oy

Metsä-Serla Corporation

Rauma Corporation

Raisio Chemicals Ltd

Tamfelt Corporation

UPM-Kymmene Corporation

We are confident that this series of books will find its way into the hands of numerous students, paper engineers, production and mill managers and even professors. For those who prefer the use of electronic media, the CD-ROM form will provide all that is contained in the printed version. We anticipate they will soon make paper copies of most of the material.

List of Contributors

Asikainen, Antti – Dr., Project Leader in the Wood Science and Industry Project,
Faculty of Forestry, University of Joensuu

Hakkila, Pentti – Dr., Professor,
Finnish Forest Research Institute, Vantaa Research Center

Harstela, Pertti – Dr., Professor,
Faculty of Forestry, University of Joensuu

Karjalainen, Timo – Dr., Senior Researcher,
European Forest Institute, Joensuu

Kellomäki, Seppo – Dr., Professor,
Faculty of Forestry, University of Joensuu

Koski, Veikko – Dr., Professor,
Finnish Forest Research Institute, Vantaa Research Center

Niemelä, Pekka – Dr., Professor,
Faculty of Forestry, University of Joensuu

Pukkala, Timo – Dr., Professor,
Faculty of Forestry, University of Joensuu

Toropainen, Mikko – Lic. Soc. Sc. (Econ.), Senior Researcher
Finnish Forest Research Institute, Joensuu Research Station

Preface

The sustainable social development of any society closely relates to the sustainable management of natural resources. The sustainable management of forest resources is a major concern in the world, since forests provide timber, other raw materials, nontimber benefits, and protection against natural and human induced threats.

Sustainable forestry is widely understood to mean only sustainable timber production; i.e. in long term the removal and growth of timber resources are in balance. Sustainable forestry currently requires using forest lands to maintain the biodiversity, productivity, and regeneration capacity of the forest ecosystems to fulfill the ecological, economic, and social functions of forests without damaging other ecosystems. Although the application of forestry and forest production varies globally, the need to sustain the forest ecosystem for current and future needs is a common goal.

This book is part of the series prepared under the general title "Papermaking Science and Technology" initiated by the Finnish Paper Engineer´s Association. The volume outlines the principles and methods of sustainable forestry. It describes the functioning and structure of forest ecosystems and world forest resources, the ecological basis of forest management, the principles and main practices applied in management, the main properties of timber, the procurement and measurement of timber, and the role of the forest industry in the context of a national economy. Shortly, the textbook describes the main features of forests and forestry to include the production of timber and its delivery to industry. The focus in this book is forests and forestry of the boreal and temperate zones particularly in Nordic countries. Much information is also valid outside the Nordic countries. The discussion includes the global dimensions of forests and forestry to place local findings in a larger context when necessary.

Writing this book to bridge the gaps between forestry and the forest industry was a challenge. The text includes a description of the timber resources, management of these resources, and how the properties of timber influence the manufacturing processes. There are many vital links between the properties of final products and the properties of timber that require careful consideration in the management of forest resources. These principles guided the experts who contributed to this book.

Dr. Seppo Kellomäki wrote the *Introduction, Trees, forests and the forest ecosystems* section, and the main part of *Management of forest ecosystems* section. He is a professor of silviculture at the Faculty of Forestry, University of Joensuu. He specializes in forest research and especially in modeling the possible effects of climate change. He also has considerable experience in modeling the effects of forest management practices on the growth and quality of timber.

Dr. Timo Karjalainen, a senior researcher at the European Forest Institute, Joensuu, wrote the *Global forest resources* section. His main interests are forest resources, and their role in the carbon cycle.

Dr. Pentti Hakkila wrote the *Structure and properties of wood and woody biomass* section. He works at the Finnish Forest Research Institute, Vantaa Research Center where he specializes in utilization of wood, technical properties of wood, forest energy, and forest operations.

Dr. Timo Pukkala, a professor of forest management planning at the Faculty of Forestry, University of Joensuu, wrote the *Forest inventory and planning* section. Dr. Pukkala has considerable experience in developing forest planning systems in Finland and other countries. His present research topics include multi-objective forest management, growth, and yield and integrating risk and uncertainty into forest decision-making.

Dr. Veikko Koski, a professor in forest genetics, wrote the portion *Genetic improvement* in the *Management of forest ecosystems* section. He works at the Finnish Forest Research Institute, Vantaa Research Center. His special interests are the genetic system of forest trees, adaption of forest trees to climate, and conservation of genetic resources.

Dr. Pekka Niemelä wrote the *Ecology and management of forest pests* portion in the *Management of forest ecosystems* section. Dr. Niemelä is a professor of forest protection at the Faculty of Forestry, University of Joensuu. He has especially studied the resistance mechanisms of trees against herbivores and the invasion of North American forests by European insects.

Dr. Pertti Harstela, a professor of forest technology at the Faculty of Forestry, University of Joensuu, wrote the *Timber procurement* and *The price of timber* sections. He has extensive experience in the research of timber and timber harvesting technologies. His main interests include planning of timber procurement, timber purchase systems, and the ergonomics of forest work.

Dr. Antti Asikainen from the Faculty of Forestry, University of Joensuu, wrote the *Timber measurement* section. He is a project leader in the Wood Science and Industry Project and a vice member of the national Timber Scaling Committee. He specializes in planning and simulation of timber transport and harvesting.

Lic. Soc. Sc. (Econ.) Mikko Toropainen, a senior researcher at the Finnish Forest Research Institute, Joensuu Research Station, wrote *The forest sector and national economy* section. His main interest is examining the significance of the forest sector in the national economy.

I am grateful to the Finnish Paper Engineer´s Association for inviting us to write this book. It was a great honor and provided a perspective of current forestry practices. I also thank all the authors for their marvellous work and for their disciplined writing habits that made my work easier. Thanks also to Ms. Marjoriitta Möttönen, M.Sc. in Forestry, whose expertise and tolerance were crucial components of the manuscript preparation. I also acknowledge the Graphical Center (Kuvakeskus), University of Joensuu, for their skillful preparation of the figures in this book.

Joensuu, July 1998

Seppo Kellomäki
Editor

Table of Contents

1. Introduction 12
2. Trees, forests, and the forest ecosystems 21
3. Global forest resources 89
4. Structure and properties of wood and woody biomass 117
5. Forest inventory and planning 187
6. Management of forest ecosystems 219
7. Timber procurement 311
8. Timber measurement 365
9. The price of timber 395
10. The forest sector and national economy 405
    Glossary 414
    Conversion factors 419

CHAPTER 1

Introduction

1	Sustainable management of forest resources	12
2	Can sustainable management be applied in practice?	14
3	Future expectations for forestry	16
4	Objectives and scope of this book	18
	References	19

CHAPTER 1

Seppo Kellomäki

Introduction

1 Sustainable management of forest resources

The sustainable social development of any society closely relates to the sustainable management of natural resources. Management of natural resources is sustainable whenever current use considers the needs of future generations. The sustainable management of forest resources is a major concern in the world, since forests provide timber, other raw materials, nontimber benefits, and protection against natural and human induced threats like avalanches, desertification, and climate change. Even in urban areas, the quality of life depends on trees and forests to improve the environment by absorbing air impurities and noise.

Although sustainable management has been a main principle in forestry since the early 1950s, it is still a challenge. Sustainable management in forestry initially meant only sustainable timber production – a balance of the removal and growth of timber resources. A current definition of sustained management is the use of forest lands to maintain the biodiversity, productivity, and regeneration capacity of the forest ecosystems to meet the ecological, economic, and social functions of forests without damaging other ecosystems. Using this principle, the guidelines discussed below are probably valid from generation to generation.

    Forestry is production using the forest ecosystem or forests. Following this definition, forestry obviously includes the production of tangible items like timber and wildlife as Table 1 shows. Forests can also provide intangible benefits like scenic beauty, wind reduction, and urban noise absorption. A unique feature of forest-based production is that it requires only small external input to control the production. Natural processes exert the most control. Forest-based production is normally integrated – the same forest land functions at the same time for timber production, game management, or for producing other environmental benefits such as the conservation of various environmental values.

Table 1. The functions of forests and forest ecosystems in the tangible and intangible production of various items and services.

Functions	Products and services
Timber	Timber for mechanical and chemical forest industry and manufacturing wood-based products
Fuel	Source of energy as fuel from ordinary forests or from fuel plantations or the use of by-products
Consumption of plants, animals, and derivatives	Source of herbs, medicinal plants, pharmaceuticals, gums, resins, tannin, waxes, oils, and livestock fodder
Regulation of climate and atmospheric composition	Stabilization of local, regional, and global climate and a sink for greenhouse gases
Protection	Buffer against spread of pests and diseases
Educational and scientific services	Research on ecosystems and organisms; zones to monitor ecological changes; environmental education; specimens for museums, zoos, and botanical gardens; wildstock for food; chemicals; and biological control agents
Source of land and living space	New lands for wildlife
Reindeer husbandry	Management of forest for grazing of reindeers in northernmost parts of Nordic countries and Russia
Game and fish management	Managing tree species composition and other properties of forests to provide food and shelter for game species
Management of lichens for decorative use	Management of lichen communities for harvesting for decorative purposes
Management for food products	Collecting wild berries, mushrooms, fruits, nuts, honey, syrups, and seeds for processing industry
Management of ground water resources	Conservation of forest land important in forming ground water
Outdoor recreation and other psychophysiological influences	Facilities and space for various outdoor activities such as skiing, bird watching, tourism, and inspiration for art, literature, music, etc.
Landscape management	Management of the amenity values of forest landscape
Noise abatement, absorption of air impurities, and related environmental benefits	Shelter belts for absorbing noise and air impurities in urban areas
Control of erosion, wind force, and related environmental benefits	Shelter belts for curbing water and wind erosion and reducing wind in agricultural and urban areas, stabilization of slopes, stream banks, water catchments, shelter belts, nutrient storage, distribution, and cycling
Management and conversation of biodiversity	Habitats for different organisms especially for those endangered

CHAPTER 1

In integrated forest production, a main problem is that only some forest-based items and services have a market value as that of timber. The remainder have no market value or a value that is difficult to define as that of scenic beauty. This incompatibility between different items and services remains unsolved and creates problems regarding management of the forest resources for different purposes – multiple purpose forestry, multiple use of forests, and multifunctional forestry – in a sustainable manner. The following principles of biodiversity, ecological balance, and multiple use are therefore necessary for sustainable management:

- The genetical resources of forests require conservation in whatever form they are contributing to the biodiversity of the forests and forest ecosystem (Principle of Biodiversity).
- The capacity of the forest ecosystem to intercept solar energy and subsequent cycles of material (water, carbon, nitrogen, mineral nutrients) requires maintenance (Principle of Ecological Balance).
- The capacity of the forest ecosystem to produce timber and other items and services requires conservation (Principle of Multiple Use).

Most global forests have natural tree species with large genetic variability. This is an advantage in commercial forests, since the large variability will reduce the risks of excessive damage due to pests, air impurities, and changing climate. The forests can regenerate and acclimate to changing conditions. Proper management can maintain the capacity of natural or seminatural forests to produce different items and services with smaller risks of disturbing the natural cycles of water and matter with their vital feedback mechanisms.

2 Can sustainable management be applied in practice?

Forestry is generally sustainable if the area of forest land does not decrease. This sustained management can also mean that stocking of forests is constant or the income from forestry is regular. The most common concept of sustained management in forestry is the balance between growth and removal of stem wood. This definition is operational, but it covers only the ecological dimension. It excludes the economical and social balances between the current use and future needs. In the following discussion, the long-term development of forest resources in Finland will exemplify the potential of sustainable forestry in boreal regions considering the balance between growth and removal of stem wood.

In Finland, a regular inventory of forest resources started in the early 1920s. Since that time, Fig. 1 shows that growth has exceeded removal of stem wood with few exceptions with a consequent increase of forest resources. The same trend should continue in the future even under increasing use of forest resources. This will provide time to formulate the future options to manage forests in a sustainable manner for different purposes and to conserve the unique features of forests.

Introduction

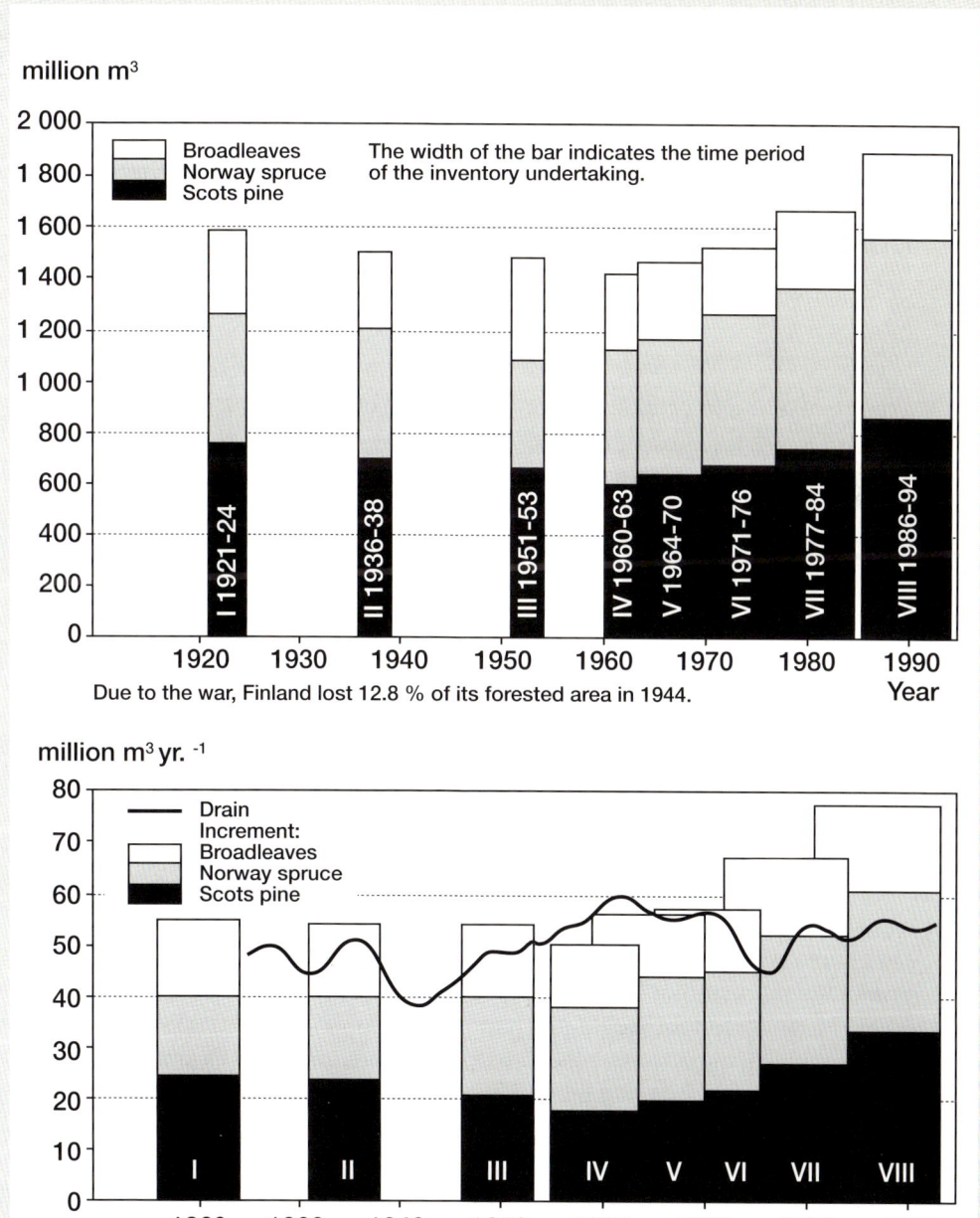

Figure 1. Total volume of forests (upper) and realized balance between growth and cuttings in Finland from 1920s to 1990 (lower)[1].

Growth exceeding cuttings since the late 1960s is mainly due to intensive management. The area of forest land increased due to the large-scale drainage and fertilization of peat lands. The growth potentials of forests increased with large-scale regeneration of old growth and understocked forests. The forests established in the 1960s and early 1970s are now having their maximum growth. Urban growth has reduced the importance of forests on the income of citizens. People often prefer their forest land as a recreation environment rather than a source of income. The management and use of forests resources are now more passive than during the 1950s and 1960s.

3 Future expectations for forestry

The development of the forests of the world closely links with the success to apply the sustainability of the social development in different parts of the world. Table 2 describes several international efforts in this regard. A common characteristic of these efforts is conservation of the world ecosystems and their sustainable management to keep the globe fit for human habitation. Sustainable forestry is a key issue, since forests are biomes that conserve the global diversity of genetic resources and ecosystems and provide resources for human welfare.

In Finland, increasing growth with greater potential for timber production should characterize the timber resources in the future. There will be many opportunities to manage forests in a sustainable manner to meet the future needs of tangible production. These include increasing consumption of forest-based products (sawed timber and related products, pulp, and paper) and energy and the increasing use of forests for recreation and producing natural products such as berries and mushrooms. The tangible production could potentially increase providing more space for the intangible production to meet all the dimensions of forest sustainability.

In the future, forests will undergo more intensive use to conserve other ecosystems. An example is watershed management to link in different ways the management of forests with the management of waters. The sequestration of carbon in forest ecosystems will also be an important tool to mitigate global warming and reduce the consequent risks for survival of current forests. In 1990, the net growth of forests in Finland with increasing timber resources absorbed nearly 60% of the carbon dioxide emissions released in energy production based on combustion of fossil fuels. Applying proper management could further enhance the carbon sequestration.

In Finland, forests are still natural or seminatural and represent a huge faunal and floral variability. In this context, even commercial forests are valuable genetic reserves. There are approximately 200 endangered species – 20% of the total of 1 000 endangered species in Finland – that the properties of forests and forestry operations affect. Natural forests still existing are therefore providing unique sanctuaries for endangered species to avoid the extinction of many species that are unable to acclimate to the management for timber production. The total land area currently set aside from commercial timber production is more than 2.7 million hectare – approximately 10% of the total area of forestry land. Balancing different dimensions of sustainability in forestry will require the balanced allocation of forest land for various purposes.

Introduction

Table 2. Intergovernmental initiatives to conserve and sustain the use of world ecosystems.

The United Nations Conference on Environment and Development (UNCED) in Rio de Janeiro in 1992 was a turning point in international discussions about forestry. UNCED adopted two important documents representing the first international consensus on forests. Forest Principles ("The Non-legally Binding Authoritative Statement on Principles for a Global Consensus on the Management, Conservation and Sustainable Development of Forests") is a tool for the conservation and sustainable development of forest resources in signatory countries. Forest Principles recognizes also the sovereign rights of countries over their forest resources. It provides a framework and gives countries flexibility to manage their forest resources according to their own goals and environmental policies. Chapter 11 of Agenda 21 is a broad and balanced foundation for the conservation, management, and development of all types of forests. It urges countries to develop forest strategies and to address deforestation problem. These documents do not provide precise guidance but propose general principles and programs for action. Development of the UNCED agreements therefore led to establishment of the Intergovernmental Panel on Forests (IPF) in 1995. IPF should progress toward an international consensus on key issues related to forests, improved national forest policies and development strategies, improved international coordination and cooperation, expanded coverage of forest resource assessments, and improved understanding of environmental implications of harvesting and trade of forest products.

The Convention of Biological Diversity (CBD) was also adopted at UNCED. The CBD is the first global attempt to address the conservation of biological diversity in natural habitats, such as forest ecosystems. The goal of the CBD is to promote the conservation and sustainable use of biological diversity and to ensure equitable distribution of the benefits derived from the use of genetic resources. CBD is working in a complementary way with IPF.

The Framework Convention on Climate Change (FCCC) was also adopted at UNCED. FCCC aims to stabilize the concentration of greenhouse gases in the atmosphere at a level that prevents dangerous human-caused disturbances of the climate system. Signatory countries of the FCCC agree to perform national inventories on greenhouse gas emissions. The FCCC has addressed the role of forests as a source and potential sink of carbon and their role in climate regulation.

The UN Convention to Combat Desertification (CCD) signed in 1994 aims to combat desertification and mitigate the effects of drought in countries experiencing serious drought, desertification, or both. Extensive efforts in Africa have involved effective action at all levels supported by international cooperation and partnership arrangements in the framework of an integrated approach that is consistent with Agenda 21 by contributing to the achievement of sustainable development in affected areas.

The Indigenous People's Convention introduced in 1989 calls upon the signatory states to take measures to protect and preserve the environment to the territories that people inhabit. It recognizes the ownership and property rights of indigenous people to the lands and the resources that they have traditionally inhabited including forests and forest resources. More diversified use of forest products and the need for a more equitable distribution of income from forest use are included as are planning, coordination, and decision-making mechanisms.

The Convention on International Trade in Endangered Species of Wild Fauna and Flora (CITES) implemented in 1973 calls for imposing restrictions on trade in threatened and endangered species including certain tree species. Species, subspecies, or populations may be components of three different lists. The first essentially prohibits commercial trade, the second requires an export permit issued by an authority in the exporting country, and the third subjects commercial trade countries.

The first International Tropical Timber Agreement (ITTA) came into force in 1985. It includes market intelligence, reforestation and forest management, additional processing in producer countries, and research and development. Originally ITTA was a commodity agreement, but it has evolved into an international framework for tropical timber development that takes into consideration environmental concerns. The term "timber producing forests" has replaced the term "tropical forests" in some places in the 1997 agreement to allow the ossible expansion of some elements of the agreement beyond the tropics.

CHAPTER 1

4 Objectives and scope of this book

This book will outline management of ecosystems and forests on a sustainable basis. The main focus will be on the timber production. Management is the process of controlling the interaction of trees and their environment to produce high quality timber. Management discussion will use an ecological context in which the physiological and ecological responses of trees link the dynamics of tree populations with the climatic and edaphic factors affecting a site. The harvest of timber links the management and forest resources with the forest-based industry and the properties of wood and other tree components to produce various items. The focus in this book is forests and forestry of the boreal and temperate zones especially in the Nordic countries. Many items are also valid outside the Nordic countries. The discussion includes global dimensions of forests and forestry whenever it is necessary to scale local findings into a larger context.

References

1. Anon., in *Statistical Yearbook of Forestry 1995* (M. Aarne, Ed.), The Finnish Forest Research Institute, Helsinki, 1995, p. 45.

CHAPTER 2

Trees, forests, and the forest ecosystems

1	**Concept of a tree**	**21**
2	**Structure of a tree**	**29**
3	**Functioning of a tree**	**34**
3.1	Concepts	34
3.2	Photosynthesis	36
	3.2.1 Description	36
	3.2.2 Biochemistry of photosynthesis	38
3.3	Transpiration	42
3.4	Respiration	44
3.5	Growth	45
	3.5.1 Components of growth	45
	3.5.2 Distribution of growth in a tree and the allometry of a tree	51
3.6	Mortality	54
3.7	Reproductive processes	56
4	**Forest ecosystem and its dynamics**	**59**
4.1	Concept of the forest ecosystem	59
4.2	Functioning and structure of the forest ecosystem	60
	4.2.1 Short-term and long-term dynamics	60
	4.2.2 Feedback through energy fixation	61
	4.2.3 Feedback through the nutrient cycle	62
	4.2.4 Feedback through the hydrological cycle	74
4.3	Productivity of the forest ecosystem	78
	4.3.1 Description	78
	4.3.2 Biomass accumulation	80
	4.3.3 Successional dynamics of the forest ecosystem	84
	References	**86**

CHAPTER 2

Seppo Kellomäki

Trees, forests, and the forest ecosystems

1 Concept of a tree

Any vascular plant is a tree if it has clearly recognizable crown, stem, and root systems. The plant therefore has a tree-like structure with a balanced relationship among different structural components – crown, stem, and roots. The concept of tree normally refers to woody plants with the above properties and appropriate size. For trees, this implies that the vascular cambium or cambium forms a continuous layer around the stem with formation of xylem (wood) inwards and phloem outwards as Fig. 1 shows. The phloem is the living part of the bark over the tree organs – branches, stem, and roots. In the boreal and temperate zones, a mature tree stem is a tapering column of wood formed by wood cones (annual growth rings) one after another enclosed in bark covering the stem. The main component of trees is therefore the wood produced by the expansion or radial growth of the woody organs. Branches, stem, and roots elongate from apical meristems. The properties of the tree structure are under the influence of genetic control but can have modification by environmental factors with a substantial plasticity in the physiological and ecological responses to the prevailing conditions.

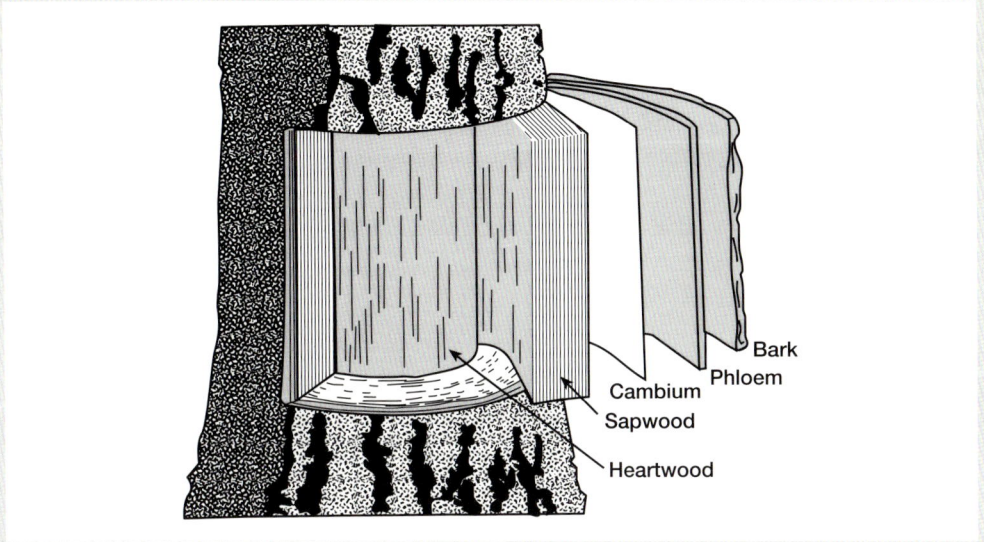

Figure 1. Main structure of the woody stem of trees in the boreal and temperate zones.

CHAPTER 2

Considering their reproductive processes, trees as seed plants are in the taxonomical classification of *Gymnosperms* and *Angiosperms*. In *Gymnosperms*, seeds develop on the surfaces or at the tip of an appendage without enclosure in it. The seed-bearing structures of *Gymnosperms* in temperate and boreal zones may occur in cones or strobili that hide seeds but do not enclose them. In *Angiosperms*, the seeds are within the structures as fruits. Examples are the broadleaved trees or flowering trees in boreal and temperate zones.

Coniferous species appeared in the plant kingdom about 180–230 million years ago, but their history dates back 400–500 million years to the times when plants with vascular system appeared in the plant kingdom. The current number of coniferous tree species in the world is more than seven hundred. They represent approximately 50 different genera. In the northern hemisphere, Table 1 shows that the most important coniferous genera are *Pinacea*, *Taxodiaceae*, and *Cupressaceae*. In the southern hemisphere, Table 2 shows that *Podocarpaceae* and *Araucariaceae* are the most important. In the boreal and temperate zones, only the species belonging to the genus *Pinaceae* have economical value. These species represent the following genera: *Pinus*, *Larix*, *Picea*, *Pseudotsuga*, *Thuga*, *and Abies*. The total number of tree species in these genera is approximately 210 species with geographical distribution from the tropics to the northern timber line. The genus *Pinaceae* represents the coniferous tree species with the most value in forestry and the forest industry.

Broadleaved tree species are substantially younger than the coniferous ones. They appeared in the plant kingdom approximately 50–150 million years ago. The current number of broadleaved species is more than 250 000. The classifications of broadleaved species cover 44 genera. In the boreal and temperate zones, Table 3 shows that the genera *Betulaceae* and *Salicaceae* are important. These include the following genera: *Betula*, *Alnus*, *Populus*, and *Salix*. The trees belonging to these genera grow throughout the boreal and temperate zones, but they are most common in the temperate zone. In the tropical zone the species richness is substantially larger as Table 4 shows.

Table 1. Selected list of coniferous species with value in forestry in boreal and temperate zones[1].

Species	Distribution or plantation area and further notes	Use
Pinus sylvestris L.	Europe, Siberia	Sawed timber, veneer, poles, pulp wood, fuel, and other uses such as resin, rubber, and food
Picea abies (L.) Karst	Europe, Siberia	Sawed timber, veneer, poles, pulp wood, and fuel
Larix sibrica (Münchh.) Ledeb.	Russia	Sawed timber, veneer, poles, and pulp wood
Larix leptolepis (Sieb. & Zucc.) Gord.	North Japan	Sawed timber, veneer, poles and other uses such as resin, rubber, and food
Abies sibirica Ledeb.	Russia	Pulp wood
Larix gmelini (Rupr.) Kuzen.	Siberia, the Far East	Sawed timber, veneer, poles, and pulp wood
Abies alba Miller	Central Europe	Sawed timber, veneer, and poles
Picea jezoensis (Sieb. & Zucc.) Carr	East Russia, Japan	Pulp wood, sawed timber, veneer, and poles
Abies balsamea Mill.	East Canada	Pulp wood, sawed timber, veneer, and poles
Abies concolor (Gord.) Ldl. Ex Hildebr.	The Rocky Mountains	Sawed timber, veneer, and poles
Pseudotsuga menziesii (Mirb.) Franco	West and North America (plantations in Central Europe, New Zealand etc.)	Sawed timber, veneer, poles, and pulp wood
Tsuga heterophylla (Rafin.) Sarg	West and North America	Sawed timber, veneer, poles, and pulp wood
Picea mariana (Mill.) Britton	Canada	Pulp wood, sawed timber, veneer, and poles
Picea glauca (Moench.) Voss	Canada	Pulp wood, sawed timber, veneer, and poles
Picea engelmannii Parry ex Engelm.	The Rocky Mountains	Pulp wood, sawed timber, veneer, and poles
Picea sitchensis (Bong.) Carriére	Alaska, West Canada, Western coast of United States	Sawed timber, veneer, poles, and pulp wood
Larix decidua Mill.	Europe	Sawed timber, veneer, poles, and pulp wood
Larix laricina (Duroi) C. Koch	Canada	Pulp wood
Pinus koraiensis Sieb. & Zucc.	North-East China, East Russia, Korea	Sawed timber, veneer, poles, and other uses such as resin, rubber, and food
Pinus strobus L.	East and North America	Sawed timber, veneer, poles
Pinus griffithii Mc Clelland	North India, Himalaya	Sawed timber, veneer, poles
Pinus nigra Arnold	Central Europe, Mediterranean countries	Pulp wood, sawed timber, veneer, and poles

CHAPTER 2

Table 1. Selected list of coniferous species with value in forestry in boreal and temperate zones[1].

Species	Distribution or plantation area and further notes	Use
Pinus densiflora Sieb. & Zucc.	Japan, China, Korea	Pulp wood, sawed timber, veneer, poles, erosion control, and fuel
Pinus thunbergii Parl.	Japan, China, Korea	Pulp wood, sawed timber, veneer, poles, erosion control, and fuel
Pinus banksiana Lamb.	Canada, Great Lakes states in United States	Pulp wood
Pinus contorta Dougl.	The Rocky Mountains (plantations in Scotland and in Nordic countries)	Pulp wood
Chamaecyparis nootkatensis (Lamb.) Spach	Alaska, British Columbia, West Coast of United States	Pulp wood, sawed timber, veneer, and poles
Thuja plicata D. Don ex Lamb.	Alaska, British Columbia, West Coast of United States	Pulp wood, sawed timber, veneer, and poles
Thuja orientalis L.	China (plantations in the Middle East and in Mediterranean countries)	Erosion control, fuel, pulpwood, sawed timber, veneer, and poles
Juniperus virginiana L.	Eastern United States	Sawed timber, veneer, and poles

Table 2. Selected list of coniferous species with value in forestry in sub-tropical and tropical zones[1].

Species	Distribution or plantation area and further notes	Use
Cryptomeria japonica L. Fil.) D. Don	Japan, China	Sawed timber, veneer, and poles
Araucaria araucana (Molina) k. Koch	Chile, Argentina	Sawed timber
Araucaria angustifolia (Bertol.) Kunze	South Brazil, Argentina	Sawed timber
Pinus elliottii Engelm.	Southeastern United States	Pulp wood, sawed timber, veneer, and poles
Pinus taeda L.	Northeastern United States	Pulp wood, sawed timber, veneer, and poles
Pinus radiata D. Don	Plantations in New Zealand, Australia, South America, and South Africa	Sawed timber, veneer, poles, and pulp wood
Pinus patula Schiede & Deppe	Mexico (plantations in Africa)	Pulp wood, sawed timber, veneer, and poles
Cupressus lusitanica Mill.	Mexico, Guatemala (plantations in Africa)	Timber
Pinus pinea L.	Mediterranean countries	Erosion control and other uses such as resin, rubber, and food
Pinus roxburghii Sarg.	North India, Himalaya	Sawed timber, veneer, and poles
Pinus merkusii Jungh. & de Vriese	Southeast Asia (a close species, *Pinus yannanensis*, in Southwest China)	Pulp wood, erosion control, and other uses such as resin, rubber, and fuel
Pinus ponderosa Laws.	The Rocky Mountains	Pulp wood, sawed timber, veneer, and poles
Pinus palustris Mill.	Southeastern United States	Pulp wood, sawed timber, veneer, and poles
Podocarpus sp.	Southern hemisphere, Asia, Africa, Australia	Sawed timber, veneer, and poles
Pinus echinata Mill.	Northeastern United States	Pulp wood, sawed timber, veneer, and poles
Pinus caribae Morelet	Central America (plantations in the countries of the southern hemisphere and Southeast Asia)	Pulp wood, sawed timber, veneer, and poles
Pinus kesiya Royle ex Cord.	Southeast Asia	Pulp wood, erosion control and other uses such as resin, rubber, and food

CHAPTER 2

Table 3. Selected list of broadleaved species with value in forestry in boreal and temperate zones[1].

Species	Distribution or plantation area and further notes	Use
Populus sp.	Plantations of many different species throughout the rich forest types	Pulp wood, sawed timber, veneer, and poles
Populus tremula L.	Europe, Siberia	Sawed timber, veneer, poles, and pulp wood
Populus tremuloides Michx	North America	Pulp wood, sawed timber, veneer, and poles
Populus balsamifera L.	Canada, Midwestern United States, Alaska	Pulp wood
Populus trichocarpa Torrey & Grey ex Hooker	The Rocky Mountains	Sawed timber, veneer, poles, and pulp wood
Salix sp.	Rich forest types (China, South America, etc.)	Pulp wood, fuel and other uses such as resin, rubber, and food
Juglans nigra L.	East and North America	Sawed timber, veneer, and poles
Juglans mandshurica Maxim.	Northeast China, East Russia (close species in Japan)	Sawed timber, veneer, and poles
Betula pendula Roth	Europe, Siberia	Sawed timber, veneer, poles, pulp wood, and fuel
Betula pubescens Ehrh.	Europe, Siberia	Pulp wood, sawed timber, veneer, poles, and fuel
Betula papyrifera Marsh.	North America	Pulp wood
Betula alleghaniensis Britt.	East and North America	Sawed timber, veneer, and poles
Alnus glutinosa (L.) Gaertner	Europe	Sawed timber, veneer, poles, and pulp wood
Alnus rubra Bong.	West and North America	Sawed timber, veneer, poles, and pulp wood
Carbinus betulus L.	Central Europe (other species in East and North America and in East Asia	Sawed timber, veneer, and poles
Fagus sylvatica L.	Central Europe (other species in East and North America and in East Asia)	Sawed timber, veneer, and poles
Quercus robur L.	Europe	Sawed timber, veneer, and poles
Quercus alba L.	East and North America (a large number of species belonging to white oaks also in Europe, Asia, and North America)	Sawed timber, veneer, and poles

Table 3. Selected list of broadleaved species with value in forestry in boreal and temperate zones[1].

Species	Distribution or plantation area and further notes	Use
Quercus rubra L.	East and North America (a number of different species in North America)	Sawed timber, veneer, and poles
Quercus mongolica Fisch.	East Asia	Sawed timber, veneer, and poles
Ulmus glabra Huds.	Europe, West Asia	Sawed timber, veneer, and poles
Ulmus americana L.	East and North America	Sawed timber, veneer, and poles
Liriotendron tulipifera L.	East and North America (one close species important in China)	Sawed timber, veneer, and poles
Robinia pseudoacacia L.	East and North America (plantations in Central and East Europe, China)	Sawed timber, veneer, and poles
Acer platanoides L.	Europe	Sawed timber, veneer, and poles
Acer saccharum L.	East and North America (also other species of rock maples)	Sawed timber, veneer, poles, and other uses such as resin, rubber, and food
Acer saccharinum L.	East and North America (also other species of silver maples)	Sawed timber, veneer, and poles
Tilia cordata Mill.	Europe, West Siberia (a number of species in Europe, Asia, East, and North America)	Sawed timber, veneer, and poles
Fraxinus excelsior L.	Europe	Sawed timber, veneer, and poles
Fraxinus pennsylvanica Marsh.	East and North America (a number of species)	Sawed timber, veneer, and poles
Fraxinus mandshurica Rupr.	Northeastern China, East Russia	Sawed timber, veneer, and poles
Nothofagus sp.	Rich forest types and heath forests in South America	Sawed timber, veneer, poles, and pulp wood

CHAPTER 2

Table 4. Selected list of broadleaved species with value in forestry in sub-tropical and tropical zones[1].

Species	Distribution or plantation area and further notes	Use
Tectona grandis L. f.	India, Southeast Asia (plantations in Africa)	Sawed timber, veneer, and poles
Eucalyptus globulus Labill.	Australia (many plantations throughout sub-tropical and tropical zones)	Sawed timber, veneer, poles, pulp wood, and other uses such as resin, rubber, and food
Eucalyptus saligna Sm.	Australia (plantations in South Africa, etc.)	Sawed timber, veneer, poles, pulp wood, and other uses such as resin, rubber, and food
Swietenia sp.	Central America, South America	Sawed timber, veneer, and poles
Shorea robusta Gaertn. f.	India, Southeast Asia	Sawed timber, veneer, poles, and other uses such as resin, rubber, and food
Leucaena leucocephala (Lamk.) de Wit	Pacific Ocean (plantations in Hawaii and Southeast Asia)	Sawed timber, veneer, poles, pulp wood, fuel, erosion control and other uses such as resin, rubber and food
Casuarina equisetifolia Forst.	Australia (many plantations throughout sub-tropical and tropical zones)	Sawed timber, veneer, and poles
Quercus sp.	(Over 200 species mainly in rich forest types but also in sub-tropical and tropical zones; often evergreens)	Sawed timber, veneer, poles, pulp wood, and fuel
Olea europaea L.	Mediterranean countries	Sawed timber, veneer, poles, and other uses such as resin, rubber, and food
Dipterocarpus alatus Roxb.	Southeast Asia	Sawed timber, veneer, and poles
Ochroma lagopus Sw.	North and South America, Central America (plantations in Africa and Asia)	Sawed timber, veneer, poles, and pulp wood
Dalbergia sisso Roxb.	India, Southeast Asia	Sawed timber, veneer, and poles
Xylia kerrii Craib. & Hutch.	Southeast Asia	Sawed timber, veneer, and poles
Gmelina arborea L.	India (many plantations in tropical zone)	Sawed timber, veneer, poles, pulp wood, fuel, and erosion control
Bambusa, Arundinaria, Phyllosthachys, Dendrocalamus and other genera	Asia, Africa, South America (planted especially in Asia)	Sawed timber, veneer, poles, pulp wood and other uses such as resin, rubber, and food

Trees, forests, and the forest ecosystems

Historically, tree species as plants are characterized with the help of the morphological properties assuming that any species was a group of individuals or populations with no change over time. This concept of species has only limited value, since the multi-dimensional variability within the populations characterizes the populations of any species. This implies that a species represented by a population can change over time and therefore represent a large intra-species variability regarding the functional and structural properties. The intra-species variability like the variability between species provides forestry with large opportunities to optimize timber production as an interaction of the genotype and environment through the proper selection of species or the local provenance (subpopulation of a tree population representing a particular region) of a species to a particular site.

2 Structure of a tree

The description of tree structure often uses the structural components of stem, branches, foliage, and roots that Fig. 2 shows. Bark covers the stem, branches and roots (coarse or woody roots). It is a separate component. The classification of tree structure into stem, branches, foliage, roots, and bark is also suitable for forestry purposes.

In forestry, *stem* is normally the prime focus due to its economical value. In the functioning of a tree, the stem connects the foliage to the soil system and with the branches provides a framework for foliage to obtain in an optimal way radiation to drive the photosynthesis and other physiological processes. The stem is normally a cylinder tapering upwards formed by annual layers of radial growth (annual rings). At any height, the radial growth decreases toward the surface of the stem. The maximum radial growth consequently shifts upwards when the tree ages, and the growth in height decreases. In many tree species, the decrease in growth of height begins earlier and more rapidly than the radial growth. The distribution of wood density follows the distribution of radial growth such that wide annual rings indicate low wood density and vice versa.

With the foliage, *branches* form the crown of the tree. Branches come from the buds around the terminal bud of

Figure 2. Schematic presentation of tree structure.

CHAPTER 2

a stem. The braches born from the same cohort of buds are a whorl. Each branch initially attached to the stem through the annual ring represents the stem growth during the season when the branch is born. Branches grow radially later to form annual layers of radial growth. The growth of branches is similar to that of the stem. Branch growth will normally cease much earlier than the stem growth. Dead branches will self-prune, but their presence is evident as knots in the wood. The inner part of the stem therefore always represents wood with green knots (innermost part of the stem) and dead knots (outside the green knots). The surface wood at the lower stem normally only contains clean wood without knots.

The relationship of the dynamics of the crown system to the formation, growth, death, and pruning of branches is clearest in the gymnosperms with branches arranged in whorls. Each branch cohort undergoes the above process as a consequence of elevation of the crown bottom with the increasing tree height. This implies that the main part of the new and most efficiently photosynthesising leaves or needles occurs at the upper part of the crown with only negligible shading from nearby trees. The lower part of the crown provides only small amounts of photosynthates for the growth of tissues other than those situated in this area. The differentiation of the growth of branches in relation to their position in the crown creates the characteristic conical shape of some tree species such as Scots pine (*Pinus sylvestris*) and Norway spruce (*Picea abies*).

In gymnosperms, the longitudinal elements of *wood* consist mainly of tracheids (90%). Figure 3 shows that these provide the pathway to transport water with dissolved nutrients from the soil to the foliage. The transverse elements primarily include ray tracheids and ray parenchyma cells. In coniferous wood, there are resin channels oriented longitudinally and transversely. In these channels, resin secretes from the epithelial cells around the channels.

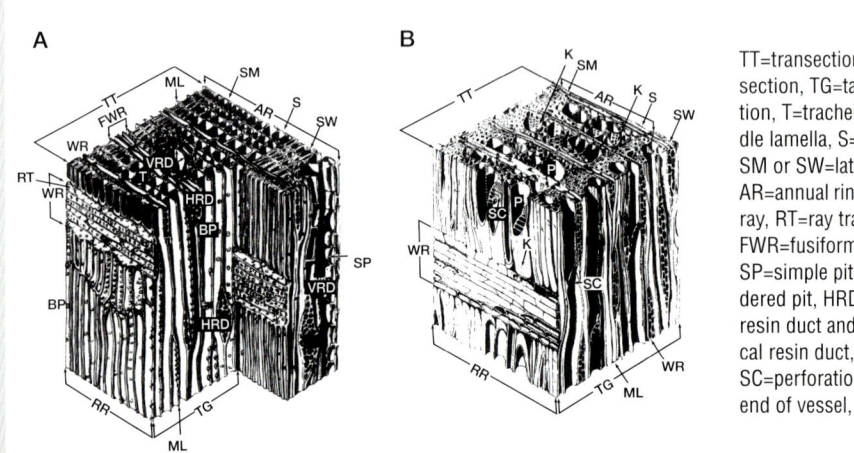

Figure 3. Schematic presentation of the structure of wood in gymnosperm (A) and angiosperm (B)[2].

Tracheids are narrow, thick-walled cells that can be 7 mm long. The tracheids connect with each other through pits that are primarily on the radial walls. The walls consist of layers with cellulose fibrils glued by lignin to each other. The fibrils include micelles consisting of cellulose molecules. Hollows between the micelles provide a capillary system for the flow of water from roots to foliage. The tracheids live only during the time necessary to form them. As a result, most tracheids in wood are dead.

In the angiosperm, the wood consists of longitudinal vessels, tracheids, and fibrils with tapering ends (prosenchymatous form). There are also longitudinal parenchyma and epithel cells. The main component of angiosperm wood consists of fibers (up to 60%) that are long, thin cells with vertical orientation. Vertical water movement occurs through vessels connected to each other through perforated walls. The vessels have a random distribution over the cross-section of wood (diffuse-porous species), but sometimes they have a ring arrangement (ring-porous species). The lateral movement of liquids occurs through pits. The transverse elements of angiosperm wood represent mainly ray parenchyma. In the boreal and temperate zones, angiosperms do not have resin channels.

Bark is tissue outside the cambium. Based on the structure and physiological functions, Fig. 4 shows that bark can be the inner bark (phloem) or outer bark. Inner bark is between the cambium and the last layer of periderm or rhytidome. Phloem provides the pathway for translocation, and rhytidome reduces water loss and provides protection from mechanical and thermal injuries. The bark tissues specifically include primary and secondary phloem, cortex formed by parenchyma cells, and periderm. Phloem or secondary phloem comes from the vertically oriented sieve tube cells and the parenchyma cells with resins, tannins, and fibers. The transversely oriented rays contain parenchyma cells.

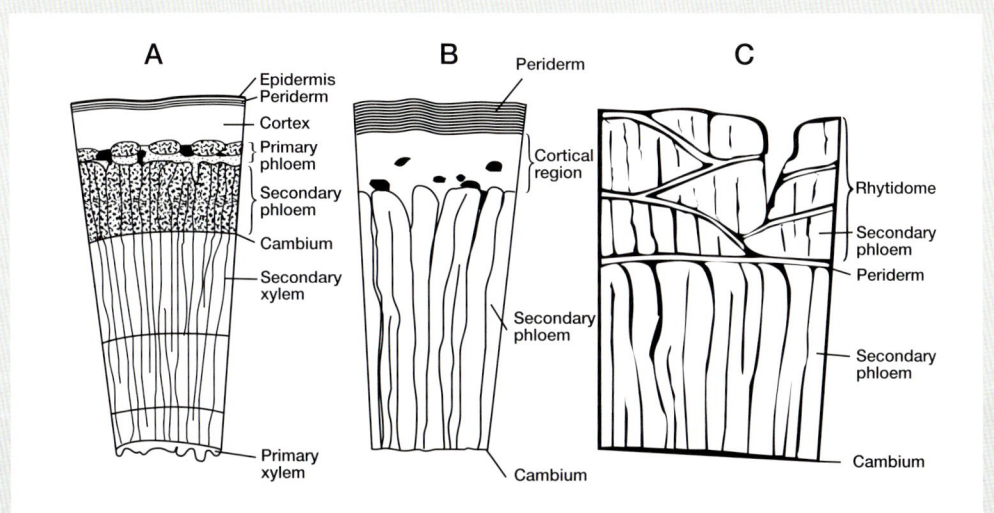

Figure 4. Structure of bark in stem of young tree (A), a mature tree without formation of rhytidome (B), and with the formation of rhytidome (C)[2].

CHAPTER 2

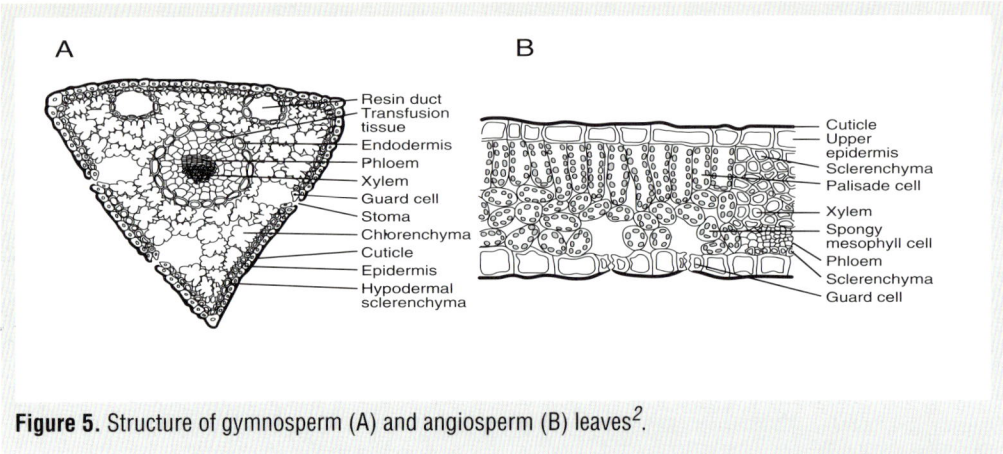

Figure 5. Structure of gymnosperm (A) and angiosperm (B) leaves[2].

The growth of tree diameter is mainly due to the cambial activity. In cambial activity, wood produces substantially more than phloem. In spring, large vessels (spring wood) form. During summer, small vessels (summer wood) form. Growth of phloem produces layers of phelloderm and phellogen that form the periderm. That replaces the epidermis during the early life of seedlings. Rhytidome represents the dead part of the outer bark. The share of the outer bark is larger at the lower part of the stem, and it decreases toward the stem apex.

Foliage of gymnosperms normally consists of evergreen leaves with few exceptions like *larches* (*Larix* sp.). Gymnosperm leaves are born from needle initials (fascicles) such that more than one needle can attach on a fascicle. The number of needles per fascicle is a common way to identify different pine species. Needle attachment has a regular pattern on shoots representing the elongation of stem and main branches or the branches attached on the main branches.

Most gymnosperms have leaves (needles) that are linear or lanceolate and bifacially flattened. Waxes normally cover the needle surfaces (epidermis) so that water vapor evaporated from needles is lost mainly through stomata as Fig. 5 shows. Stomata also provide for the intake of carbon dioxide into the needle. The stomata open and close in response to environmental factors to optimize water use for photosynthetic production.

Most needle tissue is mesophyll from the palisade and parenchyma cells. Inside the inner surface (endodermis), the vascular bundles including the xylem and phloem provide a system to supply water and nutrient (xylem) to photosynthesizing tissues of needles and the system to translocate (phloem) from the photosynthesizing tissues to growing tissues in needles or elsewhere. Resin ducts also occur in gymnosperm leaves such as in pines.

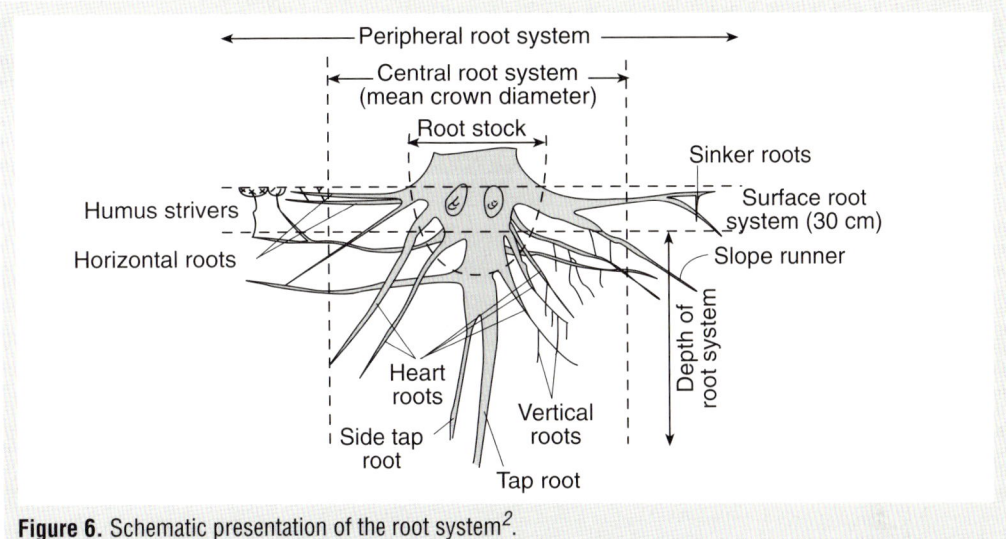

Figure 6. Schematic presentation of the root system[2].

In principle, angiosperm leaves have the same structure as gymnosperm leaves, but their leaves are normally flat and supported by petiole. The main part of the leaf tissue is mesophyll surrounded by the upper and lower epidermis. The vein system with vascular bundles supplies and removes the water and minerals. The vein arrangement in a leaf system has a net system that is characteristic for many angiosperm species. Angiosperm leaves normally have a covering of waxes, hairs, or other organs that effectively reduce the loss of water vapor from other surfaces except stomata.

Roots anchor the tree to soil and allow the system to absorb water and nutrients from the soil and supply them to living tissues in foliage and other organs for the use of the physiological process supporting regeneration and growth of the tree as Fig. 6 shows. Roots are coarse roots or fine roots to distinguish between their supporting (anchorage) and physiological (absorbing water and nutrients) functions. Modelling usually treats the root system as a crown system below the soil surface. This assumes a close functional relationship between the crown and roots in the mass flow.

The coarse roots (woody roots) form the permanent root system with long roots that can elongate and grow radially as stems or branches. The main part of the permanent roots represents long roots growing horizontally, but a part of the long roots is vertical with one or more tap roots with a high capacity to resist wind forces. The rooting zone of different species occurs at different depths in the soil. Norway spruce has its roots close to the soil surface, and Scots pine has its roots much deeper in the soil. Roots can connect neighboring trees (root graft). Branching long roots provide the short roots that form the physiologically active part of the roots. The main parts of the short roots consist of roots with mycorrhiza formation. This means that a fungal hyphae surrounds the short root and enhances the capacity of the root to absorb water and nutrients. The symbiotic relationship between trees and selected fungi is common throughout the world for all taxa of trees.

Table 5. Percentage distribution of biomass into different components for selected tree species.

Organ	Scots pine	Norway spruce	Birch
Stem	59	52	68
Roots	24	25	13
Branches	12	15	19
Foliage	5	8	
Total	100	100	100

The proportions of tree organs to the total tree biomass varies between species and also within species during the life cycle. In young trees, the proportion of foliage is larger than in older trees, and the proportion of stem increases with age. The proportion of the stem of the entire tree biomass is usually 50%–60%, that of roots is 10%–25%, that of branches is 10%–20%, and that of foliage is less than 10% as Table 5 shows.

3 Functioning of a tree

3.1 Concepts

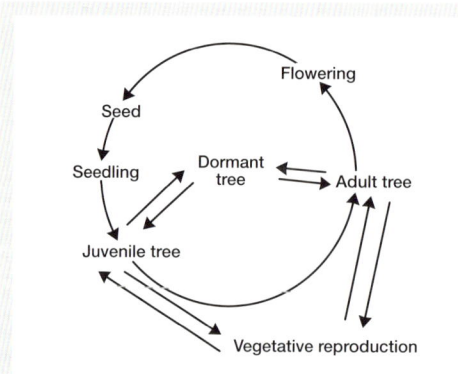

Figure 7. Annual life cycle of the tree over the life span.

In boreal and temperate zones, the functioning of trees follows the life cycle during which trees annually undergo cycling through active and dormant periods (annual cycle) and carry out physiological processes needed for regeneration, growth, and defense as Fig. 7 shows. In this context, trees can adapt to very variable conditions. In the boreal zone, some species such as Scots pine can tolerate temperatures below -50°C during dormancy. Dormancy breaks and buds burst in spring when days are sufficiently long and the temperature increases bring reduced risk of night frosts. Buds contain all that is necessary for new shoots and for the leaves and flowers that form during spring and summer. Unlike broadleaved trees, coniferous trees each year drop only the oldest part of the needles (except the *Larix* species that behaves like a broadleaved tree). Before the dormant period, deciduous trees transport all the photosynthates and chlorophyll from their foliage to the stem and roots. Before the leaves drop, new buds form.

The life span of trees varies from dozens of years to several thousands of years with no clear relation to taxonomic position or the other properties of trees. No single factor normally causes the death of a tree. The usual cause relates to a deterioration of the physiological process and attack of insect or fungi when the tree grows and matures. Both gymnosperms and angiosperms can reach an age of several hundred or a thousand years as Table 6 shows.

Table 6. Examples of how long trees can live[3].

Tree species	Lifespan, yr.
European conifers	
Abies alba	300–400
Picea abies	300–400
Pinus sylvestris	400–600
Northwest American conifers	
Picea sitchensis	800–900
Pseudotsuga menziesii	1 300–1 500
Sequoia gigantea	3 000–4 000
European broadleaved species	
Betula spp.	100–150
Ulmus spp.	500–600
Tilia spp.	800–1 000

The main characteristic in tree functioning is the growth that produces different tree structures. The common definition of tree growth is the increase of mass per time unit. The increase of mass implies also the increase of other dimensions per time unit such as the diameter, the length, and the volume of the stem. Growth is the manifestation of several metabolic processes acting under the control of the genetic properties of the tree and the environmental factors affecting the tree. The growth and the consequent structure of the tree are interactive responses of the genotype to the prevailing environmental conditions as Fig. 8 shows.

CHAPTER 2

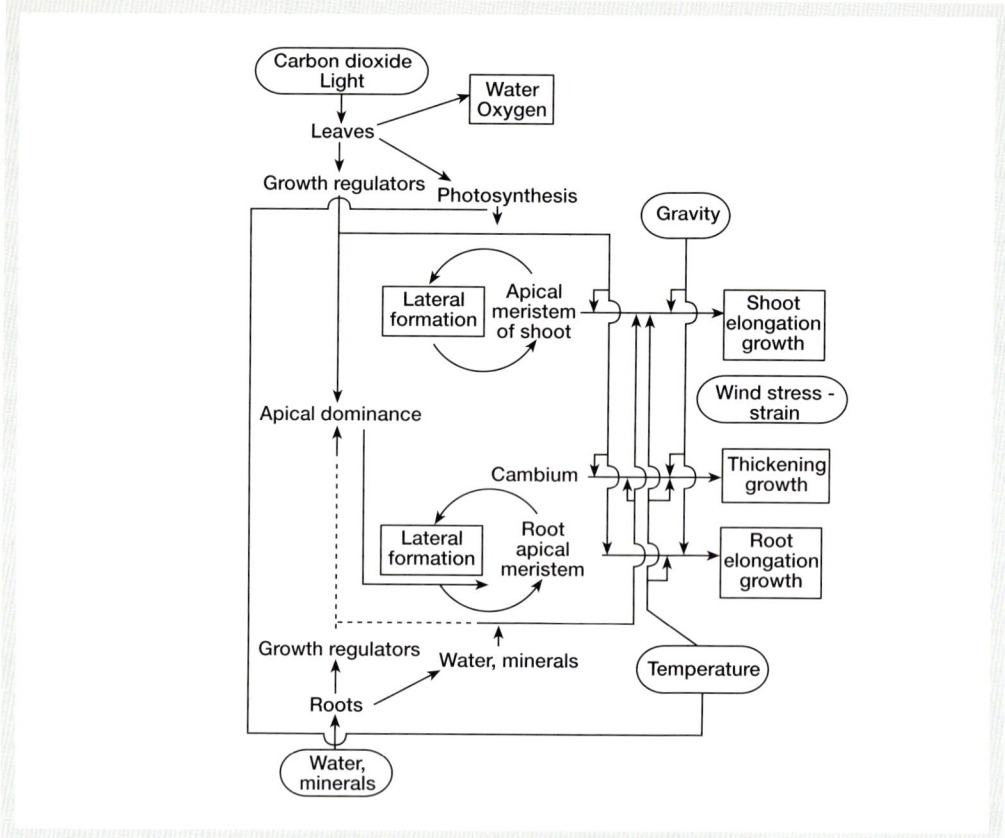

Figure 8. Some important factors and their interrelations that affect tree growth[4].

In the above context, the material factors (carbon dioxide, water, and nutrients) and the energy factors (radiation, heat, gravity, and the mechanical forces of wind and snow) affect the growth. The energy factors control the uptake and use of the material factors and the consequent metabolic processes and growth of trees. Photosynthesis, transpiration, respiration, the uptake of nutrients and water, and the translocation of metabolites to the growth and maintenance of living tissues are the main physiological processes with direct or indirect effect on the growth and regeneration of trees.

3.2 Photosynthesis

3.2.1 Description

Photosynthesis is a process where the carbon dioxide diffused in the leaves or needles is reduced to carbohydrate and other substances in the chroloplasts in the presence of water and nutrients. Solar radiation and temperature drive it. Approximately 2%–5% of the solar energy converts to chemical energy as the following equation shows:

$$6CO_2 + 6H_2O \rightarrow chlorophyl \rightarrow C_6H_{12}O_6 + 6O_2 \tag{1}$$

Trees, forests, and the forest ecosystems

Trees can store some formed carbohydrates from which they process proteins and fats. In photosynthesis, boreal and temperate trees (C_3-plants) cannot directly recycle the carbon dioxide released in leaves and needles during daytime like some tropical plants (C_4-plants). The C_3-plants can use the increase in the atmospheric carbon unlike the C_4-plants that are only slightly responsive to the elevating atmospheric carbon as Fig. 9 shows.

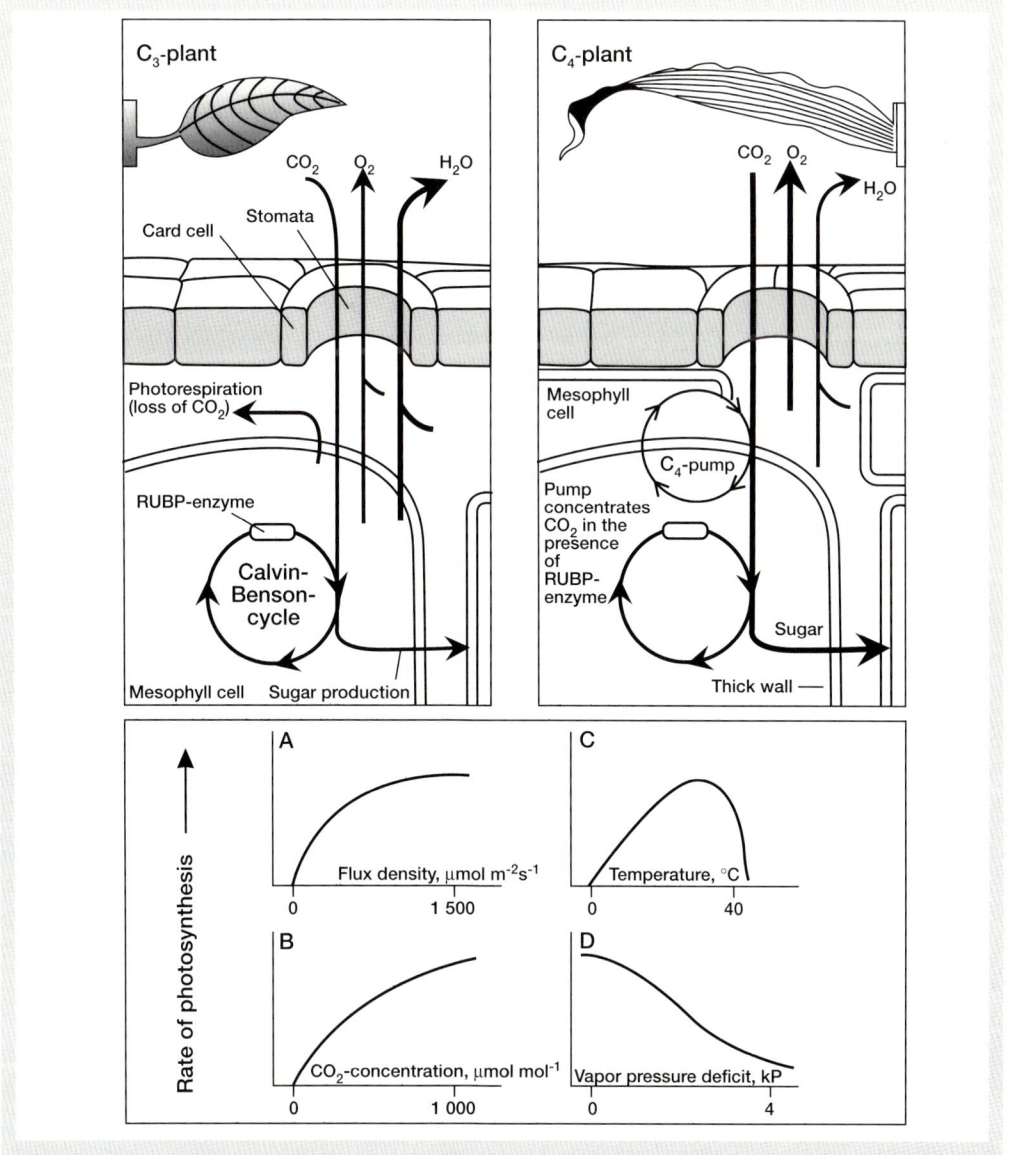

Figure 9. Outlines of the basic processes of photosynthesis in C_3-plants (above left) and C_4-plants (above right) and the response of the rate of photosynthesis to A: photon flux density, B: temperature, C: atmospheric CO_2 and D: water vapor pressure deficit (below).

CHAPTER 2

3.2.2 Biochemistry of photosynthesis

Photosynthesis harvests solar radiation and converts it into chemical energy through several reactions that need light (light reactions) or do not need light (dark reactions). In the mesophyll tissue of leaves and needles, there are many green pigments called chlorophyll. Chlorophyll pigments can absorb light energy to oxidize water to molecular oxygen and reduce carbon dioxide into primary sugars. The absorbed light drives the transfer of electrons between compounds that act as donors and acceptors of electrons. Electrons are finally accepted by nicotinamide adenine dinucleotide phosphate (NADP) that is further reduced into reduced NADP (NADPH+) ($NADP^+$ (nicotinamide adenine dinucleotide phosphate)). This is further reduced into NADPH (reduced NADP). The light reactions produce (ATP (adenosine triphosphate) and NADPH) adenosine triphosphate (ATP) and NADPH with high energy content used for synthesis of sugars shown in Fig. 10.

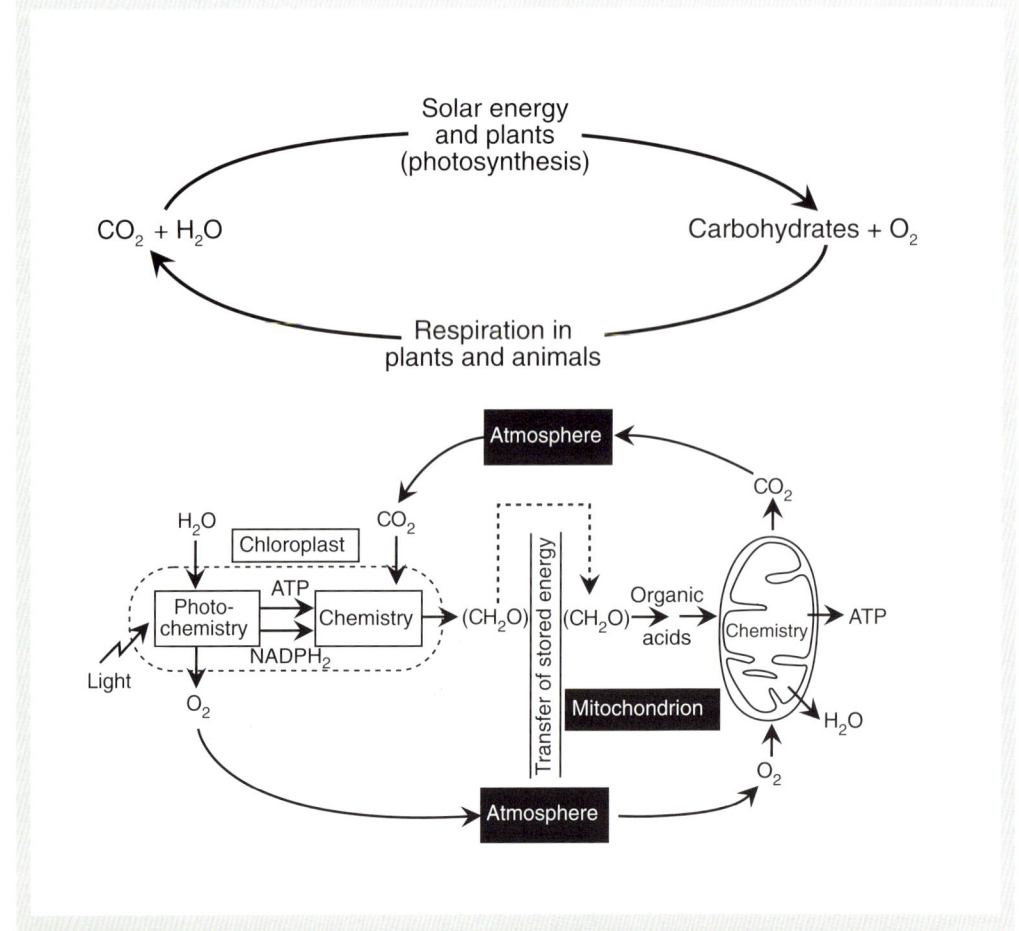

Figure 10. Schematic presentation of the fixation of CO_2 in photosynthesis and the release of CO_2 in respiration[5].

The synthesis of sugars does not need light for the photosynthetic carbon reduction cycles (PCR). In PCR, carbon and water enzymatically combine with the five-carbon acceptor molecule to generate two molecules of three-carbon intermediates that further reduce to carbohydrates using the photochemically generated ATP and NADPH. The cycle completes with the regeneration of the five-carbon acceptors. PCR of the C_3-plants proceeds in the following three phases[6] shown in Fig. 11:

- First, carboxylation of the CO_2 acceptor (ribulose 1,5-bisphosphate) forms two molecules of 3-phosphoglyserate.
- Second, 3-phosphoglyserate is reduced into glyceraldehyte 3-phosphate.
- Third, the CO_2 acceptor (ribulose 1,5-bisphosphate) is regenerated from glyceraldehyte 3-phosphate.

The enzyme (ribulose biosphosphate carboxylase/ oxygenase (Rubisco)) located in chloroplasts catalyzes the reaction between carbon and ribulose 1,5-bisphosphate. Rubisco accounts for approximately 40% of the soluble proteins in foliage.

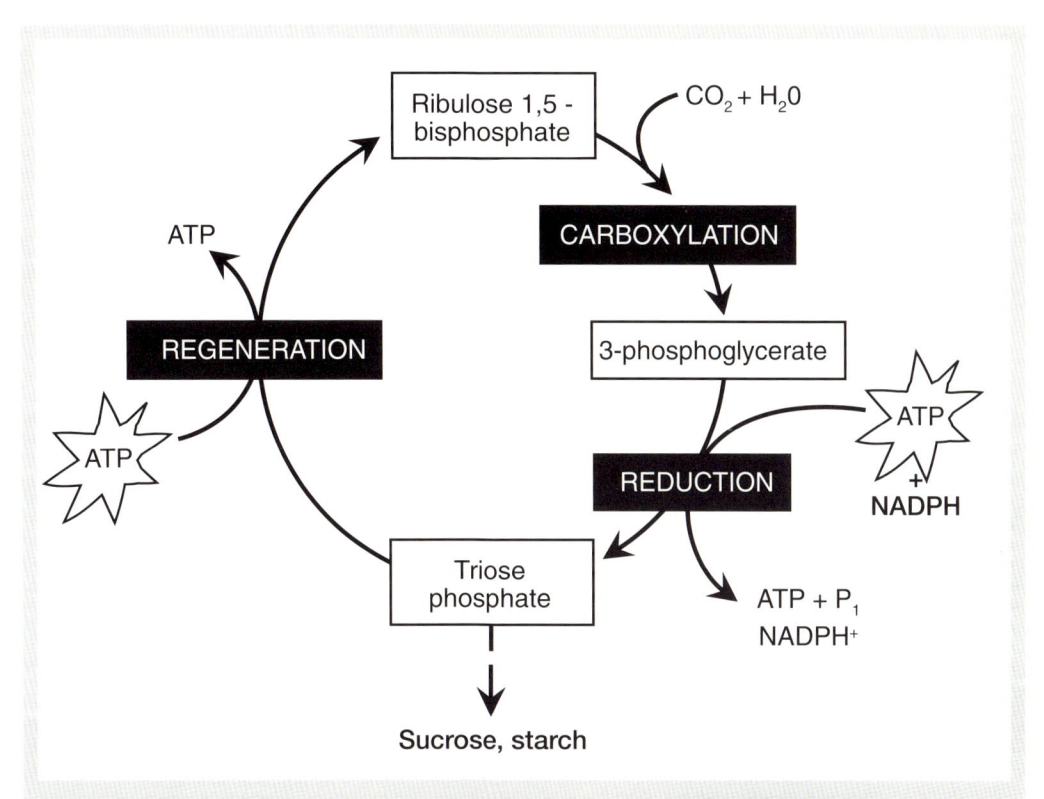

Figure 11. PCR for the C_3-plants occurs in three phases: (i) carboxylation linking CO_2 covalently with a carbon skeleton, (ii) reduction forming carbohydrates at the expense of the photochemically derived ATP, and (iii) regeneration reforming the CO_2 acceptor molecules (ribulose 1,5-bisphosphate)[6].

CHAPTER 2

In ecological modelling, the biochemical demand for CO_2 commonly uses the work of Farquhar et al.[7] They separate the diffusion of carbon dioxide into the leaves from the biochemical properties of the photosynthetic processes to give the following net photosynthetic rate:

$$A_n = min(A_c, A_q) - R_{day} \qquad (2)$$

where A_c and A_q are the gross rates of photosynthesis [µmol m^{-2} s^{-1}] limited by ribulose bisphosphate carboxylase/oxygenase (Rubisco) activity and the rate of ribulose 1,5-bisphosphate (RuP$_2$) regeneration through electron transport. R_{day} is the daytime rate of respiration [µmol m^{-2} s^{-1}]. The rates of A_c and A_q represent CO_2-limited and light-limited photosynthesis, respectively, as Fig. 12 demonstrates. This implies that under full daylight the increase of CO_2 will raise the net photosynthesis rate until the increasing supply of CO_2 no longer limits dark reactions. Under the high supply of CO_2, the increasing light will raise the net photosynthesis until the light reaction no longer limits the net photosynthesis.

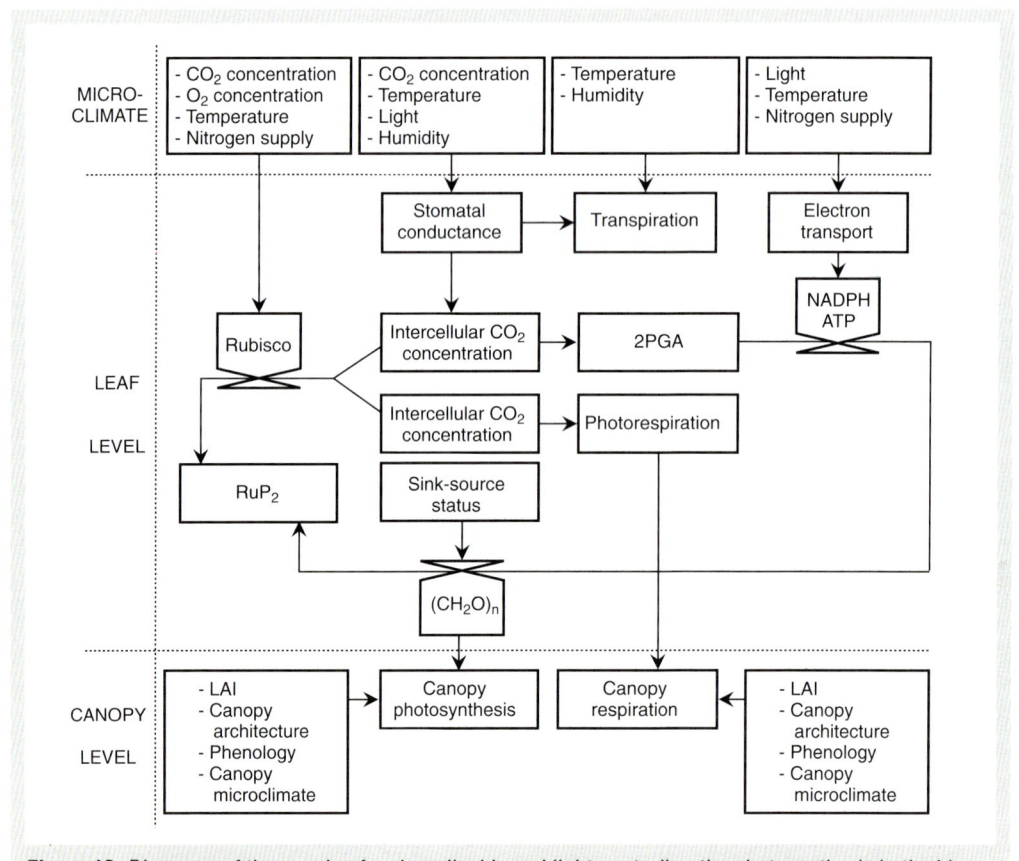

Figure 12. Diagrams of the supply of carbon dioxide and light controling the photosynthesis in the biochemical processes of photosynthesis[8].

The following equation gives the *assimilation rate limited by Rubisco activity* (A_c):

$$A_c = \frac{V_{cmax}(C_i - \Gamma_*)}{C_i + K_c(1 + Q_i/K_o)} \qquad (3)$$

where V_{cmax} [μmol m^{-2} s^{-1}] is the maximum carboxylation rate of Rubisco in the presence of saturating levels of RuP$_2$ and CO$_2$ assumed to be a linear function of the nitrogen content in the foliage[9, 10]. C_i is the CO$_2$ concentration [μmol mol^{-1}] in the inner space of a leaf, Γ_* [μmol mol^{-1}] is the CO$_2$ compensation point in the absence of daytime respiration, O_i is the intercellular oxygen concentration (210 μmol mol^{-1}), and K_c and K_o are Micháelis constants for CO$_2$ and O$_2$, respectively. The following equation gives the *assimilation rate limited by RuP$_2$ regeneration* (A_q):

$$A_q = \frac{J(C_i - \Gamma_*)}{4.5(C_i + 7/3\Gamma_*)} \qquad (4)$$

where J [μmol m^{-2} s^{-1}] is the rate of electron transport for the projected area based calculation solved as follows:

$$\theta J^2 - (Q\alpha + J_{max})J + \alpha Q J_{max} = 0 \qquad (5)$$

where J_{max} is the maximum rate of electron transport [μmol m^{-2} s^{-1}], α is the quantum requirement for electron transport, θ is the convexity factor (dimensionless with value 0.5) of the curve, and Q is the photon flux density [μmol m^{-2} s^{-1}] absorbed by photosystem II[11]. Equation 6 calculates the values of C_i at a moment t as a function of the atmospheric concentration of CO$_2$:

$$C(t)_i = C(t)_a - \frac{1.6 A_n(t-1)}{gs(t-1)} \qquad (6)$$

The factor 1.6 is the approximate ratio of binary molecular diffusion coefficients for water vapor and CO$_2$ in the air. *Stomatal conductance*, gs [m s^{-1}] can be calculated using the model developed by McMurtrie et al.[12] assuming a linear relationship to water and temperature conditions in the soil, minimum daily air temperature, vapor pressure deficit, atmospheric CO$_2$, and photon flux density.

The parameters Γ_*, K_c, K_o, and R_d depend on temperature like the values of the parameters J_{max} and V_{cmax}[13]. This implies that the net photosynthesis increases with the increasing temperature up to 20–25°C. After this, the increased level of photosynthesis decreases with increasing temperature as Fig. 13 shows. The elevated CO$_2$ level also increases the photosynthetic rate, but again the total output depends closely on the prevailing light and temperature conditions. The effect of these factors on photosynthesis is indirect, since the light and temperature also control the stomatal conductance.

CHAPTER 2

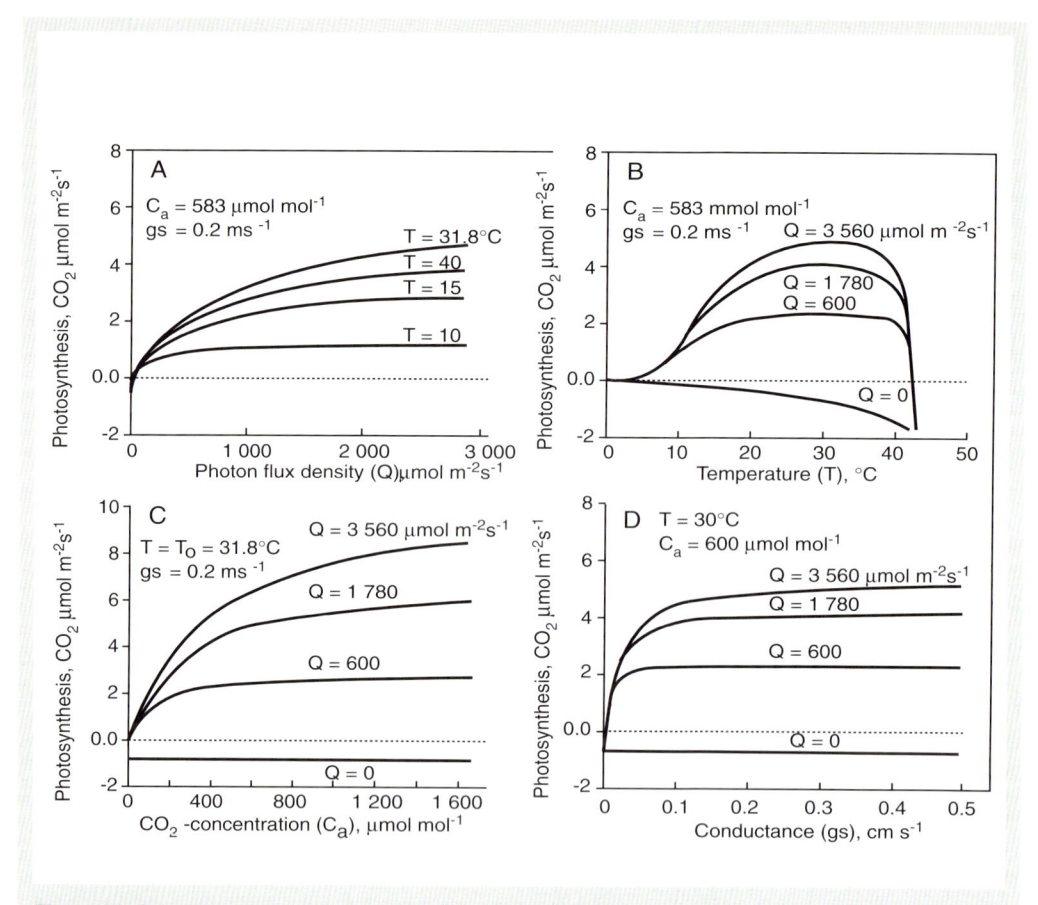

Figure 13. Schematic presentation showing the effect of selected environmental factors on the net photosynthesis rate.

3.3 Transpiration

Transpiration is a physical process where the water potential between soil and atmosphere drives the water flow from soil to atmosphere through roots, stem, and foliage as Fig. 14 shows. This flow supplies water to cells in different organs to maintain the turgor of cells and the physiological processes. The water flow simultaneously transports nutrients to foliage and other organs for use in physiological and growth processes. Equation 7 relates the transpiration rate from a unit area of foliage to the energy balance of foliage:

$$R - H - \lambda E = 0 \tag{7}$$

where R is the amount of net radiation energy per time unit [J m^{-2} s^{-1}], H is the loss of sensible heat [W m^{-2}] due to the temperature difference between air and foliage, λ is the latent heat of water [2 453 000 J kg^{-1}], and E is the transpiration rate [g m^{-2} s^{-1}]. This implies the following:

$$E = \frac{c_p \rho}{\gamma \lambda} \cdot \frac{e_s(T_l) - e_a}{r_l} \qquad (8)$$

where c_p is the specific heat of air [1 004.0 J kg^{-1} °C^{-1}], ρ is the density of air [1.220 kg m^{-3}], γ is the psychrometric constant [66.0 Pa °C^{-1}], $e_s(T_l)$ is the water vapor pressure [Pa] at the temperature of the leaf or needle, e_a is the water vapor pressure of air [Pa], and r_l is the total resistance [s m^{-1}] of the leaf and the nearby air in relation to the flow of water vapor from the foliage to the atmosphere.

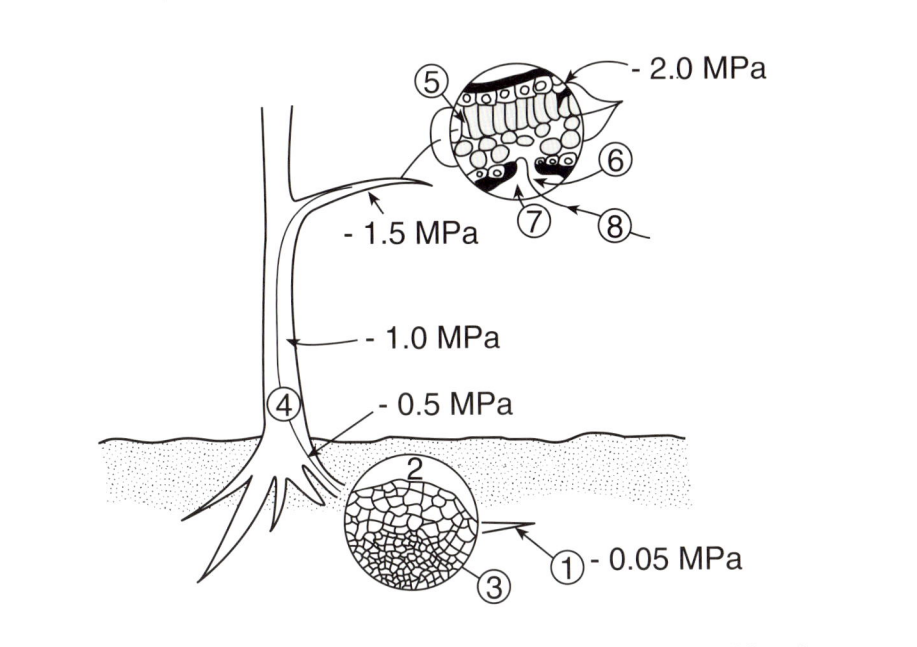

Legend: 1 = capillaric properties of soil around roots, 2 = epidermis of roots, 3 = endodermis of roots, 4 = tracheids conducting water, 5 = paremchyma tissue of leaf/needles, 6 = stomata, 7 = boundary layer of leaf/needle and 8 = difference of vapor pressure between the stomatal cavity and atmosphere.

Figure 14. Schematic presentation showing the potential difference of water in the atmosphere and the soil driving the water flow from the soil to the atmosphere through the root, stem, and foliage and showing selected structural properties of tree organs affecting the water flow.

CHAPTER 2

The previous equation shows that the rate of transpiration relates to the difference of the water vapor pressure between the space in the leaves or needles and the atmosphere as Fig. 15 shows. The stomatal conductance effectively controls the transpiration rate so that the increasing radiation increases transpiration but the decreasing water potential of foliage and the increasing water vapor deficiency in the atmosphere decreases the stomatal conductance. Since the water supply from the soil influences the water potential of foliage, the stomatal conductance will decrease whenever the amount of water around the root system decreases.

3.4 Respiration

Respiration implies chemical energy used to maintain the living tissues (maintenance respiration) and the growth of new tissues (growth respiration). The rate of *maintenance respiration* or *dark respiration* (R_m) of the existing tissues relates to the temperature conditions and the amount of the living tissues as follows:

$$R_m(t) = a \cdot \exp(k(t)) \cdot w \qquad (9)$$

In computing R_m [mol m^{-2} s^{-1}], a is a parameter, $k(t) = \ln(Q_{10}(t))/10.0$, T [°C] is the current temperature for needles, branches and stem or the temperature of the inorganic soil layer for roots and fine roots, and w is the weight [kg] of the respiring mass in a tree component. The nitrogen concentration of the mass component strongly affects the level of respiration (parameter a)[15]. The respiration rate increases exponentially with

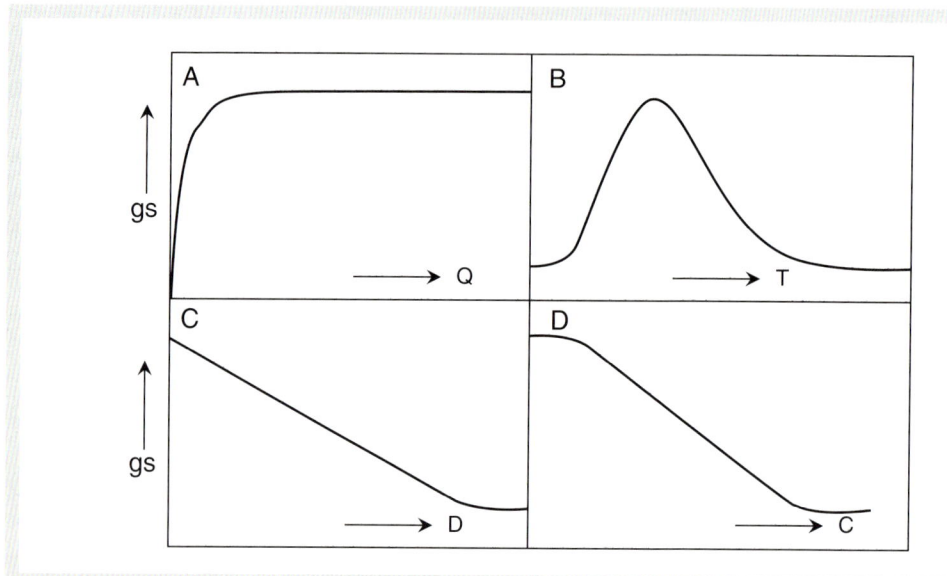

Figure 15. Schematic presentation showing stomatal conductance (g_S) responses to A: photon flux density (Q), B: leaf temperature (T), C: water vapor pressure deficit between air and stomatal cavity (D) and D: concentration of CO_2 in stomatal cavity (C)[14].

temperature as do all chemical reactions. By subtracting the values of respiration from the gross photosynthesis, one obtains the amount of photosynthates available for growth.

In *growth respiration*, glucose and other carbohydrates convert to growing cambial tissues (growing meristems) to form new tissues composed of cellulose, lignin, proteins, and fats. The synthesis of these compounds consumes energy supplied by ATP from the cell organelles (mitochondria). The different substances born in growth vary substantially in the energy needed to produce them as Table 7 demonstrates. The translocation of photosynthates and other substances to the growing tissues consumes energy, too.

Table 7. Chemical composition in % of dry weight of Douglas fir *(Pseudotsuga menziesii)* needles, branches, stem and roots, with the calculated conversion efficiency of photosynthates into dry matter (d.m.) units of various chemical compounds in growth respiration. Conversion efficiencies DWC [kg d.m. kg^{-1} CH$_2$O], the carbon production factors CPF [kg CO$_2$ kg^{-1} d.m.] and the carbon content CC of the biomass [kg C kg^{-1}d.m.] are given in terms of dry matter[16].

Mass component	Carbohydrates	Protein	Fat	Lignin	Organic acids	Minerals	DWC	CPF	CC
Needles	60	10	8	15	4	3	0.6541	0.3866	0.5061
Branches	65	3	4	24	2	2	0.6562	0.3526	0.5134
Stems	66	0.5	2	30	1	0.5	0.6443	0.3469	0.5262
Roots	62	3	5	24	4	2	0.6509	0.3644	0.5151

Conversion efficiencies in kg d.m. kg^{-1} CH$_2$O, calculated from the tissue fractions of carbohydrates (f_C), protein (f_P), fats (f_F), lignins (f_L), organic acid (f_O), and minerals (f_M), respectively.

$DWC_i = 1./(f_{C,i} \times 1.242 + f_{P,i} \times 1.704 + f_{F,i} \times 3.106 + f_{L,i} \times 2.174 + f_{O,i} \times 0.929 + f_{M,i} \times 0.050)$

$CPF_i = f_{C,i} \times 0.1701 + f_{P,i} \times 0.4620 + f_{F,i} \times 1.7200 + f_{L,i} \times 0.6589 - f_{O,i} \times 0.0110 + f_{M,i} \times 0.0730$

$CC_i = f_{C,i} \times 0.4504 + f_{P,i} \times 0.5556 + f_{F,i} \times 0.7733 + f_{L,i} \times 0.6899 + f_{O,i} \times 0.3746 + f_{M,i} \times 0.0000$

3.5 Growth

3.5.1 Components of growth

Growth is the process of converting photosynthetic products to permanent tissues of various tree organs with a specific structure. The photosynthates are translocated from foliage to the growing meristem of different organs where they are converted into the tissues of the organs as Fig. 16 shows. The translocation relates to gradients of the photosynthates with the consuming places (sinks such as shoots, or roots) and production places (source such as photosynthetising needles). The flow of photosynthates from sources to sinks occurs in phloem. This also provides a pathway for the translocation of nitrogen and mineral nutrients from the senescesing organs to the growing organs where they are bound in growth.

CHAPTER 2

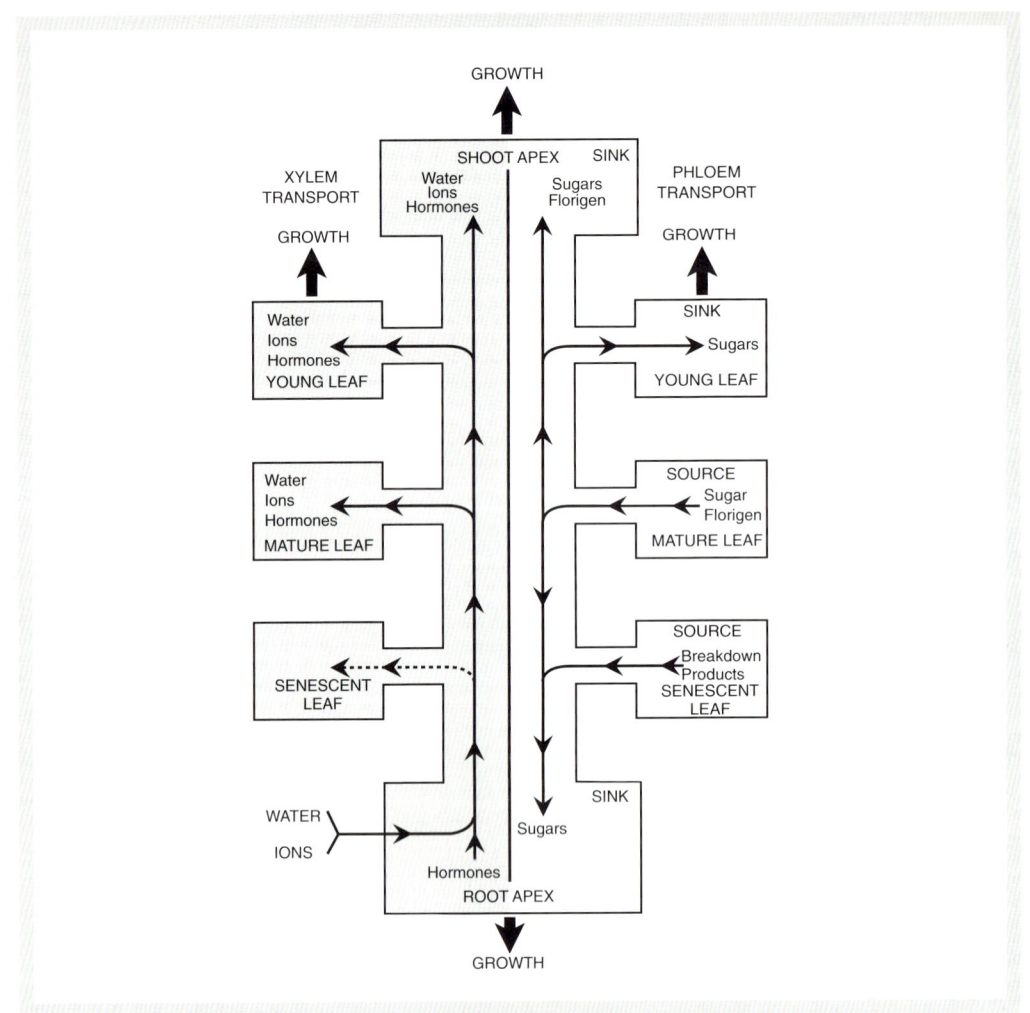

Figure 16. Schematic presentation of the mass flow in a tree and the translocation of photosynthetic products and other substances from sources to sinks.

Growth normally implies changes of mass and dimensions of tree organs. This is especially true for the stem that consists of the living surface wood (sapwood) and the dead inner wood (heartwood). The wood has a limited life, but it remains in the stem after its death with no turnover. Regarding other organs, their mass is turning over with a specific retention in the tree structure. In Scots pine, the life of needles varies 2–7 years. The shorter time is for southern locations and the longer time is for northern locations. The life of branches can range from several years to several decades. The life of fine roots is only a few weeks. This means the mass of fine roots will regenerate several times during a growing season.

Common divisions for the growth of trees are growth in height and growth in diameter. These imply activities of apical and lateral meristems. The height growth also refers to elongation of any tree organ – stem, branches, and roots – from the apical meristem. In the tree structure, a bud represents a miniature shoot as Fig. 17 shows. It contains the initial components of the leaves or needles with the nodes and internodes. After the bud opens, the leaves or needles expand and the internodes elongate. The entire shoot elongates, and the growing tissues start to perform the functions of the shoot. The main part of the shoot growth is the elongation and expanding of tissues for internodes.

In the boreal and temperate zones, the cycles of summer (high temperature) and winter (low temperature) control the growth. The *height growth* of many coniferous trees is monopodial and predetermined. This refers to the elongation of stem based on the terminal bud. Height growth ends with the formation of a bud that will burst during the following spring after a winter dormancy. In boreal conditions, the bud will very seldom flush during the same growing season (lammas growth) and form another new bud. The growth of many broadleaved trees is sympodial and continuous. This means that the height growth can start from the terminal bud, but then the growth can come from other buds. Onset of growth also initiates the formation of new buds that will break during the same growing season with full-scale growth as in the *Betula* species.

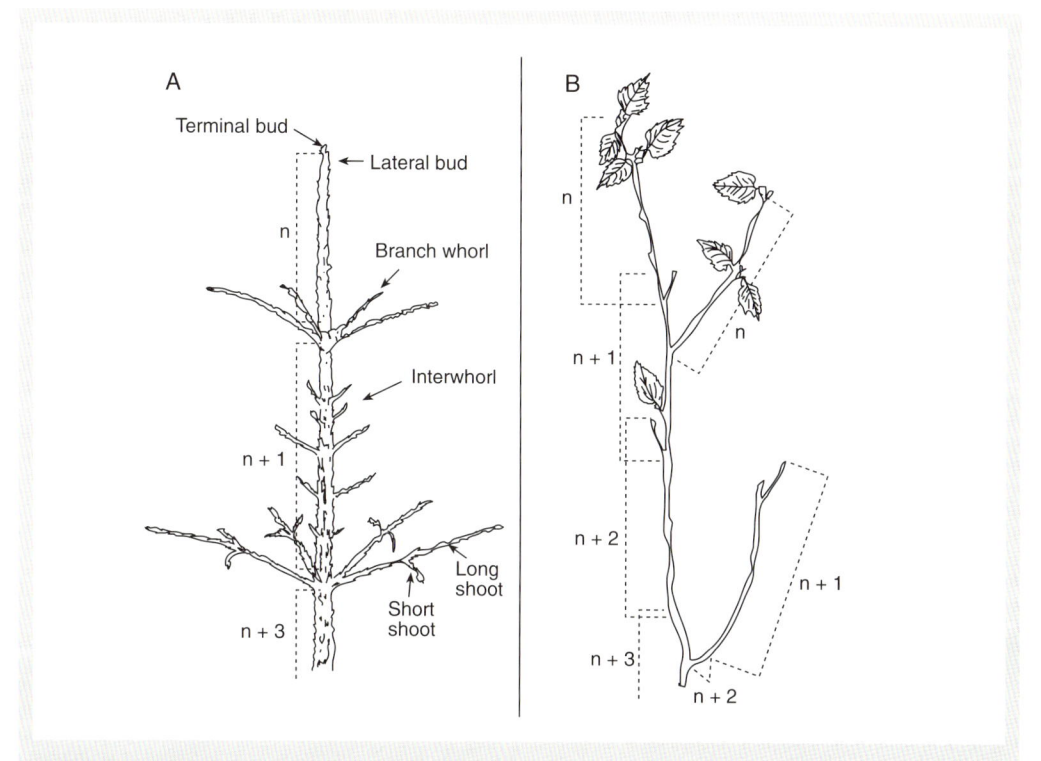

Figure 17. Schematic presentation of the height growth pattern of a coniferous (A) and broadleaved tree (B). (n=the year of birth of the shoot)

CHAPTER 2

The height growth of trees in the boreal and temperate zone closely relates to the temperature conditions during the current spring and early summer and also to the temperature conditions during the previous summer. The properties and therefore the growth potential of the bud to grow during the current growing season form during the previous growing season. The properties of the bud such as its size reflect the effects of various environmental factors on the height growth. The ability of a cell to expand initially relates to the total amount of photosynthates stored in the buds during the previous season.

As an example, the height growth of Scots pine will start when the temperature sum (a sum of daily mean temperatures higher than +5°C with the dimension of degree days (d.d.)) exceeds approximately 10 d.d. as Fig. 18 shows. Then the total elongation follows the accumulation of the temperature sum until the elongation stops at a temperature sum of approximately 500 d.d. The growth of Norway spruce will start later at 60 d.d. but continue later with completion when the temperature sum is approximately 800 d.d. Compared to Scots pine, the growth of Pendula birch and Pubescent birch will also start late at 20–30 d.d. but also continue late to 850–860 d.d. In many cases, the height growth of trees with continuous growth occurs over a longer period than the growth of trees with predetermined growth. Regardless of the annual pattern, the availability of nutrients, soil moisture, light, and temperature limit the total amount of height growth.

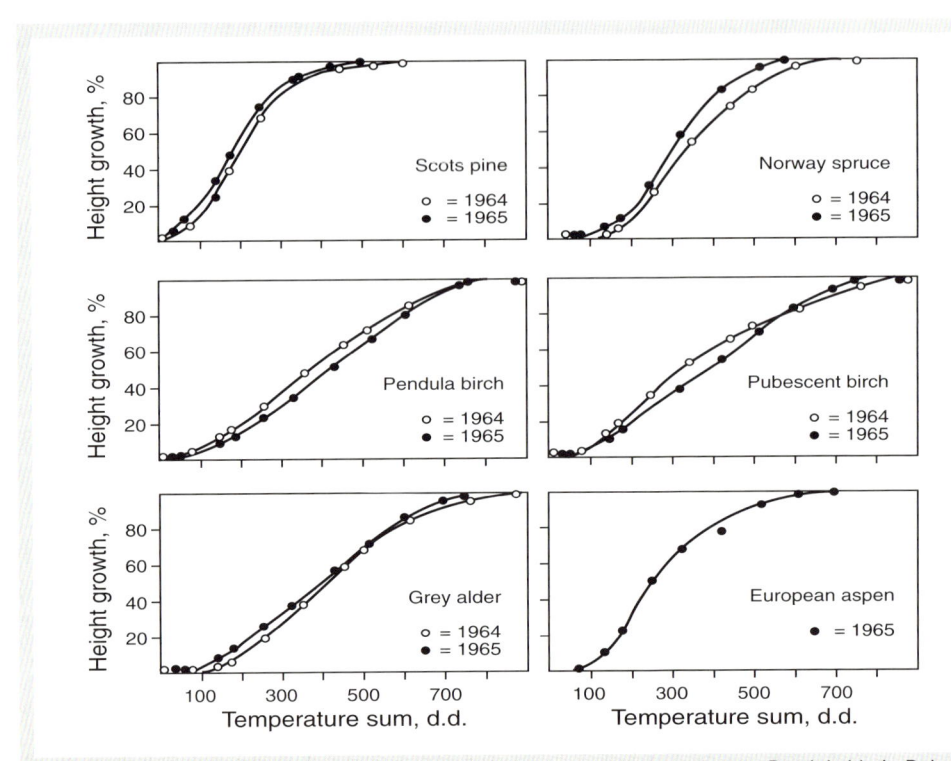

Figure 18. Seasonal course of the height growth of Scots pine, Norway spruce, Pendula birch, Pubescent birch, grey alder, and European aspen as a function of the temperature sum[17].

Trees, forests, and the forest ecosystems

In boreal and temperate conditions, radial growth starts after the bud breaks. The growth rate of the cambial cells depends on the temperature conditions similar to height growth as Fig. 19 shows. Figure 20 shows that *diameter growth* is radial and tangential. The radial growth starts from the upper part of the stem and gradually expands to the lower stem such that the formation of phloem starts earlier than that of xylem.

The radial growth starts and finishes later than the height growth. This implies that the maturation of xylem cells starts when the height growth ceases. The formation of xylem shows the shift from the formation of early wood (spring wood) with thin cell walls to the formation of late wood (summer wood) with thick cell walls. Simultaneously, the needle growth of boreal conifers such as Scots pine is at the maximum, and the new needles gradually achieve full maturation. The completion of radial growth depends on the length of the photoperiod and the reduction of day length. After the cessation of radial growth, new cells no longer form, but the cell walls can still grow thicker.

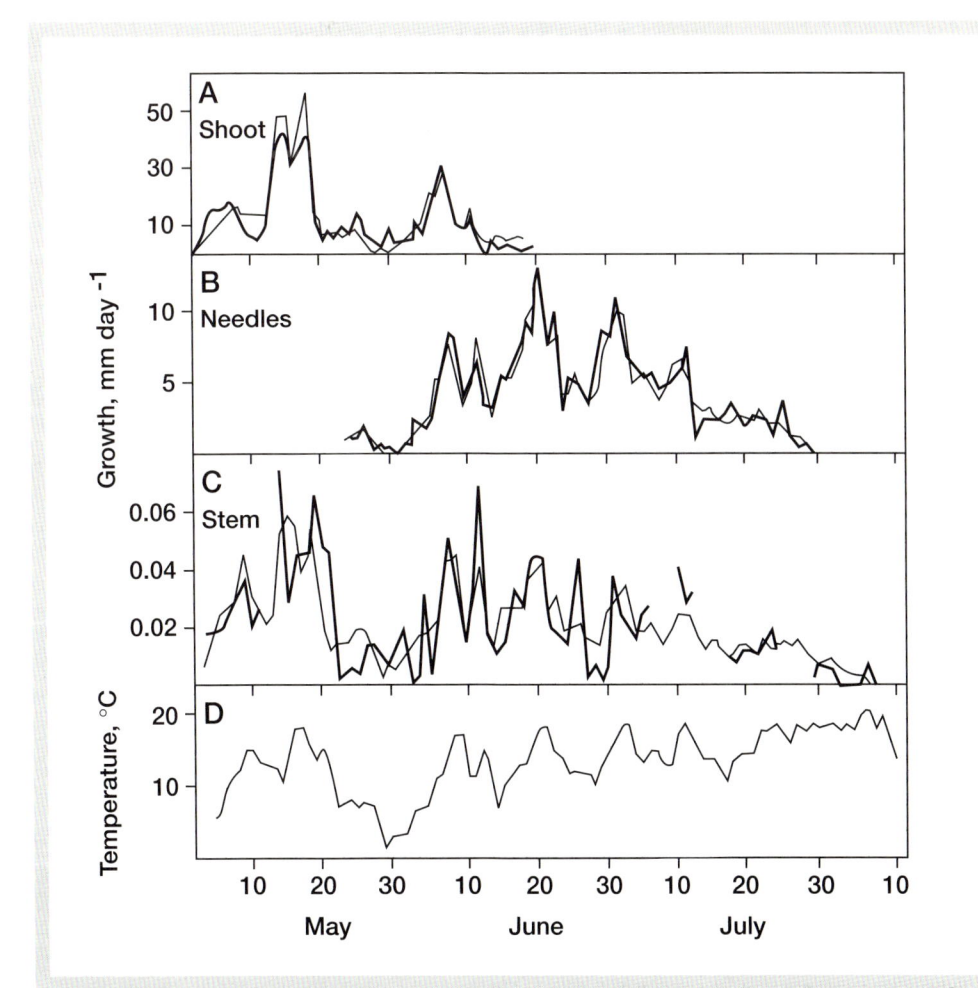

Figure 19. An example showing the seasonal effects on A: height growth, B: needle growth, and C: radial growth of Scots pine with D: air temperature[18].

CHAPTER 2

SECONDARY GROWTH

Mature phloem		
Dividing phloem	Maturing phloem	
	Phloem growing in the direction of radius	
	Dividing of phloem (basic phloem cells)	
Cambium	Dividing of cambium cells	
Dividing xylem	Dividing of xylem (basic xylem cells)	
	Xylem growing in the direction of radius	
	Maturing xylem	
Mature xylem		

Figure 20. Schematic presentation showing how xylem and phloem form from the initial cells of cambium.

During the growing season, the increase of height and diameter follows a logistical pattern with an increasing growth rate immediately after the initiation of growth and a reduction in growth after the culmination early in the growing season. This pattern also holds for the total growth over the entire life. Pioneer species like aspen and birch grow rapidly during the early phase of their life. Climax species like spruce grow slowly during the early phases but faster when they are older.

With radial growth, the expansion of the diameter gradually accelerates over the early phases of the total growth and decreases gradually when the tree turns old and mature as Fig. 21 shows. There is no distinct moment when growth ceases. The culmination of total growth occurs immediately after the closure of the canopy. This is usually earlier for many broadleaved species than for coniferous species.

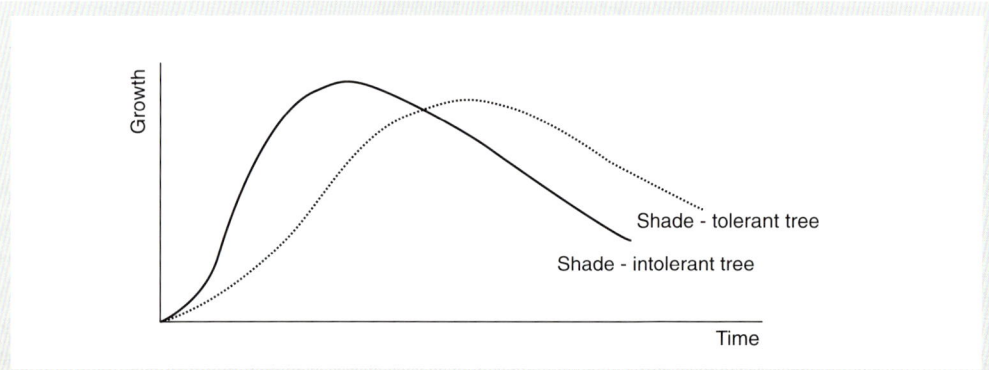

Figure 21. Schematic presentation showing the course of height and diameter growth of a shade-intolerant species and a shade-tolerant species over the life of the tree.

The following logistical function can help to describe the increase of the diameter, $dbh(t)$ (diameter at the breast height (1.3 m above soil level)), over the life of the tree

$$dbh(t) = \frac{k}{1 + a \cdot e^{b \cdot t}} \qquad (10)$$

where t is the age of the tree, k is the maximum diameter of the tree, and a and b are parameters. The first derivative of the above model gives the current annual growth of the diameter. The same model but with different parameters also holds for the mean annual growth (total growth divided by the time used to grow). The relative growth rate (annual growth divided by the total mass) decreases simulatenously as Fig. 22 shows. The pattern for radial growth also holds for the height growth.

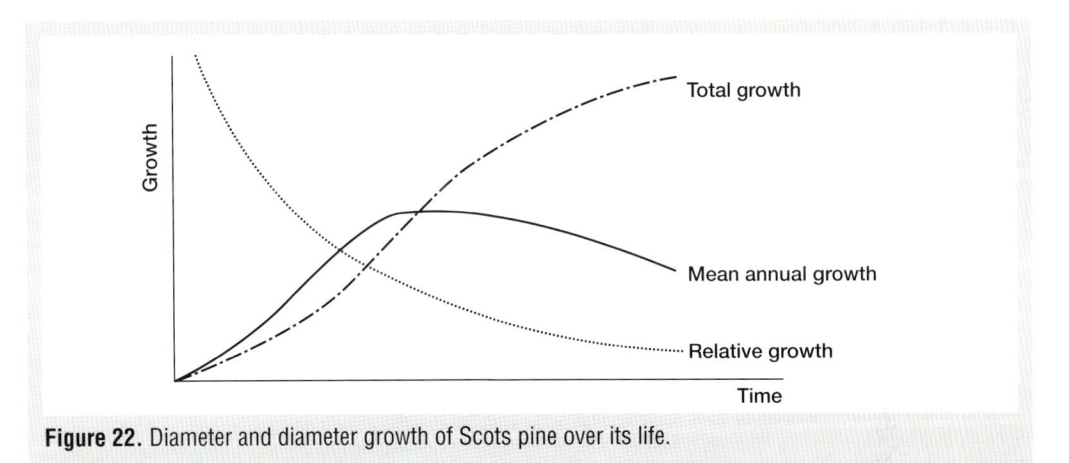

Figure 22. Diameter and diameter growth of Scots pine over its life.

3.5.2 Distribution of growth in a tree and the allometry of a tree

The close correlation of the diameter and height growth indicates a regular pattern that allocates growth over the entire tree. The annual total growth of a tree is distributed to different organs to maintain the specific mass balance between different organs. The net growth (gross growth minus litter fall) maintains the specific relationships between the masses of different organs. This implies that one can predict the existing mass of some organs such as needle mass or root mass with the help of the mass of some other organ such as stem. The allometry between different tree organs has a wide use in developing simple methods to determine indirectly the mass of different organs with the help of the stem diameter.

The allometry determines the relationship between two quantities representing the tree structure as the following equations shows:

$$y = Ax^h \qquad (11)$$

CHAPTER 2

where y is the quantity to be determined such as needle mass with the help of the quantity such as stem mass with constant A and h. This implies that the growth of the quantities x and y are relative to each other. The parameter h indicates the ratio between the relative growth rates of the quantities x and y. This leads to the following:

$$\frac{1}{y} \cdot \frac{dy}{dt} = h \frac{1}{x} \cdot \frac{dx}{dt} \tag{12}$$

Rewriting this gives the following:

$$\frac{1}{y} \cdot dy = h \frac{1}{x} \cdot dx \tag{13}$$

If y converges the upper limit of Y when $t \to \infty$, then:

$$\frac{1}{y} \cdot dy = h \frac{1}{x} \cdot \left(1 - \frac{y}{Y}\right) \tag{14}$$

Integrating this equation gives:

$$\frac{1}{y} = \left(\frac{1}{y_1} - \frac{1}{Y}\right) \cdot \frac{1}{x^h} + \frac{1}{Y} \tag{15}$$

where y_1 is the value of y when $x = 1$. Substituting with the following:

$$\frac{1}{y_1} - \frac{1}{Y} = \frac{1}{A} \tag{16}$$

gives:

$$\frac{1}{y_1} = \frac{1}{Ax^h} + \frac{1}{Y} \tag{17}$$

Then:

$$\frac{1}{y_1} = \frac{1}{Ax^h} + \frac{1}{Y} \tag{18}$$

When $Y \to \infty$, one obtains the allometric equation $y = Ax^h$ that describes the between-tree variability in a tree stand but does not include the between-stand variability. Introducing several measures of the tree structure into the calculation one can avoid this problem. A widely used form of the allometric equation is the following:

$$w = A(dbh^2 H)^h \tag{19}$$

where w is, for example, the foliage mass of tree, dbh is the diameter of the tree at the breast height, and H is the height of tree. In its implicit form, this equation includes information on the structure of the tree stand. The equation is therefore more widely applicable than the equation using only the stem diameter as an independent variable. Table 8 presents a collection of parameter values for selected tree species.

Table 8. Parameter values for the allometric equation for Scots pine[19] and birches[20] in Finland.

Tree species and mass component	Equation	Parameters A	h	r^2	S_e
Scots pine					
-Stem (w_T)	$w_T = A(dbh^2 H)^h$	0.0220	0.9403	0.995	2.43
-Bark (w_B)	$w_B = A(dbh^2 H)^h$	0.0148	0.7160	0.980	0.58
-Live branches (w_{BL})	$w_{BL} = A(dbh^2 H)^h$	0.00001	1.0540	0.956	2.75
-Dead branches (w_{BD})	$w_{BD} = A(dbh^2 H)^h$	0.0044	0.7695	0.884	0.98
-Needles (w_N)	$w_N = A(dbh^2 H)^h$	0.0002	1.056	0.928	1.06
-Stump (w_S)	$w_S = Adbh^h$	0.0095	2.1623	0.966	0.63
-Roots ($0 \geq 1cm$)(w_R)	$w_R = Adbh^h$	0.0038	2.8248	0.970	2.50
Birch					
-Stem (w_T)	$w_T = A(dbh^2 H)^h$	0.0164	0.9910	0.990	4.90
-Bark (w_B)	$w_B = A(dbh^2 H)^h$	0.0049	0.9256	0.966	3.21
-Live branches (w_{BL})	$w_{BL} = A(dbh^2 H)^h$	0.0002	1.2911	0.901	4.66
-Dead branches (w_{BD})	$w_{BD} = A(dbh^2 H)^h$	0.00001	1.5593	0.297	0.77
-Leaves (w_F)	$w_F = A(dbh^2 H)^h$	0.0004	1.0961	0.906	0.93

CHAPTER 2

3.6 Mortality

For trees, mortality can refer to tree organs or to the entire tree. The mortality of tree organs depends on the gradual senescence of foliage, branches, and roots that die and form litter with organic compounds and nutrients bound in them. Figure 23 and Table 9 show that litter represents primarily the oldest cohort of the mass accumulation in the organ. The formation of litter in tree organs therefore represents discarding of the mass of the organ following the life expectation specific to particular organs.

In Scots pines with four needle age classes, only a small amount of current-year (new) needles die compared to the one-year-old needles of which approximately 80% survive further. The mortality of the two-year-old needles is still low with a survival of 60% to the next year. From the three-year-old needles, only 40% survive further and form the four-year-old cohort. The maximum age of the needle cohort of Scots pine varies 2–7 years. The formation of litter and dead trees is the phase of the nutrient cycle where the nutrients return to the soil system and again become available for tree growth with the decomposition of organic matter.

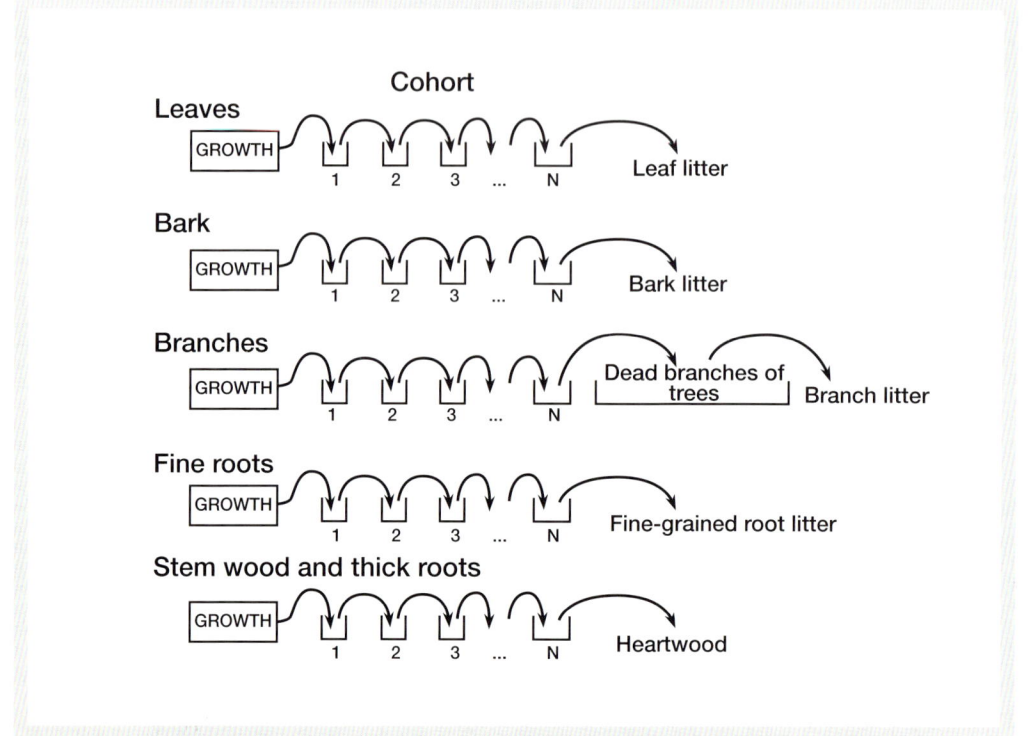

Figure 23. Schematic presentation of the formation of litter showing different organs of the tree for cohorts of the mass of an organ[21].

Table 9. Estimates for the maximum age of pine mass components before the component dies and becomes litter as a function of site fertility[19, 21].

Component	Age at death, years				
	Extremely poor site	Very poor site	Poor site	Rich site	Very rich site
Sapwood	20	15	13	12	10
Bark	15	12	11	10	8
Branches	35	30	25	20	20

Similar patterns to that for foliage also hold for living tissues of other organs, but the total life span and the survival from year to year is specific for different organs. In forestry, the death of wood in living trees is especially important, since this process controls the formation of heartwood. The death of branches and pruning of dead branches that determine the occurrence and properties of knots in wood in different parts of the stem substantially affect the wood quality. The retention of dead braches on the stem especially has a large effect on the wood quality.

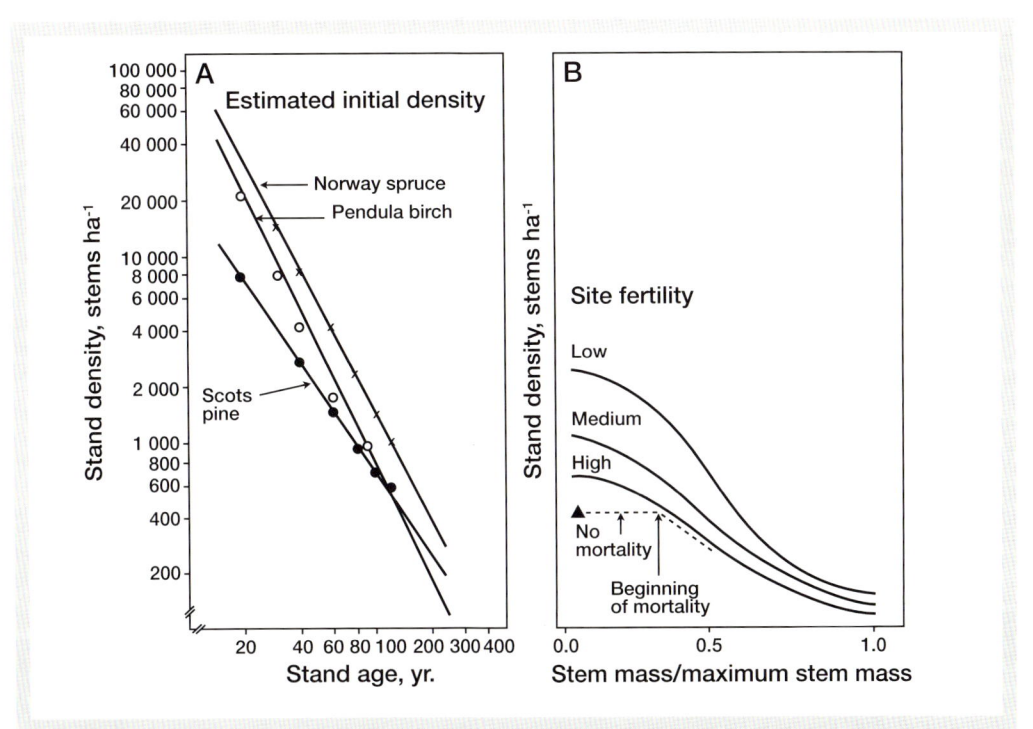

Figure 24. A: Density of Scots pine, Norway spruce, and birch as a function of stand age in the boreal zone and B: the elimination principles controlling the stand density during succession as a function of site fertility[22].

CHAPTER 2

The death of the whole tree depends on the allocation of space for growing trees with increasing phytomass. A tree with increasing phytomass needs more resources so that some trees belonging to the original cohorts will be eliminated or die during the course of the succession of tree stands as Fig. 24 shows. A particular site can support only a specific amount of phytomass that will be available to fewer and increasingly larger trees during succession. In the early phase of succession before the canopy closure, even the narrow spacing causes no excessive death of trees. The canopy closure substantially enhances the mortality due to increasingly limited space. The mortality decreases later when the phytomass of the existing trees convergences to the maximum phytomass supported by the site. The pattern holds for sites of different fertility, but the elimination process is slower when the site is less fertile.

3.7 Reproductive processes

Table 10. Examples of the age when trees are producing substantial amounts of seeds in the boreal zone.

Tree species	Age, yr.
Scots pine	60–70
Norway spruce	60–120
Pendula birch	50–60
Betula birch	40–60
Grey alder	20–30
European alder	30–40
European aspen	40–50
Willows	4–10

Regeneration refers to the capacity of trees to produce offsprings. Regeneration uses sexual or vegetative reproduction. In sexual regeneration, trees produce seeds that germinate under favorable conditions to provide established seedlings that can carry out the vital physiological processes needed for growth. Trees start to flower and produce seeds when they pass their juvenile stage and reach sexual maturity. In the boreal zone, trees start to produce a substantial amount of seeds after approximately 20–60 years. Table 10 shows that angiosperms mature earlier than gymnosperms. The female and male flowers can be in the same tree or in separate trees depending on tree species.

Sexual reproduction includes the formation of buds for flowers, flowering, formation of seeds, shedding of seed, germination of seeds, and establishment of seedlings. In angiosperms, these phases occur during a two-year period, but in gymnosperms the regeneration cycle can extend over several years. For Scots pine, the regeneration cycle covers four years as Fig. 25 shows. The sexual regeneration of Scots pine has substantial uncertainty generated by weather conditions that have considerable effect on the formation of flower buds, flowering, maturation of seeds, and on the germination of seeds and establishment of seedlings. At high latitudes and the timber line especially, the conditions for successful regeneration are minimal due to the low incidence of warm summers.

Trees, forests, and the forest ecosystems

A. Regeneration cycle of Norway spruce

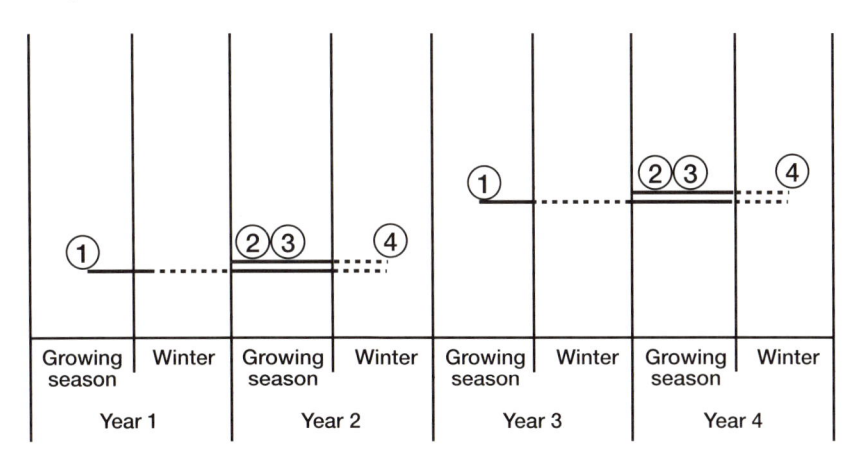

B. Regeneration cycle of Scots pine

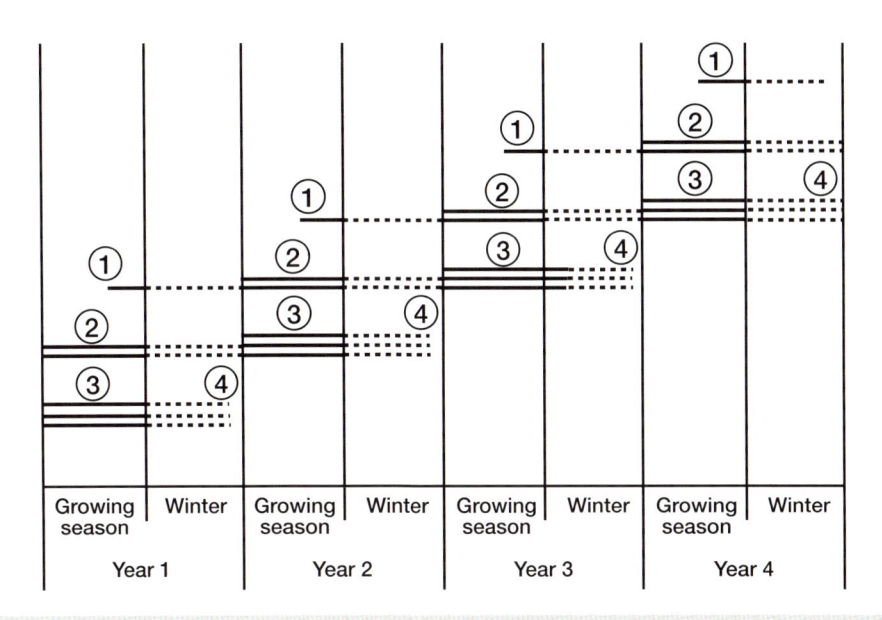

Figure 25. Schematic presentation of the regeneration cycle of Norway spruce (A) and Scots pine (B) in four-year periods. In the figure: 1: formation of flower buds, 2: flowering and pollination, 3: actual fertilization, and 4: seeds dispersal with one line marking the development during the first year, two lines marking the second year, three lines marking the third year, a solid line showing an active period, and a broken line indicating a dormant period.

CHAPTER 2

In Finland, Scots pine can only produce small amounts of mature seeds, if the long-term mean of temperature sum is less than 800 d.d. Figure 26 shows that the area from the Arctic Circle (66° N) in northern Finland to the timber line (70° N) has low summer temperatures that severely limit the regeneration of Scots pine. If the temperature sum exceeds 900–1 000 d.d. (below 66° N in southern Finland), seeds of Scots pine mature regularly and seldom does the poor maturation of seeds prolong the regeneration time. Even in southern Finland, the annual temperature variation affects regeneration, but the extent is less than in northern Finland.

Vegetative reproduction mainly occurs through the activation of dormant buds in roots and stems. This type of reproduction is particularly effective through a root system that provides a vehicle to expand the area of the site occupied by a tree. This regeneration is common among angiosperms like European aspen (*Populus tremula*). It is almost unknown among gymnosperms. In some cases, the branches will root when attached to soil. In timber line conditions, Norway spruce can regenerate through rooted branches. Even in these conditions, the vegetative regeneration of a gymnosperm is rare.

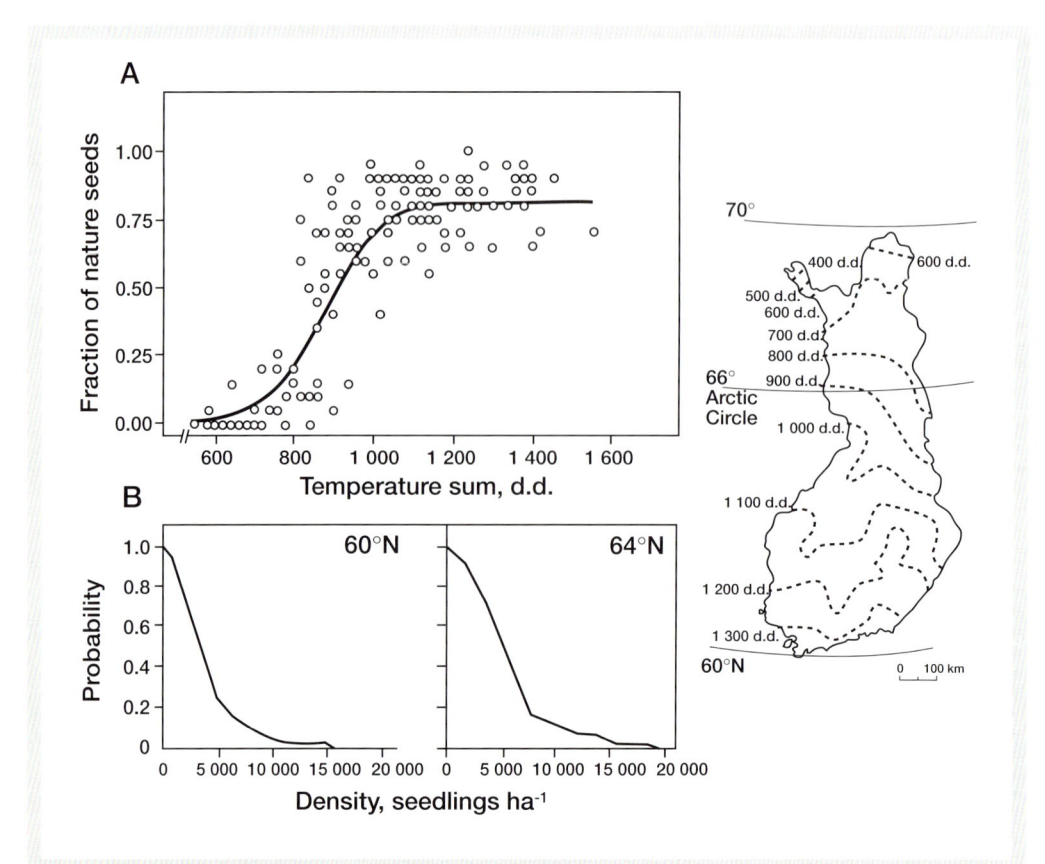

Figure 26. A: Effect of temperature conditions on the maturation of Scots pine seeds and B: probability function for selected densities of seedling stand for latitude 60° N and 64° N in Finland[23].

4 Forest ecosystem and its dynamics

4.1 Concept of the forest ecosystem

Figure 27 shows that forests are ecosystems where trees and other green organisms occupy sites and intercept solar energy under the control of climatic and edaphic factors. The solar energy flows from producers (green plants) to consumers (organisms other than green plants). In the forest ecosystem, different organisms form complex food webs. The links between different organisms are dynamic. They are the keys to the management of the forest ecosystem. Proper manipulation of forest dynamics allows production of timber or other items and maintenance of the environmental values available from forests.

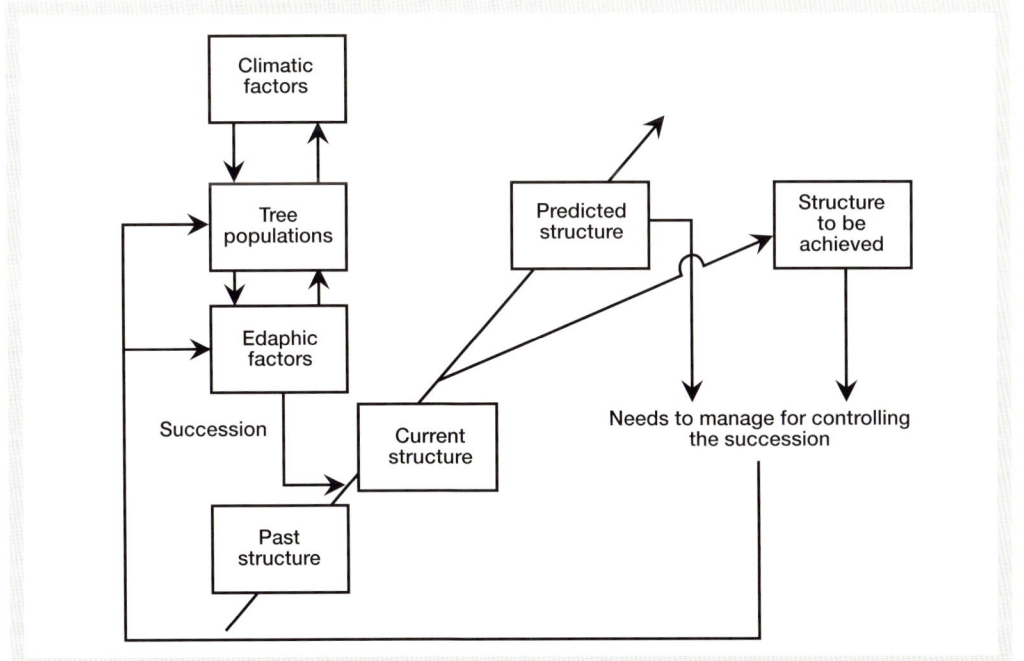

Figure 27. Schematic presentation of the structure and functioning of a forest ecosystem as an interaction between climatic and edaphic factors and the populations of organisms with the implications for management.

CHAPTER 2

4.2 Functioning and structure of the forest ecosystem

4.2.1 Short-term and long-term dynamics

Functioning and structure are concepts frequently applied to analyze the dynamics of forest ecosystems. At the physiological level, functioning refers to the metabolic processes (photosynthesis, respiration, nutrient and water uptake) and to their role in controlling regeneration, growth, and mortality of populations and communities. At the ecological level, functioning covers how regeneration, growth, and mortality modify the structure of the populations and the consequent structure of populations and communities. The long-term functioning of the forest ecosystem (succession) gradually changes the structure of the forests. This is evident in changes in the composition of tree species and the accumulation of organic matter in trees and soils. The structure of forests is therefore a result of functioning over a selected time period.

The dynamics of the forest ecosystem depend on the regeneration, growth, and mortality of trees. These factors determine the balance of tree populations and the resulting flow of energy and cycling of water and nutrients in the ecosystem. The physiological processes controlling the regeneration, growth, and mortality link the dynamics of the tree populations and therefore the dynamics of the forest ecosystem with the climatic and edaphic properties of the site as Fig. 28 shows. The structure of trees and the tree stand substantially effect the feedback to the atmosphere and the soil. Canopy closure affected by composition of tree species and the maturity of the stand controls the energy flow, nutrient cycle, and hydrological cycle.

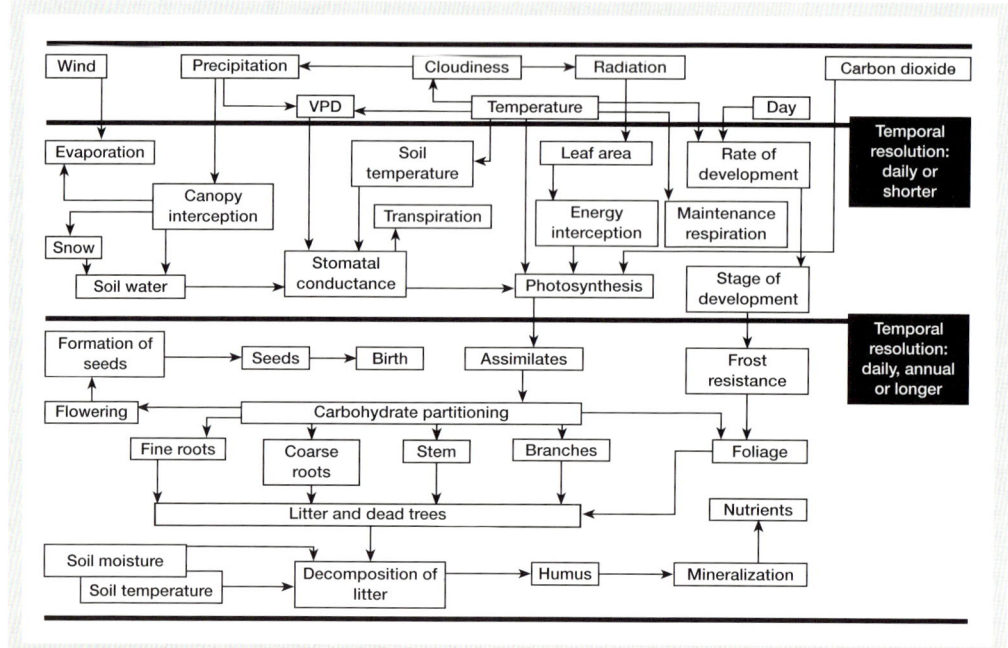

Figure 28. Schematic presentation of the structure and functioning of the forest ecosystem controlled by climatic and edaphic factors[24].

4.2.2 Feedback through energy fixation

The fixation of energy into the forest ecosystem in terms of plant growth provides the energy source for herbivores that graze in the forest – game animals or insects. Consumers and decomposers representing different levels of trophia in the energetics of the forest ecosystem use the energy bound in herbivores in several phases. The chemical energy is continually lost as heat in respiration or in the fermentation processes of micro-organisms until it reaches total exhaustion. Cycling of nutrients parallels the energy flow. The nutrients bound in plant material release during the decomposition of litter and humus. The ecosystem food webs provide the feedback controlling the photosynthesis through soil factors that depend on nutrient availability.

The physiological processes of trees occur in single trees, but they mediate the effects of trees on each other when trees are growing in a stand. A stand of trees is an assemblage of trees occupying the same site. Tree crowns and the resultant canopy in the stand modify the climate around the trees (microclimate) and the amount of precipitation falling on soil. Feedback to the physiological processes (photosynthesis, respiration, and transpiration) link the tree growth with the atmospheric and soil factors. The total foliage biomass in the crown and the canopy structure are the main determinants influencing the total energy fixation and the cycles of water and nutrients in the forested ecosystem. The density of the tree stand or the number of trees per unit area therefore has a major effect on the rate of use of resources and the total growth of the stand. This is especially true for the effect of light in the stand.

In the stand, trees shade each other from diffuse and direct radiation. Assume that tree crowns are ellipsoids over the life of the tree and that the foliage area has uniform distribution within the crown layer. Assume also that the trees have a Poisson distribution within the stand. At a given point (x,y,z) within the crown, the *transmitted fraction* (p_0) of direct solar radiation (gap probability) results from the following calculation by Oker-Blom[25, 26]:

$$p_0 = \exp(\lambda \cdot T(z, \alpha) \cdot \exp(-k \cdot LAD \cdot t(\alpha, x, y, z))) \qquad (20)$$

where the first exponential expresses the probability of a gap through neighboring tree crowns and the second exponential expresses the probability of a gap within a particular tree's crown. In addition, λ is the stand density [trees m^{-2}], $T(z,a)$ is the crown shadow area [m^2] on the horizontal plane at depth z [m, distance of crown layer from the apex of the stem] when the sun's altitude is α [radian], LAD is the leaf area density of the tree crown [m^2 m^{-3}], k is the light extinction coefficient [dimensionless], and $t(a,x,y,z)$ is the *length of the path* [m] within the crown. The following equation provides calculation of the *crown shadow area*:

$$T(z, \alpha) = \int_0^z \int_{C(z)} \exp(k \cdot LAD \cdot t(\alpha, x, y, z) \cdot k \cdot LAD/\sin(\alpha)) dx dy dz \qquad (21)$$

CHAPTER 2

where $C(z)$ is the cross-section of the crown and the horizontal plane at depth z. The radiation incidental to a crown layer is the sum of direct and diffuse radiation. The following equation provides the *mean* direct irradiance on the leaf surface at (x,y,z):

$$I_{ID} = k \cdot I_D \cdot p_0 \tag{22}$$

where I_D is current direct radiation [W m^{-2}]. *Mean diffuse irradiance* [W m^{-2}] results from integrating the relative mean direct irradiance weighted by the sky radiance over the upper hemisphere, respectively. To simplify computations, diffuse irradiance can be averaged over the horizontal plane. The following equation then provides the mean diffuse irradiance (I_{ID}) on the foliage at depth z:

$$I_{ID} = 2 \cdot k \cdot I_d \int_0^{\pi/z} \exp(-l \cdot T(z,a)) \cos(\alpha) \cdot \frac{1}{C(z)} \cdot \ldots$$
$$\ldots \int_{C(x)} \exp(-k \cdot LADt(\alpha, x, y, z)) dx dy) \alpha \tag{23}$$

where I_d [W m^{-2}] is the current diffuse radiation.

The above equations show that the properties of crown (foliage density in crown envelope, crown width, crown foliage area, and the spatial arrangement of foliage or light extinction) and the stand density (or path length) determine the radiation coming onto the foliage in different layers of the crown. The position of the sun and therefore the latitude and season have a major effect on the light conditions in the stand as Fig. 29 demonstrates, i.e., in northern conditions, the total amount of annual radiation is much smaller than in southern conditions. The same spacing of the tree stand allows much less radiation to reach into the lower crown. This limits the potential productivity of tree stands in northern conditions compared to southern ones.

4.2.3 Feedback through the nutrient cycle

Sources and cycles of nutrients

Figure 30 shows that nutrients are the result of several sources or cycles. *Geochemical cycles* represent the exchange of chemicals between ecosystems. In this case, wind or rain water carries nutrients onto the site as a dry or wet deposition. Another possibility is that calcium, magnesium, potassium, or phosphorous may weather from the parent soil. The geochemical cycles providing nutrient flow through the forest ecosystem are global.

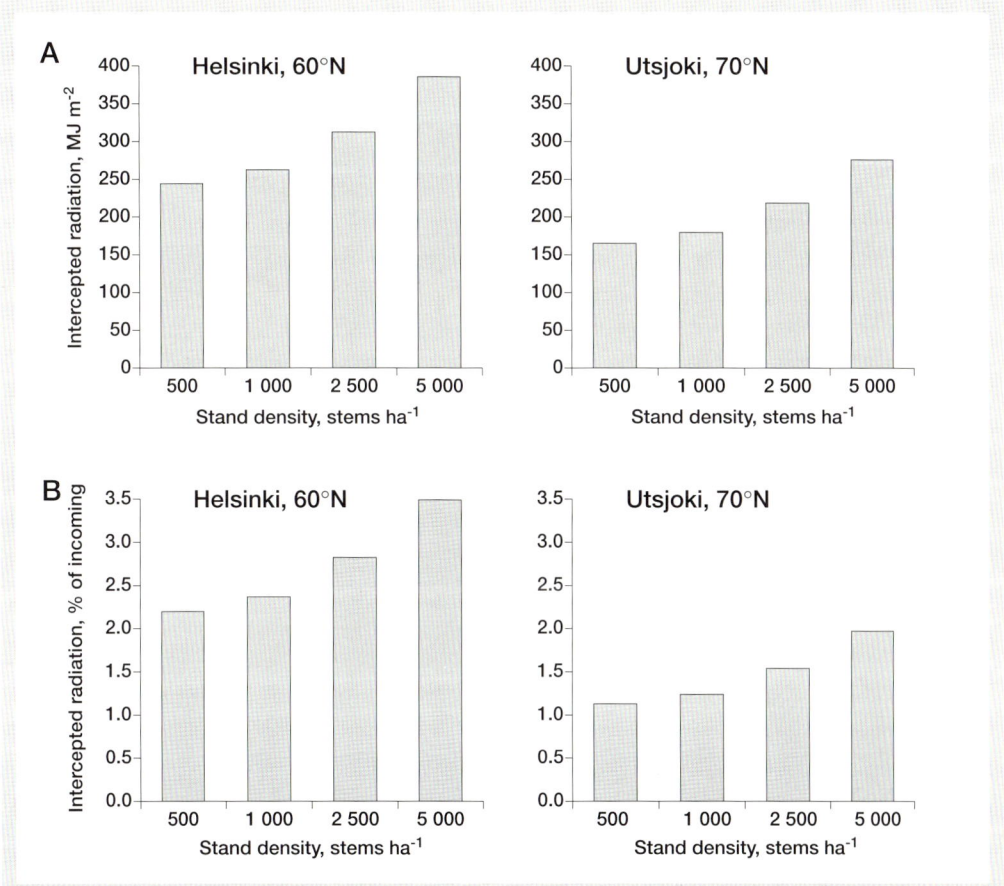

Figure 29. Example showing how the stand density and latitude affect the light conditions in a stand with A: absolute values and B: relative values as a percentage from incoming radiation.

The nutrients are lost in water flowing through the ecosystem. They cycle through the sedimentary cycle, since they will finally deposit in ocean sediment. Geological uplifts return these element for the use of plants in the form of new land surface. The temporal scale of this cycle is millions of years as is the spatial scale of thousands of kilometers. This is especially true for the sedimentary cycle that is under the control of geological processes.

Biogeochemical cycles are the exchange of chemicals within an ecosystem. An example is the interception of nitrogen by tree roots from decaying litter, the translocation to growing needles, and the return to the forest floor in falling litter. The losses of nutrient from biogeochemical cycles to geochemical cycles are small. The nutrients normally cycle in a particular ecosystem with effective retention and accumulation in the site. This implies that the temporal and spatial scales of biogeochemical cycles are much more limited than geochemical cycles.

CHAPTER 2

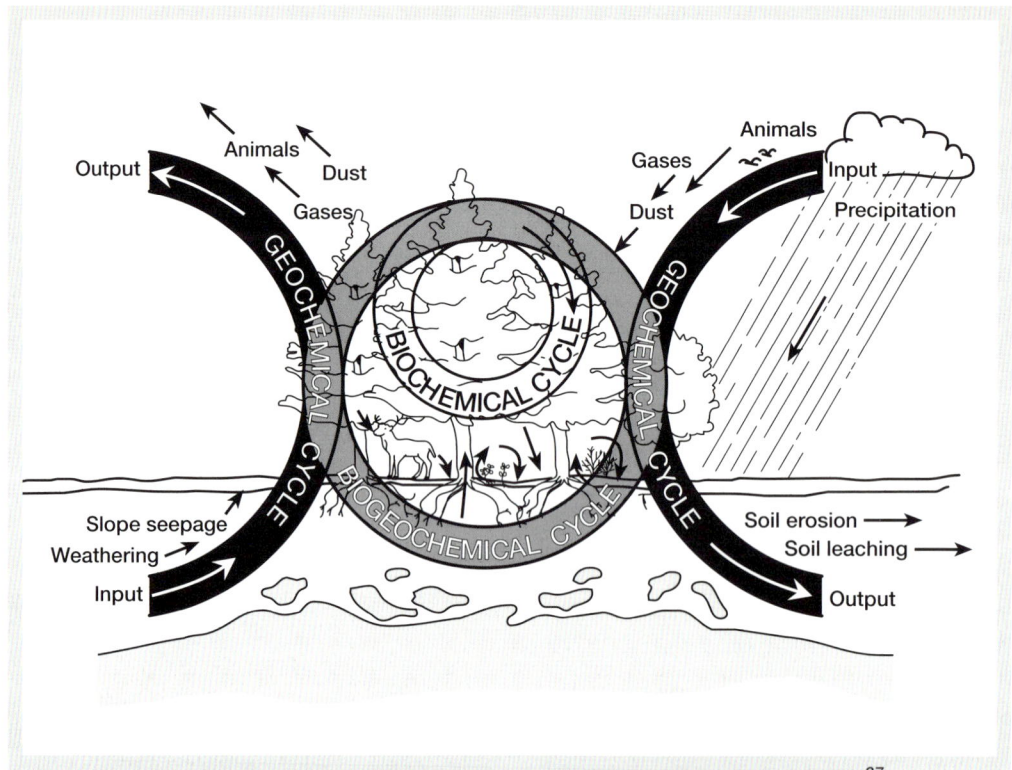

Figure 30. Schematic presentation of three major cycles of nutrient supplying plant growth[27].

Biochemical cycles provide the redistribution of chemicals within individual organisms. Plants conserve nutrients by removing or translocating nutrients from the short-lived tissues such as needles before they are shed. This means that the nutrients translocate to younger and actively growing tissues or they are stored for future use in the overall metabolic performance of the plant. The spatial and temporal scales of the biochemical cycles are very small compared with other cycles supplying nutrients for plant growth.

The geochemical cycles are typically open, and the communities representing the local ecosystem have minimum effect. This is partly true for the biogeochemical cycles, but the biogeochemical cycles also have a close relationship to the dynamics of the ecosystems. By definition, biochemical cycles represent the nutrient flow that is under the control of the eco-physiological process of plants. The biogeochemical cycles represent the source of nutrients that control the dynamics of the forest ecosystem in the temporal and spatial scale applicable in management. The nutrients in timber are those bound in the biogeochemical cycles with substantial effect on the nutrient balance of the forest ecosystem.

Decay of soil organic matter and biogeochemical cycles of nutrients

Characterization of the biogeochemical cycling of nutrients uses the within-community exchange of nutrients between trees and soil. The nutrients available in soil are bound in plant material in growth and returned in litter into the soil where they become available through the decay of the soil organic matter (litter and humus). This means that the decay of soil organic matter and therefore the return of nutrients for tree growth are the keys for the management of nutrients in the site. In natural conditions, the nutrient addition from the geochemical cycles (weathering, dry and wet deposition) is small compared to the nutrients supplied by the biogeochemical cycles. The atmospheric deposition today may be substantial as air impurities.

The decay of soil organic matter refers to the processes where litter (dead organic matter on soil with recognizable origin such as organs in the tree structure) and humus (dead organic matter on soil without recognizable origin) decomposes into CO_2, water, and nutrients. This is the mineralization of nutrients. The decay includes the three parallel processes in Fig. 31 – decomposition of organic matter through leaching, weathering, and biological decay. In leaching and weathering (physical and chemical decomposition), several organic and inorganic substances release with an increase of nutrient content of the soil organic matter. The leaching and weathering are partly associated with enzymatic processes. The biological decay represents the fungal and bacterial activity with the energy supply to these organisms. Several invertebrates graze on the soil organic matter, too. Gradually the mass of litter decreases with the conversion of the litter into humus. Finally, the litter converts into humus colloids with a life over hundreds and thousands of years.

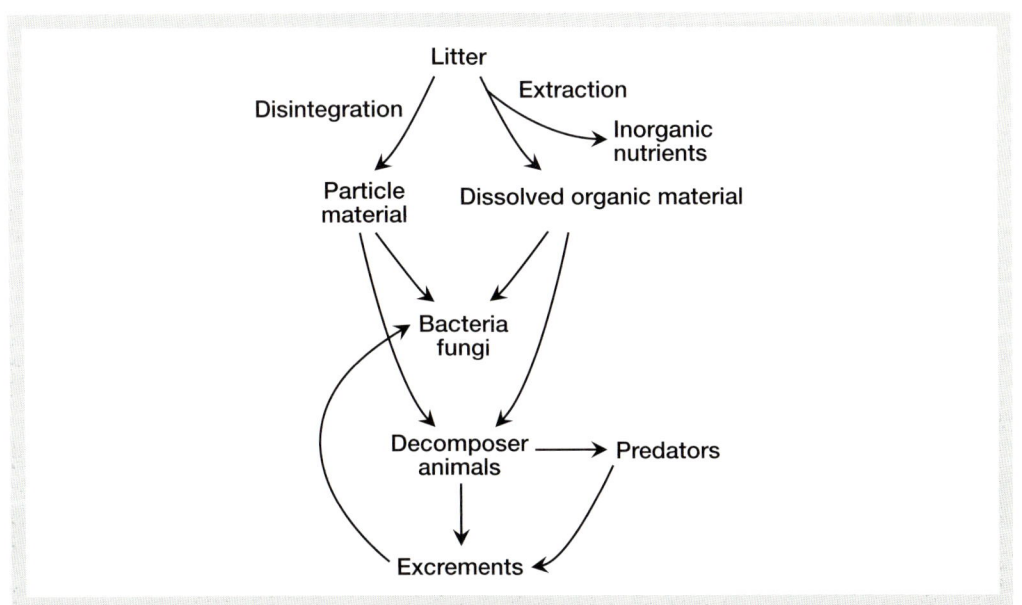

Figure 31. Schematic presentation of litter conversion into humus and final decomposition into CO_2, water, and nutrients from chemical, physical, and biological decay.

CHAPTER 2

If X is the amount of the soil organic matter per area unit at the moment t, then the change of X is the following[28]:

$$\frac{\Delta x}{\Delta t} = \text{Increase of matter} - \text{Decrease of matter} \qquad (24)$$

If L indicates the constant flow of litter on the soil as is typical for old forests with stabile conditions of the forest ecosystem, then the change of the soil organic takes the following form:

$$\frac{dx}{dt} = L - kX \qquad (25)$$

The parameter k is the rate of decay. In stable conditions, the increase and decrease of organic matter on soil are equal. This means that $L = kX_s$, where X_s is the amount of organic matter per unit area assuming a stabile ecosystem. Consequently, $k = L/X_s$.

All the organic matter on soil originates from the litter formed annually. The decay of a litter cohort (the amount of litter formed annually) can be a special case of the above equation when assuming no litter fall ($L = 0$). This implies the following:

$$\frac{dx}{dt} = -kX \qquad (26)$$

If X_0 ($t = 0$) is the original mass of the litter cohort, then the share of the original mass still remaining at the moment t ($t > 0$) is the following:

$$\frac{X}{X_0} = e^{-kt} \qquad (27)$$

This is a function of the decay rate (k) and time elapsed from the formation of the litter cohort. The quantity, quality, and timing of the litter crop have a substantial effect on how the organic matter will accumulate on soil over a long period. Consider two cases: the decay under a continuous litter fall (e.g. fall of pollen, dead bark, other minor particles, the needles in boreal conditions to a smaller extent, and a weed community with short life) and the decay under the litter fall at regular intervals (e.g. the fall of leaves and dead needles in boreal forests). In the former case, $dX/dt = 0$ and $k = L/X$, assuming $t \to \infty$. This gives the following:

$$\frac{dX}{\frac{L}{k} - X} = -kdt \qquad (28)$$

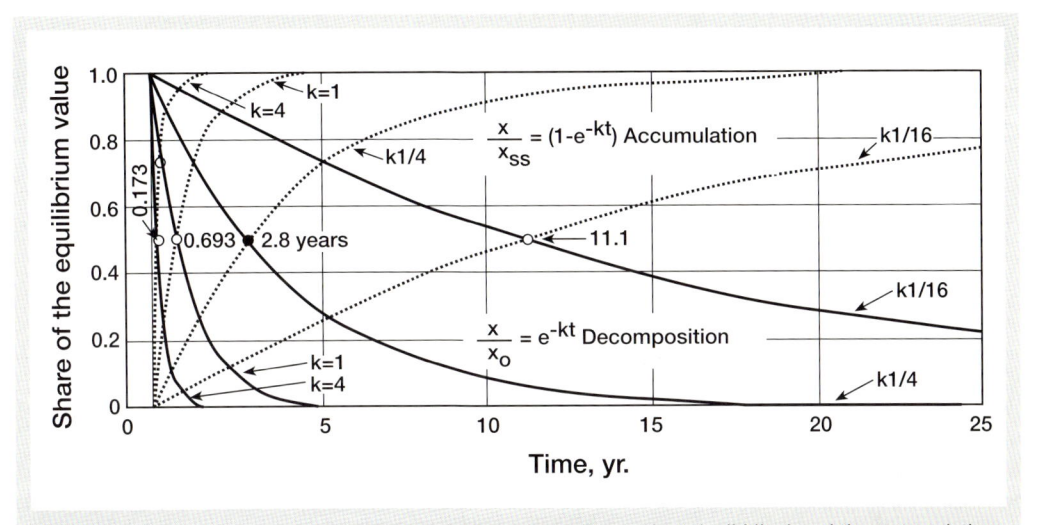

Figure 32. The remaining amount of the original mass of a litter cohort (solid line) and the accumulation of the organic matter on soil assuming the continuous litter fall (dotted line) as a function of decay rate (k) and time (t)[28].

Integration of the previous equation gives:

$$\ln\left(\frac{L}{k} - X\right) = kt - constant \tag{29}$$

Now assume that there is no organic matter on the forest soil after an intensive forest fire ($X = 0$, $t = 0$). The amount of the organic matter on the soil at a moment t is the following:

$$X = \left(\frac{L}{k}\right)(1 - e^{-kt}) \tag{30}$$

This equation implies that the accumulation of organic matter will be enhanced rapidly if the decay rate decreases as Fig. 32 shows. For example, the mass of litter cohort will decrease by half in 2.8 years with $k = 0.25$ but in 11.1 years with $k = 0.125$. The same time is necessary to achieve the balance between the litter fall and the decay of litter — the stabilised amount of organic matter on the soil.

With the litter fall at regular intervals, Eq. 31 gives the amount of the organic matter on soil (J_n) at the year n indicated in Fig. 33:

$$X = \left(\frac{L}{k}\right)(1 - e^{-kn}) \tag{31}$$

CHAPTER 2

where $k' = X/X_0 = 1 - X/X_0 = 1 - e^{-kt}$ with $k = \ln(X/X_0) = -\ln(1 - k')$. During the time between two preceding crops of litter, the amount of the matter on soil will achieve the following value:

$$X = \left(\frac{L(1-k')}{k'}\right)(1 - e^{-kn})$$

(32)

The value of the parameter k' can be estimated when the value of the litter crop and the amount of the organic matter on soil approaches the value T; i.e., $k' = L/T$.

The decay rate of soil organic matter (k) is a function of the site quality and the quality of the litter. Table 11 shows that the decay rate will increase under higher temperatures and a larger supply of water and nutrients. The decomposition of litter with a higher amount of nitrogen in relation to lignin will be higher compared to the litter with a low nitrogen/lignin ratio. These effects show the influence of the amount of the decomposing organisms and the structure of the communities of the decomposing organisms on the decomposing rate. The micro-flora has the dominant role in the boreal conditions, and the mesofauna and bacteria have the dominant roles in the temperate conditions. In the tropical conditions, many vertebrates also have a substantial effect on the decay of soil organic matter.

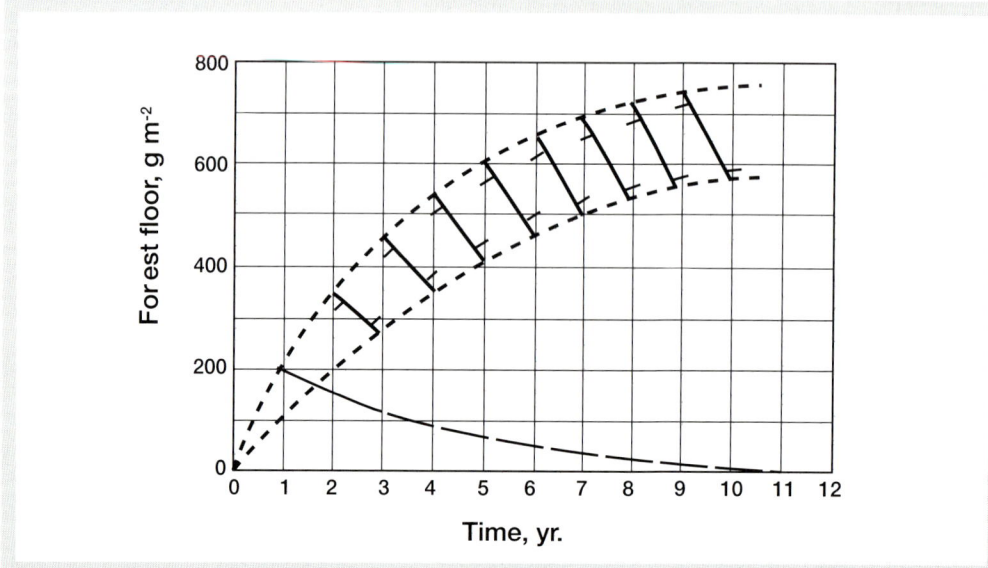

Figure 33. Example showing how litter will decompose and organic matter will accumulate on the forest soil under the litter crop at regular intervals with solid line indicating the mass of a single litter cohort and the dotted lines indicating the accumulation of organic matter at the values k=0.25 and k=0.288 with the value of L=200 g m^{-2}. The line down to the right indicates the weight of a single cohort of litter.

Table 11. Selected factors affecting the decay rate of soil organic matter in general.

Factor	Effect on decay
Environment	
Higher temperature	Increasing
Higher moisture	Increasing
Higher nitrogen supply	Increasing
Quality of litter (tree species)	
Higher lignin (L)	Decreasing
Higher nitrogen (N)	Increasing
High L/N ratio	Decreasing
Decomposing organisms	
Higher ratio of fungi	Decreasing
Higher ratio of bacteria	Increasing
Higher ratio of invertebrates and other grazing animals	Increasing

The variability of the decomposing organisms primarily depends on the effects of *temperature*. This is the most dominant factor controlling the decay of soil organic matter. The accumulation of organic matter in soil in the vegetation zones depends closely on the temperature conditions. The energy supply and the physical conditions of soil therefore have substantial influence on the decay of organic matter as the following equation for soil temperature shows:

$$\frac{\delta E}{\delta t} = L_f \cdot r_i \cdot \frac{\delta w_i}{\delta z} = \frac{\delta}{\delta z}\left[k_h \cdot \frac{\delta T}{\delta t}\right] - \left(C_w \cdot \frac{\delta(Tq)}{\delta t}\right) + s_h \qquad (33)$$

where E is the heat capacity [J m^{-3}] of the inorganic layer of soil, C_w is the volumetric heat capacity of water [J m^{-3} °C^{-1}], k_h is the thermal conductivity [W m^{-1} °C^{-1}], L_f is the latent heat of freezing [J kg^{-1}], q is the water flow [mm day^{-1}], s_h is the sink/source term for heat [J m^{-3} s^{-1}], T is the temperature [°C] of the inorganic soil layer, and z is the thickness [m] of the inorganic soil layer.

In the vegetation zones, Fig. 34 shows the existence of an inverse correlation between the content of nutrients in falling litter and the mass and nutrient content of the forest floor. In moist tropical forests, high litter fall with large nutrient content equates to a small accumulation of organic matter on the soil. This is contrary to the boreal forests where large amounts of organic matter and nutrients will accumulate on soil even under small litter fall. In the moist tropics, most nutrients are bound in the phytomass of trees. In the boreal conditions, the soil organic matter provides a large reserve of nutrients that can be available for tree growth with the help of proper soil management.

CHAPTER 2

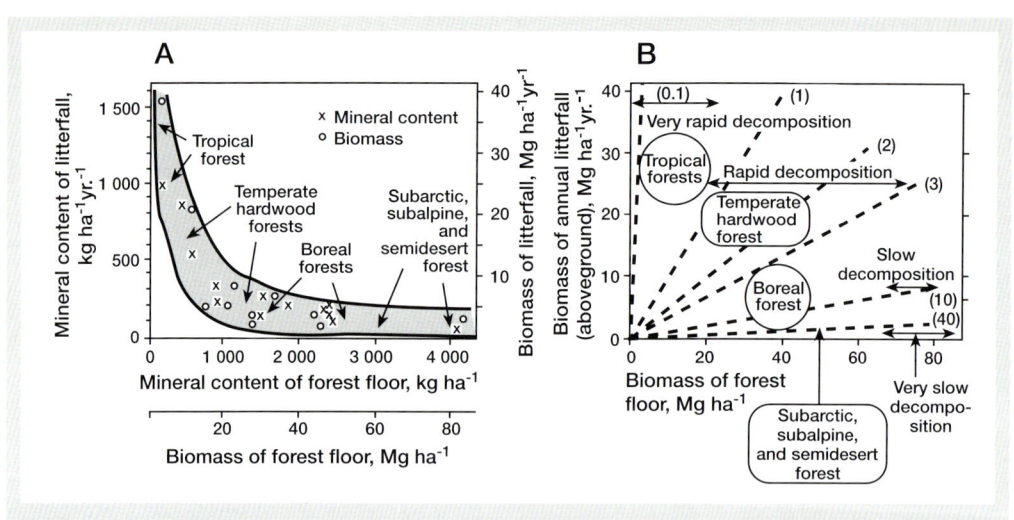

Figure 34. Accumulation of organic matter in soils over different vegetation zones showing A: relationship between litter fall mass and nutrient content, and forest floor mass and nutrient content and B: rate of decay of litter in terms of the ratio between the litter fall and the mass of forest floor with the numbers in the brackets indicating the time in years needed for the complete decay of a litter cohort[27].

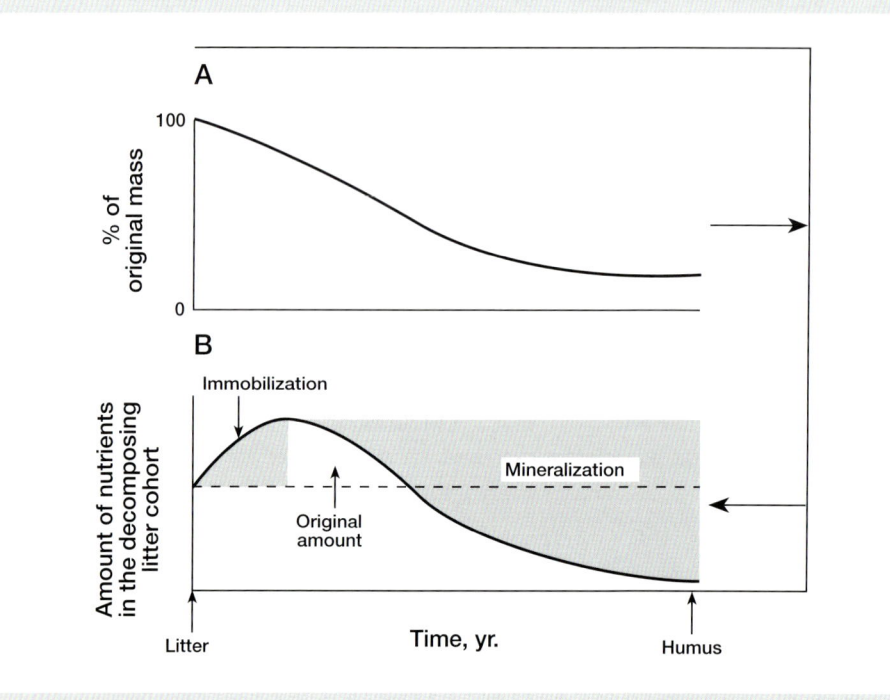

Figure 35. Schematic presentation of the weight loss of litter and humus in decay with implications on the content of nutrients in organic matter and the amount of nutrients to be released in the decay[20] with A: the weight loss of a litter cohort and B: mineralization of nutrients during the weight loss of the litter cohort.

Decomposition is a process where the nutrients bound in the organic matter of soil become available again for plant growth. The same nutrients bound in litter go through the different phases of the decay. This means that the content of nutrient in decaying matter will increase during the course of the decay due to the decreasing mass of the litter cohort. Figure 35 shows that the availability of nitrogen drives the decay – immobilisation of nitrogen. During the final phases of decay, the nutrients bound in organic matter will be released. This means that nutrients will be mineralised earlier when the decay rate is higher. The minerals in herb and grass litter will become available earlier than those bound in needle and woody litter.

Nitrogen has a close association with the biochemical processes that drive photosynthesis. The photosynthetic activity of trees is linearly proportional to the nitrogen content of the foliage. This links the productivity of trees to the within-stand nitrogen cycle and the availability of nitrogen. The nitrogen content of foliage increases with increasing site fertility. The increasing availability of nitrogen from 0.8% to 4% of dry weight as is typical for Scots pine is an example of the doubling of the maximum photosynthesis provided no other factors limit photosynthesis. The foliage mass of trees will simultaneously increase with further increase of the productivity. The site can support more biomass and therefore grow with the increasing availability of nitrogen as discussed in the section on nutrient management.

Balance of nutrients in the forest ecosystem

In the *biogeochemical cycle*, the amount of nutrients in the forest ecosystem (storage) balances the input and output of nutrients from several cycles of nutrients as follows:

$$Storage(t) = Storage(t-1) + Inputs(t) - Outputs(t) \qquad (34)$$

or

$$Inputs(t) - Outputs(t) = \Delta\, Storage(t) \qquad (35)$$

where *Storage(t)* indicates the change of the amount of the nutrients at the moment *t*. Table 12 shows that the integration of the change of the amount of nutrients over a selected period provides the amount of nutrients currently existing in the site.

Table 12. Input and output processes of nutrient and the storage of nutrients in the forest ecosystem.

Nutrient stores as defined by the balance between increasing and decreasing processes	Processes and explanations
Nutrients bound in vegetation - Increasing stores of nutrients	Growth, nutrient uptake, biological nitrogen fixation, and internal cycle of nutrients
- Decreasing stores of nutrients	Formation of litter, mortality, burning, predation, harvesting
Litter and humus - Increasing stores of nutrients	Litter coming to the ground, microbial activity, fertilization
- Decreasing stores of nutrients	Erosion, burning, decomposition and mineralization, harvesting
Available nutrients of the soil - Increasing stores of nutrients	Decomposition of litter and humus and mineralization of nutrients, burning, precipitation
- Decreasing stores of nutrients	Nutrient uptake, erosion, dentrification, nutrient immobilization, leaching

The *inputs* to the ecosystem include the weathering of the parent material in the site, dry and wet deposition of different compounds from the atmosphere, biological fixation of nitrogen, nutrients carried to the site by the surface flow, and fertilization whenever applied. The *outputs* of nutrients from the ecosystem similarly include leaching of nutrients in surface flow and in percolation (water flow across the soil profile down to ground water) and removal of nutrient in the harvested biomass outside the site. The nutrient *storage* in the forest ecosystem includes the nutrients bound in the biomass of organisms (trees and other green plants, other organisms) occupying the site, nutrients in dead organic matter on the soil surface (litter and humus), and soil profile (humus). In natural conditions, the nutrients annually enter during weathering, and dry and wet precipitation into the forest ecosystem amount to a few kilograms per hectare that the output in leaching and instream waters balances.

In the *biogeochemical cycles,* the nutrient balance consists primarily of the within-stand balance between the nutrient uptake of trees and the return of nutrients to forest soils in litter fall. The *translocation of nutrients* from senecesing tissues to living tissues substantially affects this balance, since the internal nutrient cycle will satisfy a substantial part of the nutrient requirements of the growing trees as Table 13 shows. In natural Scots pine stands, the internal cycle of nitrogen could comprise nearly the same amount of nitrogen supply as the uptake through the root system. An indication of this is the large amount of litter fall compared to the total growth of tree stand. This means that the harvest of timber could have a substantial effect on the nutrient balance by removing the nutrients bound in timber from the site.

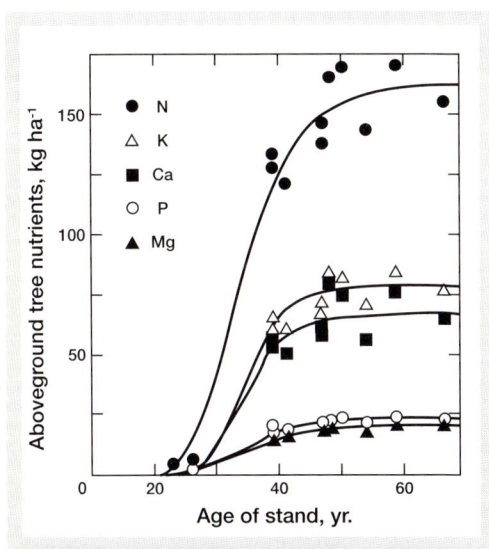

Figure 36. Accumulation of selected nutrients during the postfire development of Jack pine stands in New Brunswick, Canada[29].

In the example presented in Table 13, the nitrogen balance is positive mainly due to the management regime that includes no thinning but clear cutting at the end of the selected rotation. If using regular thinnings, the balance over the rotation would most probably be negative as Fig. 36 shows. The uptake of nitrogen bound in the tree biomass closely depends on the growth rate of tree population with the implication that the early thinnings will remove relatively more nitrogen from the site than the final cutting. This is because the thinnings will substantially decrease the return in litter fall. The mass of younger trees contains relatively more nitrogen than the mature wood of older trees. This will further enhance the loss of nutrients outside the forest ecosystem when applying thinnings.

Table 13. Productivity, litter fall, and nitrogen cycle in a Scots pine stand over 100 years with the harvest of the bole wood as clear cut at the end of the rotation using the model of Kimmins and Scoullars[21].

Characteristic	Total over 100 years, Mg ha^{-1}
Productivity of tree stand	
- Stemwood	1 170
- Harvest yield	195
- Litter fall	840
Dynamics of nitrogen	
Increasing processes	
- Dry and wet deposition	0.3
- Biological fixation	0.1
- Return to soil in litter	2.5
- Translocation from dying tissues and reuse	2.8
Decreasing processes	
- Uptake to tree biomass	3
- Leaching from site	0.3
- Timber harvest	0.2
Balance	0.2
Productivity per unit of nitrogen	0.4

4.2.4 Feedback through the hydrological cycle

The hydrological cycle encompasses the feedback where soil factors control the energy fixation and the consequent dynamics of the forest ecosystem. The hydrological cycle is a physical phenomenon with links to tree functions through the effects of water availability on photosynthesis and other physiological processes. The hydrological cycle also controls the nutrient supply when providing the vehicle to carry nutrients to different tree organs.

The within-stand hydrological cycle is a part of the global hydrological cycle. At the local level, it is therefore more open than the nutrient cycle. It is comparable to the geochemical cycle, since it is open with short-term retention of water within a particular forest ecosystem. In this context, one or several ecosystems form watersheds where some precipitation will evaporate into the atmosphere and some will percolate into the ground water or flow over land (surface flow) through river systems to the oceans. The retained precipitation indicated by the balance of water in the watershed defines the moisture of the soil and the water available for tree growth. The properties of the ecosystem forming the watershed determine the properties of the watershed. The water balance of the separate forest ecosystems therefore determines the properties of the entire watershed as Fig. 37 shows.

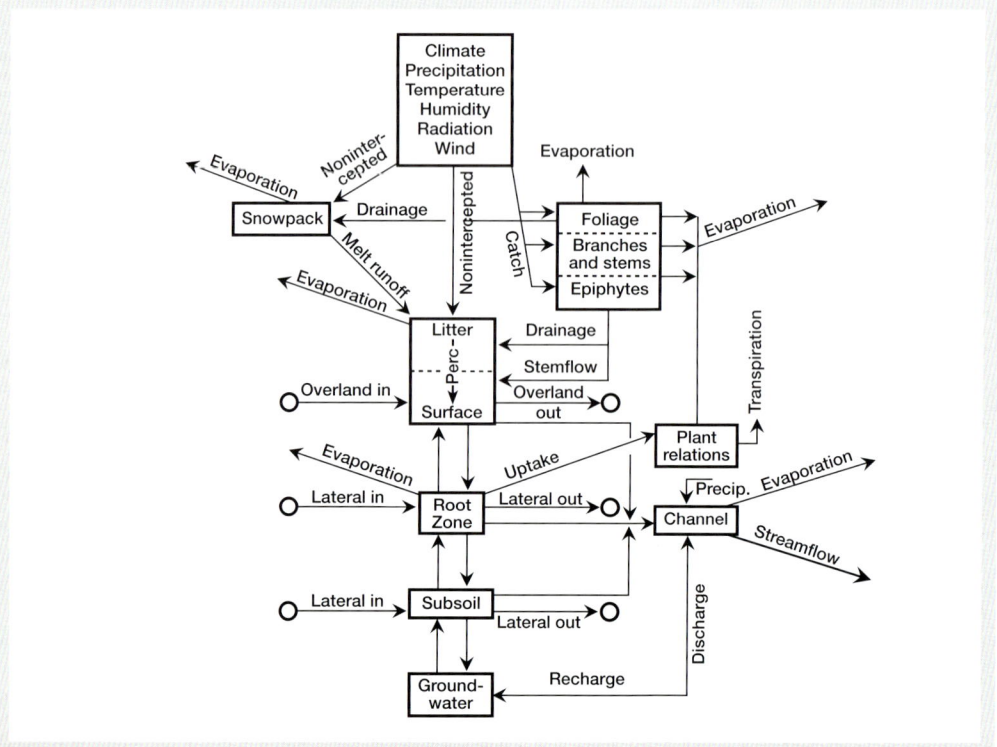

Figure 37. Structure of the water fow through a forest ecosystem with the lateral and vertical movement of water through storage into stream channels and into the atmosphere by evaporation and transpiration[30].

In a forest ecosystem, the foliage and other components of the above-ground biomass retain water. Water also comes from snow cover, soil organic matter, soil profile, and ground water. The amount of water in all these storages is a balance between precipitation, evapotranspiration, and inflow and outflow of water to the ecosystem. The water balance may be present in the whole ecosystem or as the separate storage of water as part of the entire water balance as follows:

$$WaterStorage(t) = WaterStorage(t-1) + Precipitation(t) + ... \qquad (36)$$
$$... + Inflow(t) - Evapotranspitation(t) - Ourflow(t)$$

where *t* is time. In the long term, the inflow and outflow balance. The precipitation entering the soil surface and the evaporative and transpirational water losses (evapotranspiration) with the percolation (downflow through the soil profile) will therefore determine how much water will be present in different layers of the soil profile and available for tree growth as the following shows:

$$AW = \frac{W - W_{min}}{W_{max} - W_{min}} \qquad (37)$$

where AW (0–1) is the fraction of the field capacity, W is the volumetric water content [m^3 m^{-3}] in the soil layer, W_{max} is the water content [m^3 m^{-3}] at field capacity – indicating the storage is full with water available for plants, and W_{min} is the water content [m^3 m^{-3}] at the wilting point – indicating the water content with no water available for plant growth.

The effects of precipitation and the feedback through the water conditions on the tree growth interact in several ways with the effects of the climatic factors and the properties of the tree populations on the productivity of the forest ecosystem. The water entering the soil surface is equal to the precipitation, if no trees are on the site. When trees occupy the site, the tree crowns or the canopy of the tree stand – the total cover of the soil formed by the foliage of the tree crowns – intercepts some precipitation. When the precipitation exceeds the interception capacity of the canopy and the evaporative loss of water, water is stored or passes through. The throughfall is the amount of water entering the soil surface and the water potentially available to recharge the water in the soil profile.

Each day the *amount of water in vegetation* (AWV, mm) is the daily interception ($INTERCEP$, mm day^{-1}) minus daily evaporation (EVA, mm day^{-1}):

$$AWV = [INTERCEP - EVA] \cdot s \qquad (38)$$

CHAPTER 2

where s is the integration step and *daily evaporation* indicates the amount of water evaporating from the canopies as follows:

$$EVA = \min\left(\frac{INTERCEP}{s}, PTR \cdot r\right) \tag{39}$$

where r is the maximum value of (1, PEV/PTR where PEV = the daily potential evaporation, and PTR = the daily potential transpiration) and s is the integration step. Some precipitation falling on the canopy will simultaneously infiltrate through it ($INFIL$, mm) as follows:

$$INFIL = \max\left(0, \frac{D \cdot s + AWV - i}{s}\right) \tag{40}$$

where D is the daily precipitation [mm day^{-1}], i is the interception capacity of the tree canopies [mm] ($i = c \cdot LAI$, where c is the specific interception storage capacity [mm], LAI is the leaf area index [m^2 m^{-2}]), and s is integration step. *The daily interception* ($INTERCEP$, mm) is the following:

$$INTERCEP = \min(i, D \cdot s + AWV) \tag{41}$$

The *daily potential evaporation* (PEV) is the potential evaporation from intercepted water calculated using the Penman-Monteith equation[31] that integrates the effects of the climatic factors and the properties of plants in terms of the energy and the resistance of plant surfaces to the air flow as follows:

$$PEV = \frac{d \cdot R + Da \cdot Ha \cdot VPD/Ra}{d + G \cdot (1 + ini/Ra)} / HV \tag{42}$$

where d is the derivative of saturated vapor pressure, R is the daily net shortwave radiation [J m^{-2}], Da is the density of air [1.220 kg m^{-3}], Ha is the specific heat of the air [1004.0 J kg^{-1} °C^{-1}], VPD is the current vapor pressure deficit [Pa], G is the psycrometric constant [66.0 Pa °C^{-1}], ini is the surface resistance when intercepted water occurs [s m^{-1}], Ra is the aerodynamic resistance [s m^{-1}], and Hv is the heat of vaporization for water [2 453 000 J kg^{-1}].

The interception capacity and the total surface area of the crown primarily indicated by the foliage area determine the total water intercepted on the tree crowns. The interception capacity is the amount of water intercepted on a unit area of canopy. In Scots pine, the amount of rain retained on the surfaces of needles is 0.3 mm. In boreal conditions, the leaf area index [LAI] in Scots pine stand may be up to 5–6 m^2 m^{-2} (m^2 of total foliage area per m^2 of land). The total foliage area in natural Scots pine stands has maximum values at the canopy closure at the age of 40–60 years. After this, the foliage area gradually reduces with a reduction in tree number as a consequence of the enhanced mortality of the trees. At the maximum foliage area, the canopy of the Scots pine may intercept 15%–20% of the precipitation as Fig. 38 shows.

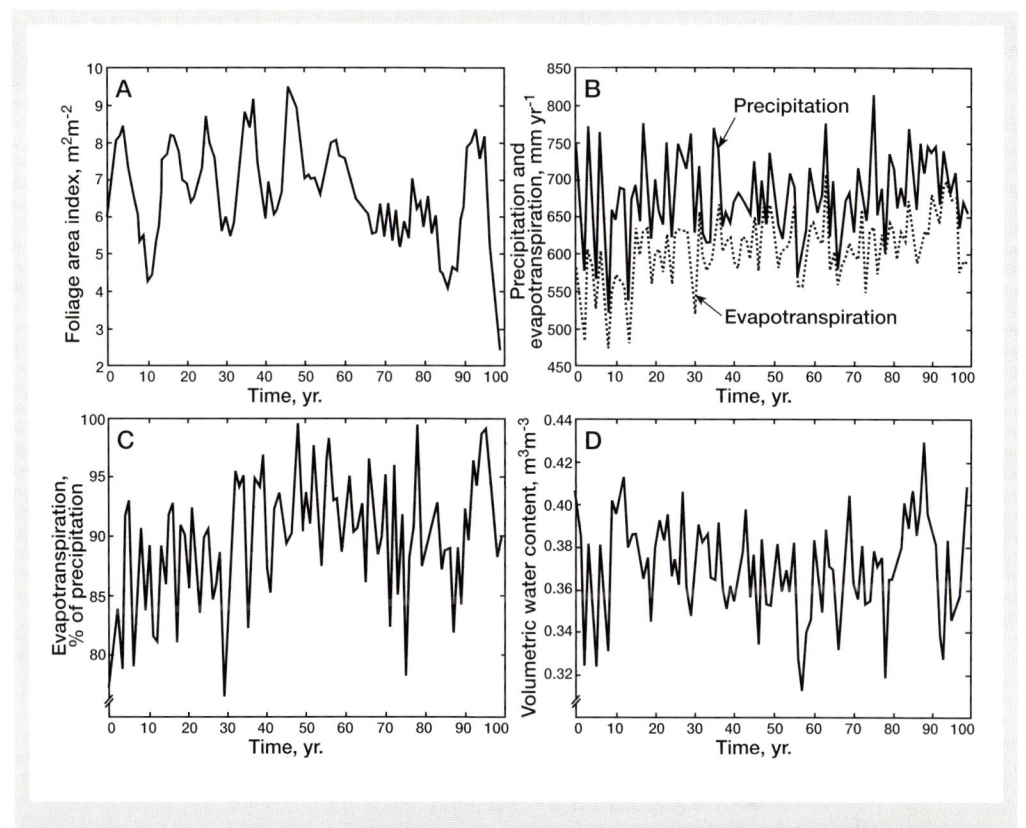

Figure 38. Balance of water in a natural Scots pine stand over a rotation of 100 years with A: foliage area, B: amount of precipitation and amount of evapotranspiration, C: evapotranspiration as a percentage of precipitation, and D: moisture content of the uppermost layer of soil.

The foliage area is a primary driver for the evaporative and transpirational loss of water from a tree stand, since it provides the surface for both processses. The evaporation and transpiration are physical processes related to the energy balance of the foliage that drive the water loss in both cases. Stomatal processes also affect the transpirational water loss through which the physiological proceses of the plant effect on the transpirational water loss. The calculation of the transpiration water loss from the foliage of a tree uses the same technique as the evaporative water loss. The *daily potential transpiration* (*PTR*) is the following:

$$PTR = \frac{d \cdot R + Da \cdot Ha \cdot VPD/Ra}{d + G \cdot (1 + ef/Ra)} / HV \tag{43}$$

where *ef* is the effective surface resistance (a parameter) and the other symbols are as noted above. The evaporative and transpirational water losses in boreal conditions may be 80%–90% of the precipitation entering the stand with the largest loss during the culmination of the foliage area.

CHAPTER 2

The amount of water infiltratred into the soil (precipitation not evaporated from the canopy and ground), the water holding capacity of the soil, the amount of water infiltrated in the ground water, and the water uptake of trees through roots (transpirational water loss) affect the amount of water stored in the soil profile. The infiltration rate and the water holding capacity of soil depend primarily on the soil texture. The annual cycle of freezing and melting of soil water also controls the infiltration and the amount of water in the soil profile. Snow cover provides shelter from freezing and the storage of water to recharge the soil water in the spring. This increases the soil moisture content higher than without the recharging from snow cover as the following equation for moisture of the soil profile indicates:

$$\frac{\delta W}{\delta t} = \frac{\delta}{\delta z} \cdot \left(k_w \cdot \left(\frac{\delta p}{\delta z} + 1 \right) \right) + s_w \qquad (44)$$

where p is the water tension [Pa], k_w is the hydraulic conductivity [mm day^{-1}], s_w is the sink or source term for water [s^{-1}], t is the time, W is the volumetric water content [m^3 m^{-3}], and z is the thickness [m] of the inorganic soil layer.

Precipitation, climatic factors, and the properties of soil determine the moisture content of soil, but tree cover also has a substantial effect on soil moisture especially of the surface soil. Moisture of the surface soil decreases following the growth of the foliage area to increase the evaporative and transpirational loss of water. The accumulation of snow decreases with an increase in the canopy, too. The soil may therefore freeze deeper and the duration of the freeze may be longer with less infiltration of water into the soil profile and a lower moisture content in the soil. Under a prevailing climate, this tendency could reduce the photosynthesis and growth of trees less than potentially possible.

4.3 Productivity of the forest ecosystem

4.3.1 Description

One can summarize the short-term and long-term functioning of the forest ecosystem in the terms of the productivity. This is a key concept in forestry used to analyze the applicability of different tree species and management practices in management. The *productivity* or *primary productivity* is the rate at which the photosynthetic and chemosynthetic activity of the producer organisms store radiation energy in the form of organic substances further usable for food material of other organisms[32]. The concept of productivity therefore applies to an individual or to a community of plants or an ecosystem that the interaction of the community with the environment is forming. Following the energy flow, Table 14 shows that several successive steps in the productive processes can distinguish the application and interpretation of the concept of productivity.

Table 14. Concepts used in the analysis of productivity of the forest ecosystem and the growth and yield of tree stands.

Concept	Definition
Gross primary production, GPP	The amount of carbon fixed through the process of photosynthesis by the ecosystem's green plants in a year
Autotrophic respiration, R_A	The amount of carbon released to the atmosphere as CO_2 by the ecosystem's green plants through respiration in a year
Heterotrophic respiration, R_H	The amount of carbon released to the atmosphere as CO_2 by the ecosystem's animals and micro-organisms through respiration in a year
Net primary production, NPP	The difference between gross primary production and plant respiration in a year, NPP=GPP-R_A
Net ecosystem production, NEP	The yearly rate of change in carbon storage in an ecosystem: A positive NEP indicates that the ecosystem has accumulated carbon during the year, while negative NEP indicates that it has lost carbon during the year, NEP=GPP-R_A-R_H=NPP-R_H

The *gross primary productivity* (GPP) is the total rate of the photosynthesis including the organic matter used in the respiration (R_A). The concepts of the total photosynthesis and the total assimilations are parallel to the gross primary productivity with the same content. When excluding the organic matter used for respiration, one obtains the *net primary productivity* (NPP) that is equal to the gross primary productivity minus the respiration (or autotrophic respiration) – NPP = GPP - R_A. The concepts for gross primary productivity and net primary productivity are applicable at the level of individuals, populations, communities, and the ecosystem.

The concepts of gross ecosystem and net ecosystem productivity are commonly used to emphasize the application of the functioning of the ecosystem. The *gross ecosystem productivity* (GEP) is the amount of carbon fixed in the ecosystem through photosynthesis occurring in communities of green plants in the ecosystem. One can also use the concept of the *net ecosystem productivity* (NEP) obtained by subtracting the ecosystem autotrophic and heterotrophic respiration (R_H) from the gross ecosystem productivity (NEP = GEP - R_A - R_H). The heterotrophic respiration indicates the rate at which the carbon releases from the ecosystem through the functioning of the heterotrophic organisms.

These concepts have wide use to analyze the functioning of the ecosystem. The net primary productivity of the ecosystem compares how different ecosystems perform in relation to each other or how they are contributing to the processes of the biosphere. In forestry, the net ecosystem productivity as applied in tree populations directly indicates the growth rate of trees in terms of mass or volume. Over a longer period, the accumulation of the net ecosystem productivity indicates the stocking or the amount of tree biomass standing on the site.

4.3.2 Biomass accumulation

Figure 39 shows that the structural change of the forest ecosystem with time is directional with an accumulation of organic matter – mass of trees and mass of humus in the soil. The difference between total mass growth and mortality (mass of dead trees and the mass of dead tree organs in the form of litter) initially increases indicating enhanced growth of individual trees. It begins to decrease after the culmination of the total mass growth (approximate age of 50–60 years in fertile sites in the boreal zone) with the closure of canopy with full use of resources for tree growth. The rate of litter fall and tree mortality depend on the growth rate of a tree with a few years delay. The above model applies over the entire range of sites with varying fertility, but the maximum accumulation of mass in a fertile site is several times larger than at a poor site.

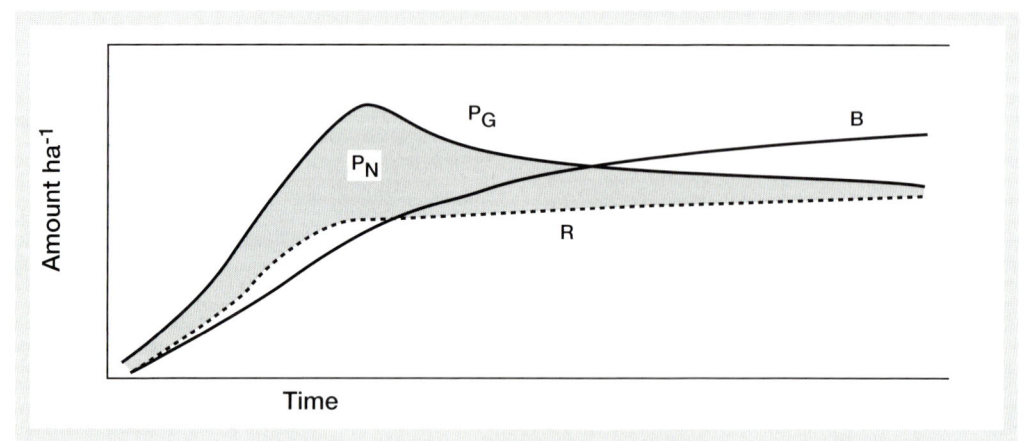

Figure 39. Schematic presentation on the course of the total growth (PG), net growth (PN), respiration loss (R), and the accumulation of mass (B) in trees during succession[33].

Whenever the space and availability of resources (energy, water, and nutrients) become sparse and the death of trees enhances, growth of individual tree reduces. This means that the limits set by the site conditions will influence the total growth so that the energy intercepted in the forest ecosystem balances the energy used in the forest ecosystem with a subsequent levelling of the mass accumulation. The growth that increases the mass and the mortality that decreases the mass balance with a resultant convergence of mass to the level determined by the local climatic and edaphic conditions. Figure 40 shows that fertile sites enhance the accumulation of mass more than poor sites. The following equation describes the growth of trees in different sites as a function of the age of the tree stand:

$$M(Age) = B_1 \cdot (1 - e^{B_2 \cdot Age \cdot B_3}) \tag{45}$$

where M is the mass and Age is the age of the stand, B_1 (maximum value of mass existing in the stand), B_2 (time for the start of enhancement of the accumulation of mass), and B_3 (time when the maximum accumulation is obtained) are parameters with the values given in Table 15.

Table 15. Values of the parameters in Eq. 45 used to model the stemwood accumulation in different sites.

Site		B_1 $m^3\ ha^{-1}$	B_2 yr.	B_3 yr.
OMT	Very rich	600	10	135
MT	Rich	520	15	160
VT	Poor	400	25	175
VT	Very poor	300	40	185
CIT	Extremely poor	200	65	190

The gross productivity of a tree stand over a selected period is much larger than the standing mass indicates at the end of the period due to the mortality through succession. Table 15 shows that on a fertile site, the standing stem mass of Scots pine after 100 years is approximately 60% of the total stem mass produced over the same period with tree stand density of approximaely 700 trees per hectare. The mean height is approximately 25 m, and the mean diameter is approximately 26 cm. On a poor site, the standing mass covers approximately 80%. There is slower growth and development on a poor site compared to a fertile site. The tree stand density is approximately 820 trees per hectare with a mean height of approximately 21 m and a mean diameter of approximately 22 cm. An analysis of growth and yield shows a close correlation between growth and mortality. High growth rate also means a high mortality rate or a high turnover of the biomass.

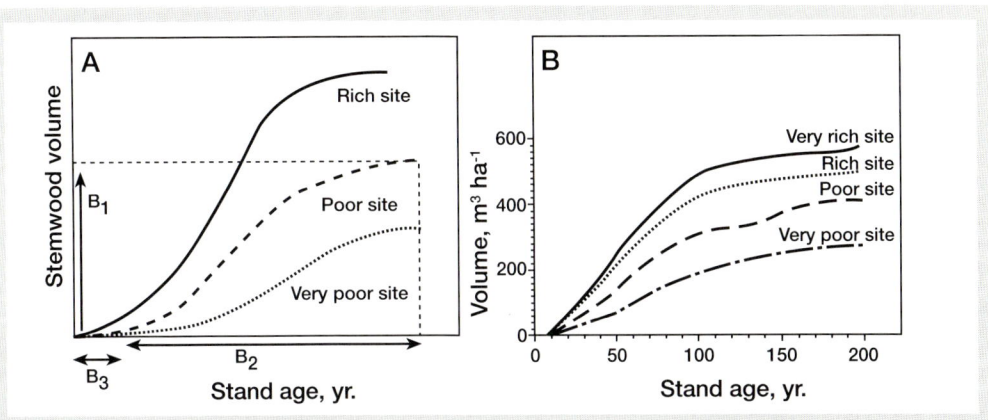

Figure 40. A: Schematic presentation of the accumulation of mass in sites representing variable fertility[21] and B: the accumulation of mass of Scots pine on different sites in Finland calculated with Eq. 45 applying the parameter values presented in Table 15.[22]

CHAPTER 2

Table 16. Growth and yield of Scots pine on fertile and poor sites in the southern boreal zone (62° N) in Finland[22].

Age, yr.	Number of stems ha^{-1}	Mean height, m	Mean diameter incl. bark, cm	Volume of the stand, m^3 ha^{-1}	Yield, incl. bark, m^3 ha^{-1}
Fertile site					
20	7 900	4.7	4.8	60	68
30	4 370	7.7	7.6	135	164
40	2 700	10.9	10.4	200	258
50	1 885	14.4	13.2	260	347
60	1 415	17.8	16.0	313	425
70	1 140	20.6	18.8	363	500
80	940	22.7	21.6	407	567
90	800	24.1	24.0	443	628
100	703	25.0	25.9	472	682
110	625	25.7	27.3	492	728
120	570	26.2	28.5	503	761
Poor site					
20	9 300	3.7	3.6	44	45
30	6 090	6.3	5.9	87	102
40	4 050	8.9	8.3	134	170
50	2 565	11.5	10.7	177	239
60	1 830	13.9	13.1	219	304
70	1 418	16.3	15.4	262	368
80	1 137	18.3	17.8	299	425
90	943	19.8	20.1	328	474
100	820	20.7	21.8	351	515
110	735	21.5	23.1	366	547
120	675	22.1	24.2	375	569

The mass of the other organs of a tree also depend on the age of the tree or the mass of the trees. During succession, Table 17 shows that the mass of each component increases following the allometry of the tree structure, but the share of needles, branches, bark, and roots from the total tree mass decreases as expected due to the longevity and the mass turnover of different tree organs. The maximum mean of mass in mature Scots pine stands is as much as 240 Mg ha^{-1} of which 75% is stem wood, 6% is bark, 7% is branches, 1% is needles, and 11% is roots.

Table 17. Density and mass of Scots pine stand during succession showing different mass components on fertile site.

Age, yr.	Density, stems ha^{-1}	\multicolumn{6}{c}{Mass component of the trees, Mg ha^{-1}}
		Stem	Bark	Branches	Needles	Roots	Sum
5		0.5	0.1	0.4	0.4	0.1	1.5
10		2.6	0.5	0.5	1.0	0.3	5.4
20	7 900	15.9	3.4	4.2	3.4	1.9	28.8
30	4 370	39.4	7.4	6.7	6.4	4.6	64.5
40	2 700	64.9	9.4	8.7	6.2	7.9	97.1
50	1 885	88.0	10.5	10.9	6.4	11.1	126.9
60	1 415	108.2	11.2	12.7	6.7	14.2	153.0
70	1 140	126.8	12.1	13.7	7.1	17.3	177.0
80	940	143.5	12.7	14.5	7.2	20.1	198.0
90	800	157.2	13.4	15.1	5.0	22.8	213.5
100	703	167.9	14.0	15.3	3.6	25.1	225.9
120	570	179.2	14.7	15.3	3.4	25.5	238.1

The mass of stocking trees also contains many nutrients bound in different organs of the trees. The leaves and needles are especially important sinks of nutrients, but Table 18 shows that nutrients also accumulate in other organs with an increase in mass. The amount of nitrogen bound in tree organs in mature Scots pine stands is of the same magnitude as that currently available in the soil. This means that timber harvesting removes large amounts of nutrients bound in tree organs from the forest ecosystem.

Table 18. Amounts of some primary nutrients bound in Scots pines in different phases of succession on fertile site.

Age, yr.	\multicolumn{4}{c}{Nutrients, kg ha^{-1}}
	Nitrogen	Phosphorus	Potassium	Calcium
5	8	<1	3	3
10	23	2	11	12
20	95	10	48	60
30	184	19	93	120
40	225	23	115	156
50	251	26	129	175
60	278	28	143	195
70	301	31	155	211
80	319	33	165	226
90	302	31	157	226
100	291	30	153	227
120	284	30	150	222

4.3.3 Successional dynamics of the forest ecosystem

Over long periods, the growth and development of tree populations occurs under disturbances due to wind, wildfire, and attack by insects and pests. In the boreal and temperate conditions especially, disturbances cause cycling in the dynamics of the forest ecosystem so that disturbances return the successional phase of the ecosystem to earlier phases. As an example, wildfire that kills old trees provides space for regeneration, enhancement of growth, and the accumulation of mass. For large forest areas, this means that forestry represents a mosaic of tree stands of varying age from seedling stand to mature trees. For maximum growth and timber yield in the forest area, the distribution of tree stands with the mean mass of trees is theoretically equal to half the mass potentially supported by the climatic and edaphic factors (site properties).

After a long period, the populations of different tree species can grow parallel in the same site with changing composition of tree species. This is mainly due to the varying physiological and ecological responses of different tree species to the prevailing environmental conditions and the length of the life span of competing tree species. Each tree species simultaneously modifies the growing conditions representing a feedback related to the functioning (differences in using nutrients or water) and structure (shading cast trees or quality of litter) of trees. The result is that the growth and development of a particular tree species can be enhanced or diminished during the course of succession. In this context, Fig. 41 shows that the disturbances have a major effect on the success of different tree species to occupy a particular site.

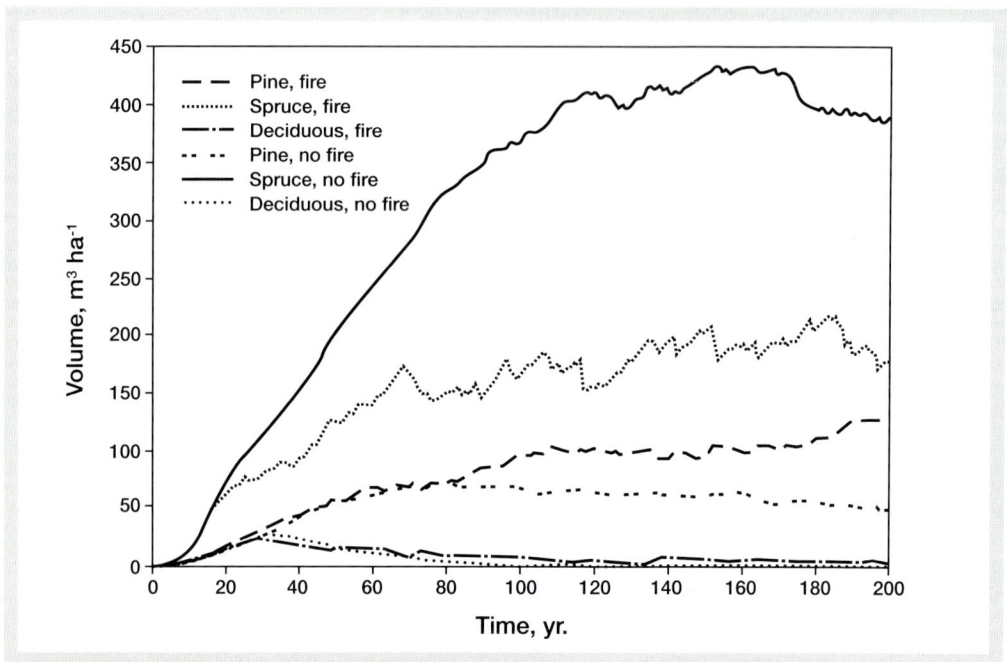

Figure 41. Successional course of a tree stand in volume per hectare for a multiple-species stand including wildfire (fire) or excluding wildfire (no fire).

In boreal conditions, many broadleaved tree species on fertile sites can more easily use the abundant supply of resources than coniferous trees. Broadleaved species typically dominate the initial phase of succession after a wildfire. Whenever the site fertility is high enough later, the changing conditions start to support shade-tolerant conifers more than broadleaved species. This brings an invasion of conifers such as Norway spruce to the site. Conifers finally replace the broadleaved species unless a disturbance causes the succession to return to an earlier phase. The disturbances and the dynamics of the tree species result in substantial variability in the mass of organic material in the forest ecosystem. This cycling determines the management of the forest ecosystem and the timber yield obtained.

CHAPTER 2

References

1. Luukkanen, O., Liite dendrologian kurssi-monisteeseen, Helsingin yliopisto, Helsinki, 1981, pp. 183–189.

2. Kramer, P. J. and Kozlowski, T. T., Physiology of woody plants, Academic Press, New York, 1979, pp. 13–57.

3. Sarvas, R., Havupuut, Werner Söderström Oy, Helsinki, 1964, pp. 48–51.

4. Wilson, B. F., The growing tree, The University of Massachusetts Press, Amherst, 1970, pp. 18–26.

5. Hall, D. O. and Rao, K. K., Photosynthesis, 5th edn., Cambridge University Press, Cambridge, 1994, pp. 1–19.

6. Taiz, L. and Zieger, E., Plant Physiology, Benjamin/Cummings Publishing Company Inc., Redwood City, 1991, pp. 219–248.

7. Farquhar, G. D., von Gaemmerer, S., and Berry, J. A., Planta 149(1):67(1980).

8. Wang, K., Effects of long-term CO_2 and temperature elevation on gas exchange of Scots pine, Research Notes 47, University of Joensuu, Joensuu, 1996, p. 8.

9. Tissue, D., Thomas, R. B., and Strain, B. R., Plant, Cell and Environment 16(7):859(1993).

10. Pettersson, R. and McDonald, J. S., Photosynthetic Research 39(3):389(1994).

11. Farquhar, G. G. and Wong, S. C., Australian Journal of Plant Physiology 11(4):191(1984).

12. McMurtrie, R. E., Rook, D. A., and Kelliher, F. M., Forest Ecology and Management 30(4):381(1990).

13. Brooks, S. and Farquhar, G. D., Planta 165(3):397(1985).

14. Jarvis, P. G., in Heat and Mass Transfer in the Biosphere (D. A. de Vries and N. H. Afgan, Ed.), Scripta Book Company, Washington, 1975, pp. 369–394.

15. Ryan, M. G., Plant, Cell and Environment 18(7):765(1995).

16. Mohren, G. M. J., Simulation of Forest Growth, Applied to Douglas Fir Stand in the Netherlands, Pudoc., Wageningen, 1987, pp. 41–72.

17. Raulo, J. and Leikola, M., Commun. Inst. For. Fenn. 81(2):1(1974).

18. Kanninen, M., Hari, P., and Kellomäki, S., J. Appl. Ecol. 19(2):465(1982).

19. Mälkönen, E., Commun. Inst. For. Fenn. 84(5):1(1974).

20. Mälkönen, E., *Commun. Inst. For. Fenn.* 91(5):1(1977).

21. Kimmins, J. P. and Scoullar, K. A., Forcyte-10, A user's manual, University of British Columbia, Vancouver, 1984, pp. 27–29.

22. Koivisto, P. *Commun. Inst. For. Fenn.* 65(8):1(1962).

23. Henttonen, H., Kanninen, M., Nygren, M., et al., *Scand. J. For. Res.* 1(2):243(1986).

24. Kellomäki, S. and Väisänen, H., *Ecological Modelling* 97(2):121(1997).

25. Oker-Blom, P., *Agricultural and Forest Meteorology* 34(1):31(1985).

26. Oker-Blom, P., *Acta For. Fenn.* 197(1):1(1986).

27. Kimmins, J. P., Forest Ecology, Macmillan Publishing Company, New York, 1987, pp. 81–96.

28. Olson, J. S., *Ecology* 44(2):322(1963).

29. MacLean, D. A. and Wein, R. W., *Canadian Journal of Forest Research* 7(3):562(1977).

30. Waring, R. H., Rogers, J. J., and Swank, W. T., in Dynamic Properties of Forest Ecosystems (D. E. Reichle, Ed.), Cambridge University Press, London, 1982, pp. 205–264.

31. Monteith, J. L. and Unsworth, M. H., Principles of Environmental Physics, Edward Arnold, London, 1990, pp. 245–263.

32. Odum, E. P., Fundamentals of Ecology, 3rd edn., W. B. Saunders Co., Toronto, 1971, pp. 43–52.

33. Kira, T. and Shidei, T., *Jap. J. Ecol.* 17(2):70(1967).

CHAPTER 3

Global forest resources

1	**Biosphere, vegetation zones, and forest formations**	**89**
1.1	Biosphere	89
1.2	Vegetation zones	89
1.3	Forest formations	91
	1.3.1 Boreal forests	91
	1.3.2 Temperate forests	92
	1.3.3 Tropical forests	92
1.4	Forest biomass and productivity	93
2	**State of global forest resources**	**95**
2.1	Forest cover	95
2.2	Amount of wood, growth, and fellings	97
2.3	Consumption of roundwood	101
2.4	Conservation of forests	106
2.5	Forest health	108
2.6	Forest fires	109
2.7	Climate change	110
2.8	Plantation forests	111
	References	**113**

CHAPTER 3

Timo Karjalainen

Global forest resources

1 Biosphere, vegetation zones, and forest formations

1.1 Biosphere

The *Biosphere* is the part of the Earth's surface that includes the upper horizons of the soil, the atmosphere near the ground, and the surface waters where living organisms exist and biological cycles occur[1]. The terrestrial ecosystems cover approximately 26% of the global area, but they contain 99% of the biomass. Given the large area and diversity of the terrestrial ecosystems, estimates of the terrestrial biomass vary from 480 Pg (Pg = petagram = 10^{15} g) of carbon (C) to 1 080 Pg of C. A recent estimate is 610 Pg of C[2]. Terrestrial biomass has a very uneven distribution. Forests represent the largest reservoir – 80%–90%. Forests include a diverse range of formations from high density tropical rain forests to less dense northern or high altitude forests.

In the terrestrial vegetation of the biosphere, large areas with plant cover of similar properties are called *vegetation formations*. Examples are coniferous forests or broad-leaved forests. The properties of vegetation change within zones of the globe from the Equator toward the poles. Thermal conditions are a prime distinguishing feature of the *vegetational zones*. Humidity conditions provide further classification in these zones into subgroups or sections. The same zonal pattern could theoretically exist on both hemispheres, but the uneven distribution of the continents may make certain areas recognizable only on one hemisphere. Figure 1 showing the thermal zones and bioclimatic regions of the world indicates the classifications using temperature and humidity.

1.2 Vegetation zones

The *Arctic zone* in the north and the *Antarctic zone* in the south are treeless areas behind the timber line where the mean temperature of the warmest summer month is less than 10°C. The numbers of moss and lichen species in this area are large, and dwarf shrubs are common. The next zone south from the arctic zone is the *boreal zone* in the northern hemisphere. The annual mean temperature is -5°C – +5°C, and the annual precipitation is 300–1 500 mm. The mean maximum temperature of the warmest summer month is more than 10°C, and the duration of summer is not longer than four months. This zone is humid and typically characterized by coniferous tree species. Common names for the zone are the coniferous zone or the taiga zone. Because of the cold winter and thin cover of snow, permafrost covers large areas in Alaska and the high-continental boreal zone in Canada and Siberia. The soil temperature regularly remains below 0°C in summer.

CHAPTER 3

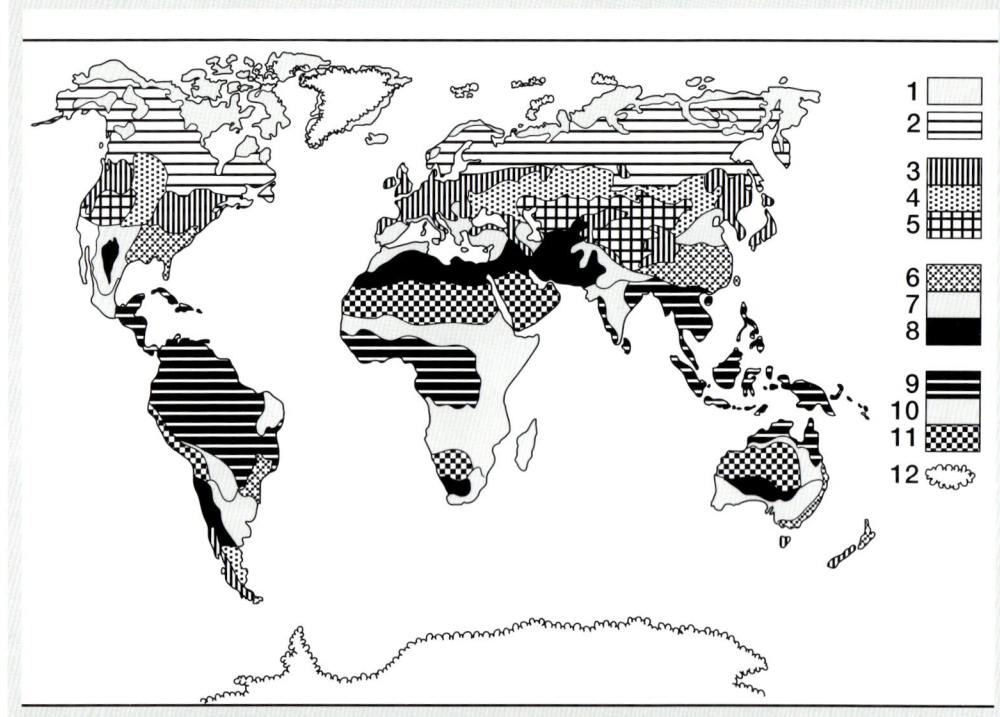

Figure 1. Thermal zones and bioclimatic regions of thermal conditions and humidity: 1=Arctic zone; 2=Boreal zone; 3–5=Temperate zone: 4=semiarid region, 5=arid region; 6–8=Subtropical zone: 6=humid region, 7=semiarid region, 8=arid region; 9–11= Tropical zone: 9=humid region, 10=semiarid region, 11=arid region; 12=Glaciers and areas of permanent snow[1].

The *Temperate zone* represents areas north and south from the Equator at the middle latitudes where the annual mean temperature is +5°C – +15°C, the duration of summer is four to seven months, and the temperature of the coldest month less than +2°C. Trees can grow here since the annual precipitation is 500–2 000 mm. The temperate zone is usually humid. The seasonal nature of the precipitation influences the vegetation. Although the temperate zone typically contains broadleaved, deciduous species, coniferous species are also common.

The *Tropical zone* covers regions near the Equator where the temperature of the coldest month is more than +18°C, and the mean annual temperature is 22°C–28°C. In the tropical zone, the existence of forests strongly depends on precipitation and its distribution. The main deserts of the world exist in the tropical zone. This indicates that precipitation is a very decisive factor influencing growing when temperature increases. Figure 2 shows the relationship for all vegetation zones to temperature and precipitation.

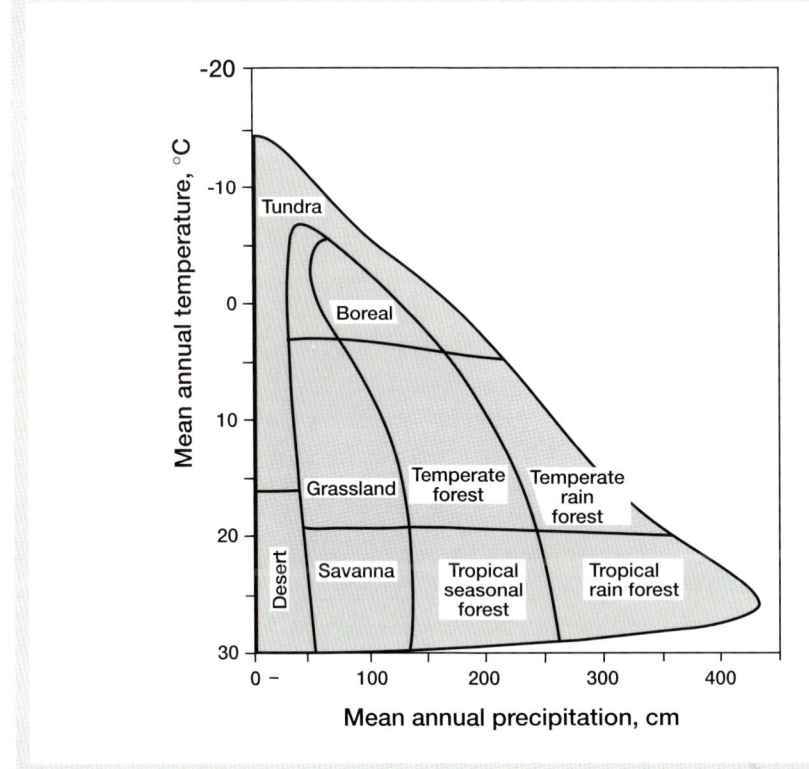

Figure 2. The distribution of major vegetation types or biomes related to the mean annual precipitation and temperature[3].

1.3 Forest formations

1.3.1 Boreal forests

The classification of the forests of the world differ according to tree species and productivity. The circumpolar belt of boreal forests covers the northern latitudes in North America, the Nordic countries, and the former USSR. The boreal forests occur between the northern tundra and the temperate mixed forests of the south and at high altitudes in the temperate zone. The boreal zone comprises almost 1 000 million ha of which approximately 600 million ha are forests. The climate is humid, and rainfall exceeds potential evaporation to give a positive water balance. Water that does not flow to the rivers increases the level of the water table to form mires. Peaty soils cover large areas of the boreal zone.

Approximately 20% of the world's industrial wood raw material comes from the boreal forests. Commercially the most important coniferous genera are pine (*Pinus*), spruce (*Picea*), fir (*Abies*), larch (*Larix*), juniper (*Juniperus*), thuja (*Thuja*), and hemlock (*Tsuga*). The most common deciduous genera are aspen (*Populus*), birch (*Betula*), willow (*Salix*), and alder (*Alnus*).

CHAPTER 3

1.3.2 Temperate forests

The location of the temperate forests is between the boreal forests and the tropical forests (from 25–50° N and S). This covers large areas of the northern hemisphere including southern Canada, the United States, non-Nordic Europe, parts of the former USSR, Japan, China, and elsewhere in East Asia. Temperate forests also occur in the southern hemisphere south of the tropics – some parts of Chile, Argentina, New Zealand, and Australia. The area of closed temperate forests has decreased to approximately 700 million ha. This is nearly half of its original area. Half is in North America, and one-quarter is in Europe. Human interests have influenced these forests for a long time with large areas used for timber production or other applications.

The temperate forests have broadleaved and coniferous species. Principal forest genera include pine (*Pinus*), juniper (*Juniperus*), fir (*Abies*), oak (*Quercus*), beech (*Fagus*), maple (*Acer*), elm (*Ulmus*), ash (*Fraxinus*), walnut (*Juglans*), hornbeam (*Carpinus*), chestnut (*Castanea*), sycamore (*Platanus*), willow (*Salix*), and alder (*Alnus*). Temperate forest trees of the southern hemisphere include southern beech (*Nothofagus*) and more than a hundred species of eucalypts (*Eucalyptus*).

In the temperate zone, dry forests occur where annual precipitation is less than 1 000 mm. The forests vary in climatic gradients from structured, closed forests to open woodlands and other areas of short and sparse, woody vegetation. Dry forests exist over large areas of southwestern North America and most of the Mediterranean Basin. Some areas of dry forest are the results of repeated burning over thousands of years.

1.3.3 Tropical forests

The tropical forests range over five continents and cover approximately 1 790 million hectares. These forests occur between 25° N and 25° S and are mainly in developing countries. The tropical forests represent over half of the world's forested area and contain approximately 60% of the global forest biomass and 25% of the soil organic matter. These forests cover approximately 15% of the world's surface, but they contain almost half of all plant and animal species.

The tropical regions contain the *evergreen rainforests* where the annual precipitation is greater than 2 000 mm with an even distribution throughout the year – more than 100 mm per month. The three major rain forest regions are the following:

- Amazonia with adjacent northern South America and the Atlantic coasts of Central America and southern Mexico
- Congo basin and adjacent western equatorial Africa
- Indo-Malayan region including the west coast of India, large areas of Southeast Asia, Papua New Guinea and other Pacific islands, and northeast coastal Australia.

The number of tree species in this forest formation is high. As many as 40–100 species can occur on one hectare – most belonging to different families. Deciduous genera that have substantial commercial value are *Eucalyptus, Pasania, Nothofagus*. Coniferous species are *Podocarpus, Dacrydium, Agathis, Libocedrus*, and *Phylocladus*.

In tropical regions where the annual precipitation is 1 000–2 000 mm and a dry season occurs for one month or more, the forests are *tropical moist deciduous forests*. Some dominant trees lose leaves toward the end of the dry season. Monsoon forests are the particular deciduous and semi-deciduous forests in South and Southeast Asia where heavy rains follow a very dry period of 2–6 months. The Asian tropical deciduous forests include timber groups significant for the forest industry such as the *dipterocarps* (lauan, Philippine mahogany, meranti, etc.), teak (*Tectona grandis*), sal (*Shorea rogusta*), *Swietenia,* and *Sideroxylon*.

The tropical zone also has low and simply structured wooded areas where the annual precipitation is less than 1 000 mm. These dry forests vary in climatic gradients from structured closed forests to open woodlands, thornlands, shrublands, savannas, and other short and sparse woody vegetation. Dry forests occur over most of sub-Saharan Africa not occupied by the equatorial rainforests. They also occur over large areas in India, Australia, South and Central America, Mexico, and parts of the Caribbean Basin. Some areas of dry forest are the results of repeated burning over thousands of years. Oaks (*Quercus*), mesquite (*Prosopis*), and *Acacia* (tropics) are common in dry forest types.

1.4 Forest biomass and productivity

The terrestrial biomass is mainly in the forest ecosystems – 80–90% of the total terrestrial biomass, although forests cover less than 30% of the global land area as Table 1 shows. Over half the forests are in the tropics – approximately 17% in the temperate zone and approximately 30% in the boreal zone. Although the total desert area is nearly 20% of the global land area, the desert vegetation represents less than 1% of the global terrestrial biomass.

Table 1. Distribution of the global land area for different vegetation types[4].

Type of vegetation	Area, million km^2	% of total
Tropical	17.9	14
Temperate	5.9	5
Boreal	10.6	8
Total forest	34.4	27
Savanna	17.2	13
Grasslands	17.0	13
Tundra	11.6	9
Deserts and scrublands	24.2	19
Cultivated lands, wetlands, etc.	25.0	19
Total	129.4	

CHAPTER 3

Figure 3 shows that the productivity of the vegetation in different vegetation zones depends on the variability of temperature and precipitation. In humid conditions, the annual dry matter production increases with increasing temperature from less than 2 Mg ha^{-1} yr.$^{-1}$ in the boreal zone to more than 20 Mg ha^{-1} yr.$^{-1}$ in the tropical zone.

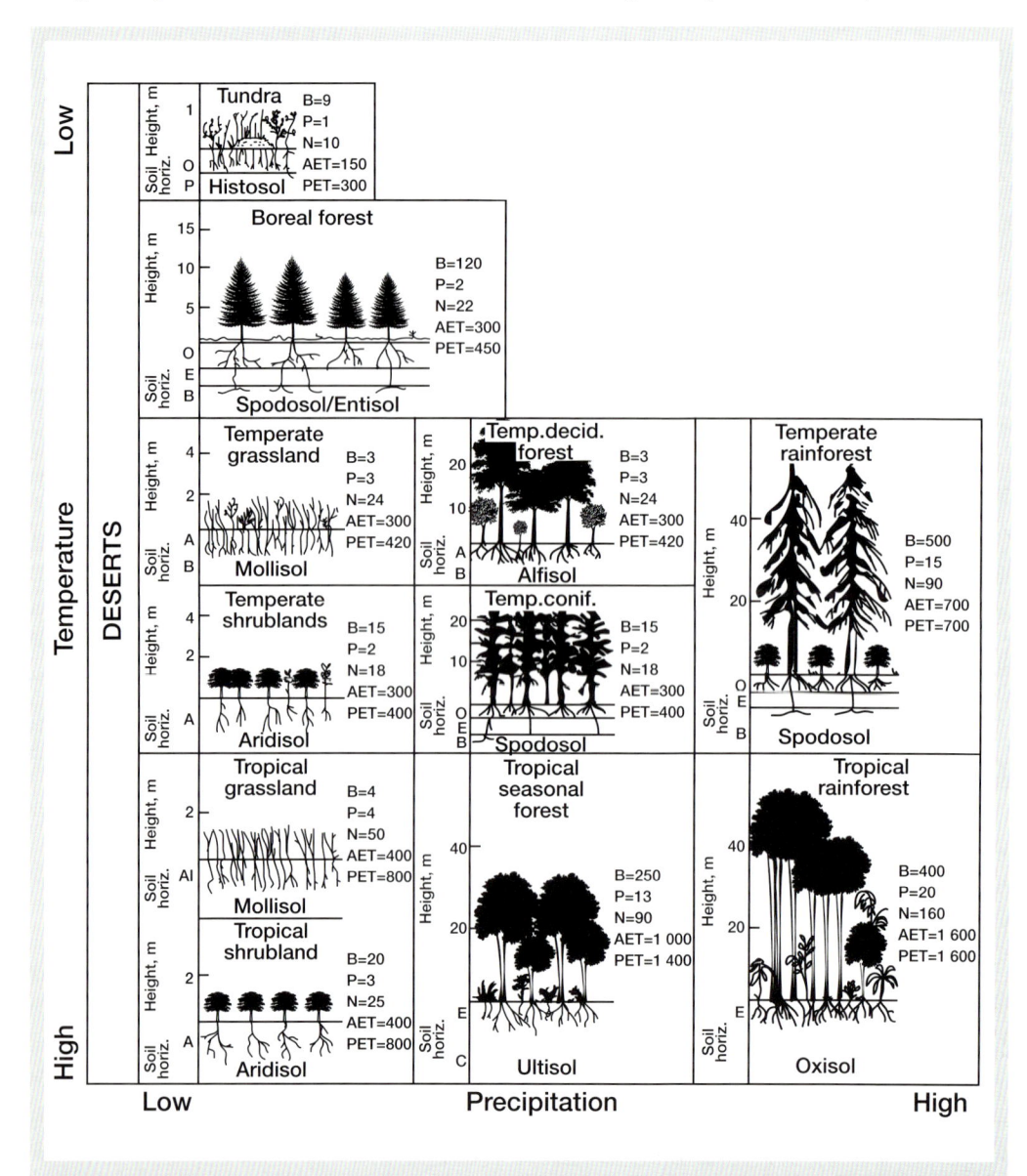

Figure 3. The relationship of the properties of major ecosystem types to precipitation and temperature with elements of structure and function used to characterize ecosystems for each type where B=total plant biomass in Mg ha^{-1}, P=total plant production above ground in Mg ha^{-1} yr.$^{-1}$, N=nitrogen uptake by plants in kg ha^{-1} yr.$^{-1}$, AET=actual evapotranspiration in mm of water per year, and PET=potential evapotranspiration in mm of water per year[5].

The biomass per unit area increases similarly with the increasingly favorable temperature conditions from approximately 120 Mg ha^{-1} in the boreal forests to 400 Mg ha^{-1} in the tropical rain forests. The amount of soil organic matter is higher in the boreal zone at 400–1 000 Mg ha^{-1} than in the tropics at 250–300 Mg ha^{-1}. This indicates a faster decay of the dead plant material in warmer conditions. The scarce supply of water substantially reduces the productivity and the biomass per unit area. The difference in productivity between the humid and semiarid regions is not as large as that for the biomass per unit area. In these conditions, the role of herbs and grasses with a short decay of the mass is more important than with woody species with a long decay of the mass.

Table 2 shows that the tropical forests have a much higher rate of gross and net photosynthesis than forests in other biomes, but the accumulation of net biomass is smaller. This means that maintenance and growth processes (respiration and litter fall) in the tropical forests consume a large proportion of the gross photosynthesis. Only a small fraction actually accumulates in the biomass. In the temperate forests, a much larger proportion accumulates in the standing biomass than in the tropical forest.

Table 2. Comparison of gross and net carbon flux through a tropical forest and a temperate forest, Mg C ha^{-1} yr.$^{-1}$ [16].

Carbon flux	Temperate deciduous forest	Tropical rainforest
Gross photosynthesis	19.6	127.5
Leaf respiration	4.6	60.1
Net photosynthesis	15.0	67.4
Other respiration	4.2	38.8
Net primary production	10.8	28.6
Litter fall	3.9	25.5
Net biomass accumulation	6.9	3.1

2 State of global forest resources

2.1 Forest cover

Table 3 contains data from the Global Forest Resources Assessment of the Food and Agriculture Organization of the United Nations (FAO)[4] and shows that forests covered nearly 27% or 3 443 million ha of the land area in 1990. This area includes land with a minimum crown cover of 20% in the developed countries and 10% in the developing countries. Approximately 58% of the land area and forests are in the developing countries. On the average, the forest area or the volume of growing stock per capita for developing countries is approximately half that for developed countries, although substantial variability exists from country to country. For example, Europe has 0.27 ha of forests per capita, and Finland has over 4 ha per capita. Nearly half the forests in the developing world are in Latin America and the Caribbean.

CHAPTER 3

Table 3. Some basic results of the global forest resources assessment for 1990[4].

Region	Number of countries	Number of inhabitants, mill.	Land area, mill. ha	Forest and other wooded land[1] mill. ha	% of the land area	Annual change 1980-90, 1 000 ha	Forest land[2] Area, mill. ha	% of the land area	Growing stock volume, mill m³ over bark	Mean growing stock volume m³ ha⁻¹	Volume per capita, m³	Area per capita, ha
Europe [3]	28	564.9	550.4	194.9	35	191	149.3	27	19 264	129	34	0.27
Former USSR	3	350.5	2 139.0	941.5	44	51	755.0	35	84 324	112	240	2.15
North America	2	276.5	1 835.2	749.3	41	-317	456.7	25	53 401	117	193	1.65
Developed Asia and Oceania	3	144.0	818.6	177.8	22	-4	71.5	9	6 553	92	46	0.50
Developed regions	36	1 335.9	5 342.2	2 063.6	39	-79	1 432.5	27	163 451	114	122	1.07
Africa	53	642.1	2 964.0	1 136.7	38	-2 828	545.1	18	55 655	103	87	0.85
Asia and Pacific	46	2 921.8	2 613.3	660.3	25	-999	497.4	19	55 200	125	19	0.17
Latin America and Caribbean	44	447.8	2 016.4	1 259.7	62	-6 047	967.5	48	109 421	114	244	2.16
Developing regions	143	4 011.7	7 593.7	3 056.7	40	-9 874	2 009.9	26	220 276	113	55	0.50
All regions	179	5 347.6	12 935.9	5 120.2	40	-9 953	3 442.4	27	383 727	114	72	0.64

[1] Forest and other wooded land (FOWL): includes also other formations of woody vegetation such as open woodland, scrub and brushland, forests under shifting cultivation, etc.
[2] Forest land: lands with a minimum crown cover of 20% in the developed countries and 10% in the developing countries
[3] Including Israel, Cyprus and Turkey

In the future, the global forest cover will probably decrease due to the increasing population. The population is likely to increase in the developing countries from 5 billion in 1990 to 8 billion in 2020. In the developed countries, the increase seems to be much less, from 1.1 billion to 1.2 billion. Increasing population will raise the pressure for agricultural land and built-up areas. This is particularly true in the developing countries where the forested area decreased by 16.3 million ha per year in 1980–1990[4]. Deforested area in 1980–1990 was nearly one-third of the land area of Europe. At the same time, the annual planting was 3.2 million ha and about 2.1 million ha per year underwent conversion into other wooded land. These resulted in a net loss of forested and other wooded land of 11 million ha per year. The area of forested and other wooded land expanded in Europe by 1.9 million ha in the 1980s.

2.2 Amount of wood, growth, and fellings

Table 4 shows that the area of forest land in the tropics is 1 790 million ha, in the temperate regions is 590 million ha, and in the boreal regions is 1 060 million ha. Note that some forest land in Canada and the former USSR belongs to temperate forest, and some forest land in the United States belongs to boreal forests. The figures in Table 4 are therefore probably overestimates for boreal regions and underestimates for temperate regions. The growing stemwood volume of global forests in 1990 was 384 billion m^3 – 53% in the tropics, 16% in the temperate zone, and 31% in the boreal zone. When averaging growing stock volume over the land area, differences between boreal, temperate, and tropical regions are small. The highest volume per forest area is in Switzerland with 329 m^3 ha^{-1}. This is much larger than the average for Europe. When comparing the total tree biomass, the biomass density in tropical regions is clearly higher than in the temperate or boreal regions.

Table 5 shows that fellings were less than the net annual increment in most developed regions. In Europe, fellings were 71% of the net annual increment in 1990. They were 74% in the former USSR and 79% in North America. In Greece, however, fellings were equal to the net annual increment, but in Albania fellings substantially exceeded the net annual increment. In Europe, annual fellings were approximately 2% of the growing stock. In the developing countries, deforestation and forest degradation led to a decrease in the total growing stock. In addition, the volume per hectare of the remaining forest may have decreased when removals exceeded growth. Unfortunately, similar statistics for the developing regions are not available.

The annual net increment was approximately 4.3 m^3 ha^{-1} in Europe with a range from 3.2 m^3 ha^{-1} in Southern Europe to 6.3 m^3 ha^{-1} in Central Europe. In the former USSR, the net annual increment was much lower at 1.7 m^3 ha^{-1}. In North America, the average net annual increment was lower than in Europe with 3.2 m^3 ha^{-1} – 1.9 m^3 ha^{-1} in Canada and 3.9 m^3 ha^{-1} in the United States. This variation is partly due to the differences in age class structure affected by different forest management practices – intensity in management and proportion of managed and unmanaged forests, fire control, etc. and to the differences in climate, site productivity, and in the occurrence of disturbances such as fire, insect, fungi, snow, wind, etc.

CHAPTER 3

Table 4. Forestry statistics for tropical, temperate, and boreal regions calculated from the FAO statistics[4].

Region	Land area, mill. ha	FOWL[1], mill. ha	Forest, mill. ha	Exploitable, mill. ha	Un-exploitable, mill. ha	Natural, mill. ha	Plantations, mill. ha	Average volume, $m^3\,ha^{-1}$	Total volume, $M\,m^3$	Forest Average biomass, $Mg\,ha^{-1}$	Total biomass, Pg
Tropical	4 788.1	2 728.0	1 792.0			1 761.2	30.8	115.3	203 077	168.7	297 197
Asia	900.9	452.9	338.0			315.4	22.6	140.0	44 106	181.0	57 149
S. America	1 650.4	1 191.3	924.2			918.1	6.0	114.0	104 273	185.0	169 847
Africa	2 236.8	1 083.8	529.8			527.7	2.1	104.0	54 698	133.0	70 201
Temperate	4 985.3	936.4	595.0	323.3	53.8			105.5	62 786	106.8	63 525
Europe	2 179.7	607.8	377.1	323.3	53.8			120.9	45 587	87.1	32 840
USA											
Oceanic											
Asia	1 712.5	207.4	159.3			125.7	33.6	88.0	11 094	144.0	18 120
S. America	366.0	68.5	43.3			41.6	1.7	124.0	5 148	252.0	10 460
Africa	727.2	52.3	15.3			13.0	2.3	74.0	957	162.0	2 105
Boreal	3 162.5	1 455.8	1 055.4	574.3	481.1			111.7	117 847	75.6	79 757
Nordic	102.0	61.0	53.3	48.2	5.1			93.0	4 942	50.0	2 651
Canada	921.5	453.3	247.2	112.1	135.1			116.0	28 671	103.0	25 458
Former USSR	2 139.0	941.5	755.0	414.0	340.9			112.0	84 234	68.0	51 648
TOTAL	12 935.9	5 120.2	3 442.4					111.5	383 710	128.0	440 479

[1]) FOWL = Forest and other wooded land; includes also other formations of woody vegetation such as open woodland, scrub and brushland, forests under shifting cultivation, etc.

Global forest resources

Table 5. Information for exploitable forests in the developed regions in 1990[4].

Region and country	Forest area million ha	Forest area per capita, ha	Growing stock	Net annual increment (NAI)	Fellings	Removals, million m³ under bark	Proportion of fellings from NAI, %
			million m³ over bark				
Europe total	133.0	0.24	18 509	576.7	408.3	342.7	71
The European Union	41.9	0.21	6 374	209.0	144.8	129.4	69
Belgium	0.6	0.06	90	4.5	3.3	3.0	75
Denmark	0.5	0.09	54	3.5	2.3	1.8	65
France	12.5	0.22	1 742	65.9	48.0	43.2	73
Germany	9.9	0.12	2 674	57.6	42.6	42.6	74
Greece	2.3	0.23	149	3.3	3.4	2.5	102
Ireland	0.4	0.11	30	3.3	1.6	1.4	48
Italy	4.4	0.08	743	17.8	8.0	7.3	45
Luxembourg	0.1	0.22	20	0.7	0.4	0.3	57
The Netherlands	0.3	0.02	52	2.4	1.3	1.1	54
Portugal	2.3	0.22	167	11.3	10.9	7.8	96
Spain	6.5	0.17	450	27.8	15.0	12.1	54
United Kingdom	2.2	0.04	203	11.1	8.1	6.4	73
Nordic countries	48.2	2.67	4 721	178.3	125.2	102.7	70
Finland	19.5	3.91	1 679	69.7	55.9	44.6	80
Iceland							
Norway	6.6	1.57	571	17.6	11.8	10.1	67
Sweden	22.0	2.58	2 471	91.0	57.5	48.0	63
Central Europe	4.4	0.31	1 313	27.8	22.6	19.5	81
Austria	3.3	0.43	953	22.0	17.3	15.0	79
Switzerland	1.1	0.16	360	5.8	5.3	4.5	91
Southern Europe	15.5	0.17	1 895	49.7	41.4	28.1	83
Albania	0.9	0.28	73	1.0	1.6	1.5	163
Cyprus	0.1	0.13	3	0.0	0.6	0.5	
Israel	0.1	0.02	4	0.2	0.1	0.1	29
Turkey	6.6	0.11	759	20.8	17.2	11.1	82
Former Yugoslavia	7.8	0.33	1 056	27.7	22.0	15.0	79
Eastern Europe	22.9	0.24	4 207	111.9	74.2	62.9	66
Bulgaria	3.2	0.36	405	10.6	4.8	3.5	45
Former Czechoslovakia	4.5	0.29	991	31.0	20.2	18.1	65
Hungary							
Poland	1.3	0.13	229	8.2	6.1	4.8	74
Romania	8.5	0.22	1 380	30.5	27.3	22.1	90
	5.4	0.23	1202	31.6	16.0	14.2	50
Former USSR	414.0	1.43	50 310	699.9	517.6	416.4	74
Belarus	5.4	0.53	720	13.0	11.6	9.2	90
Ukraine	5.8	0.11	895	20.8	14.6	11.3	70
North America	307.7	1.11	37 947	971.2	771.3		79
Canada	112.1	4.23	14 855	207.5	151.7		73
USA	195.6	0.78	23 092	763.7	619.6	497.9	81
Australia	17.0	1.00	1 796	35.8	20.0	16.8	56
Japan	23.8	0.19	2 861				
New Zealand	2.1	0.61	351			13.5	
Grand total	897.5	0.70	111 774				

CHAPTER 3

In the developed countries, Fig. 4 shows that the exploitable coniferous forests cover a larger area and have a larger growing stock than the exploitable deciduous forests. The exploitable forests here refer to forest and other wooded land having no legal, economic, or technical restrictions on wood production. They also include areas where harvesting is not currently taking place – areas included in long-term use plans or intentions. The area and volume of the coniferous forests are larger in the former USSR than any other country or region. Even in Europe, the area of the coniferous forests is larger than that of the deciduous forests.

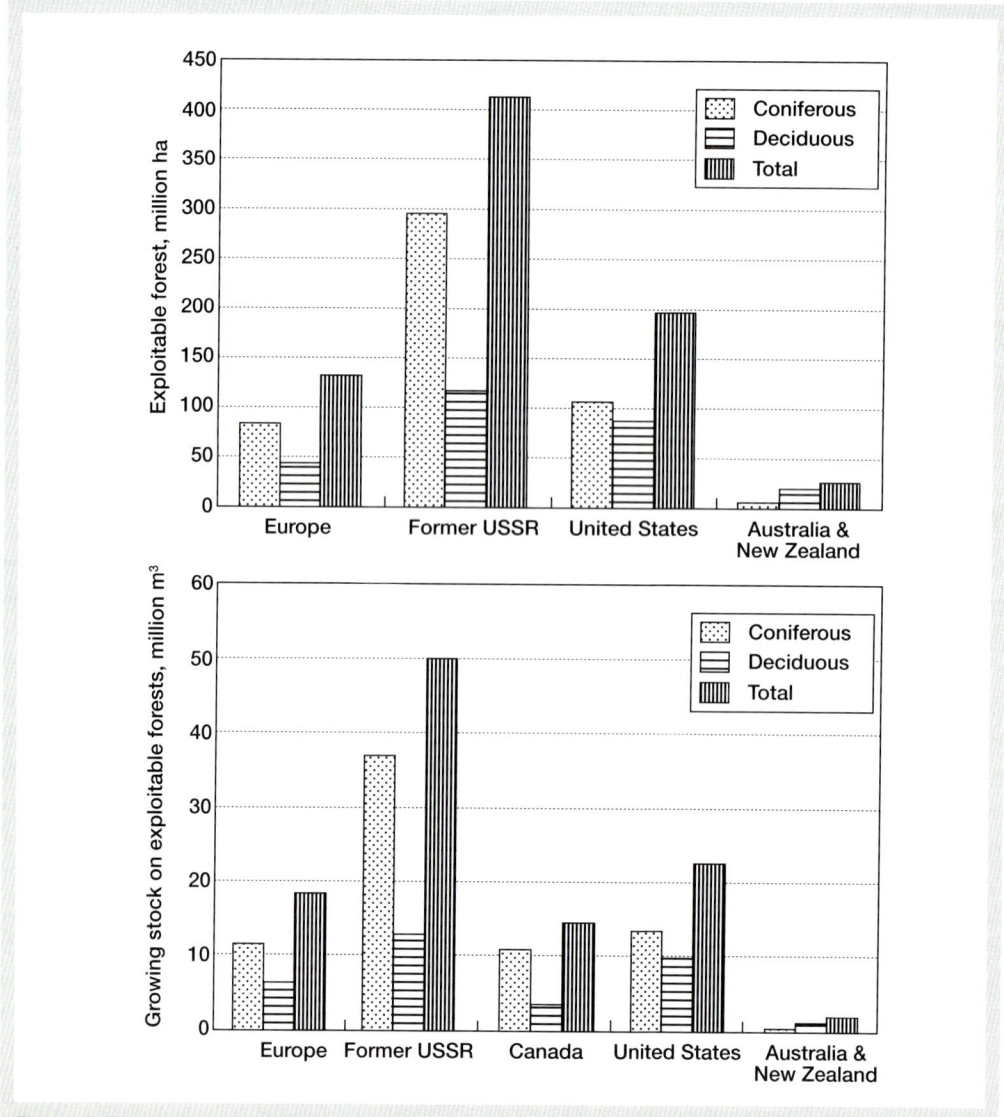

Figure 4. Distribution of exploitable coniferous and deciduous forest area and growing stock in the developed regions[7].

A recent report of the European Forest Institute shows that the growth of European forests has increased on many sites[8]. There is an increasing growth trend in the southern regions of Northern Europe, in most regions of Central Europe, and in some parts of Southern Europe. No clear trend exists in the most northern part of Europe. A decreasing trend occurs in exceptional cases having extreme growth conditions such as intense exposure to pollutants or exceptional climatic conditions. Several factors such as changes in land use, forest management, natural disturbances, climatic conditions, nitrogen deposition, and CO_2 concentration in the atmosphere may be causing the increasing trend. Other factors or a combination of several factors may also be occurring. As a result, past growth observations or growth and yield tables may not adequately describe current growth.

2.3 Consumption of roundwood

Figure 5 shows that roundwood consumption increased by 80% between 1961 and 1991. The consumption of fuelwood and charcoal has increased even more, by 108%. This is a result of the increasing population in developing countries where these items are common fuels for food preparation. The consumption of industrial roundwood increased by only 57%. In 1991, the amount of fuelwood and charcoal of the total roundwood consumption was 16% in developed countries and 80% in developing countries. The amount of fuelwood and charcoal of the total consumption of roundwood was previously larger in the developed countries – 23% in 1961. The consumption of roundwood will probably increase in the future from 3 430 million m^3 in 1991 to 5 070 million m^3 by 2010. In relative terms the consumption of industrial roundwood is predicted to increase more (67%) than that of fuelwood and charcoal (30%).

CHAPTER 3

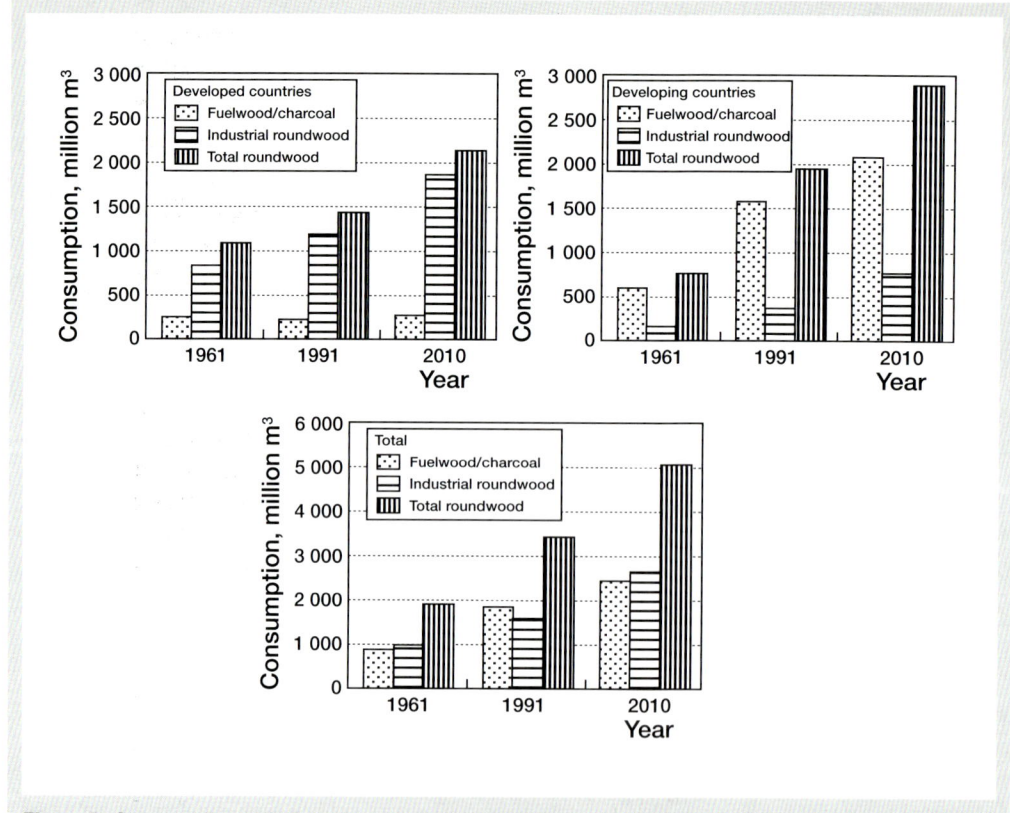

Figure 5. Consumption of fuelwood and industrial roundwood in developed and developing countries in 1961–2010[4].

A closer look at roundwood production in Table 6 shows that production of coniferous industrial roundwood is larger than that of nonconiferous industrial roundwood. The largest producers of coniferous saw and veneer logs in 1994 were the United States and Canada with half the total amount. The largest producers of coniferous pulpwood were the United States, Canada, Sweden, and Finland. The developing countries were larger producers of nonconiferous saw and veneer logs than the developed countries, but the developed countries were larger producers of nonconiferous pulpwood. Most production of fuelwood and charcoal was in developing countries.

Table 7 shows that the United States with 17%, the Russian Federation, and Malaysia were the largest exporters of roundwood in 1994. The European countries in transition were also large exporters. The global export of roundwood was mostly from developed countries at 70%. Japan was clearly the largest importer of roundwood, since Table 8 indicates that its share was 36% of the global total. The developed countries had 82% of the imports and developing countries had 18%. Wood supply will probably be inadequate during the next century compared to consumption. The annual need for timber will exceed the current annual growth increment by 2% in 2050 assuming a constant per capita wood use[9].

Table 6. Roundwood production by regions with some sample countries in 1994, million m^3 under bark[4].

Region / country	Industrial roundwood, coniferous — Saw and veneer logs	Pulpwood	Total	Industrial roundwood, nonconiferous — Saw and veneer logs	Pulpwood	Total	Fuelwood and charcoal	Roundwood total
Developed countries	519	228	778	125	123	273	191	1 318
The EU	48	22	73	17	17	36	26	149
France	13	6	19	8	5	14	10	50
Germany	18	7	26	3	4	7	4	37
Other Western Europe	65	43	110	3	8	11	13	145
Finland	21	17	39	1	4	5	4	48
Sweden	29	20	49	0	3	4	4	66
Countries in transition	66	28	109	27	15	54	43	248
Russian Federation	43	15	66	16	8	29	26	161
North America	302	120	434	73	67	147	99	680
Canada	137	32	173	5	4	8	7	188
USA	165	88	261	68	63	139	92	492
Australia	5	3	9	4	5	10	3	22
New Zealand	11	5	17	0	0	0	0	17
South Africa	4	3	8	0	7	9	7	24
Japan	16	2	19	1	5	7	0	33
Developing countries	82	35	146	169	28	271	1 700	2 122
Latin America	39	25	64	34	23	65	320	451
Brazil	22	11	33	19	20	45	197	275
Near East Asia	3	2	7	3	1	9	17	33
Turkey	3	2	6	2	0	3	8	17
Far East	38	7	70	112	4	150	862	1 085
China	33	6	64	19	2	36	204	306
India	3	0	3	16	1	22	269	294
Indonesia	0	0	1	35	-	38	149	187
Malaysia	0	-	0	35	1	36	10	46
Whole world	601	263	923	294	151	544	1 891	3 440
1990	707	298	1 138	273	142	516	1 661	3 315
1985	668	256	1 054	254	122	470	1 554	3 078
1980	614	256	992	264	115	460	1 376	2 828

CHAPTER 3

Table 7. Exports of roundwood in some regions and in the ten largest export countries, 1 000 m^3 under bark in 1994[4].

Exports	Industrial roundwood				Fuelwood and charcoal	Roundwood total
	Coniferous	Non-coniferous	Chips, particles, and wood residues	Total		
Developed countries	43.090	11.961	26.654	81.704	2.905	84.609
The EU	6.019	3.125	4.954	14.098	767	14.867
Other Western countries	4.115	532	2.209	6.856	75	6.928
Countries in transition	15.680	6.384	1.785	23.851	1.158	25.009
North America	11.816	1.502	8.628	21.946	731	22.677
Developing countries	3.289	19.831	8.604	31.725	2.627	34.352
Africa	53	4.799	1	4.854	182	5.036
Latin America	2.495	1.227	5.463	9.185	280	9.465
Far East	711	10.131	2.790	13.633	2.159	15.792
World	46.379	31.792	35.258	113.429	5.531	118.961
1. USA	10.961	1.195	7.523	19.679	322	20.001
2. Russian Federation	10.100	3.700	746	14.546	70	14.617
3. Malaysia	71	8.561	47	8.678	131	8.809
4. Australia	415	3	6.917	7.335	7	7.341
5. Germany	4.135	547	2.365	7.047	74	7.121
6. Chile	1.671	171	4.689	6.531	-	6.531
7. New Zealand	4.837	204	472	5.513	-	5.513
8. France	412	1.423	994	2.829	443	3.272
9. Latvia	1.544	885	115	2.544	135	2.679
10. Canada	855	307	1.105	2.267	410	2.677

Global forest resources

Table 8. Imports of roundwood in some regions and in the ten largest import countries, 1 000 m³ under bark in 1994[4].

Imports	Industrial roundwood				Fuelwood and charcoal	Roundwood total
	Coniferous	Non-coniferous	Chips, particles, and wood residues	Total		
Developed countries	36.818	28.859	35.928	101.605	3.486	105.091
The EU	5.831	9.596	4.761	20.188	1.600	21.789
Other Western countries	11.749	9.556	5.660	26.964	850	27.813
Countries in transition	558	273	110	941	17	958
North America	4.234	1.280	2.184	7.698	712	8.410
Developing countries	10.364	8.575	2.084	21.023	1.464	22.487
Africa	12	581	9	710	305	1.016
Latin America	114	60	24	198	44	242
Far East	9.022	7.628	1.968	18.618	958	19.576
World	47.183	37.434	38.012	122.629	4.950	127.579
1. Japan	14.438	7.947	23.201	45.586	302	45.887
2. Republic of Korea	7.321	2.243	938	10.503	304	10.807
3. Austria	4.037	791	3.023	7.851	295	8.146
4. Finland	1.925	4.848	745	7.518	45	7.563
5. Sweden	3.626	3.034	769	7.429	95	7.524
6. Italy	2.485	3.583	1.088	7.156	363	7.519
7. Canada	3.807	1.174	1.191	6.172	97	6.269
8. China	1.588	2.164	834	4.587	232	4.818
9. Belgium-Luxembourg	866	1.951	1.134	3.951	136	4.087
10. Norway	1.877	740	453	3.070	344	3.414

CHAPTER 3

2.4 Conservation of forests

Table 9 shows that currently protected areas total 790 million ha[10]. Totally protected areas maintained in the natural state and closed to extractive uses cover almost 60% of the protected area. Over 60% of the protected area is in developing countries. The average size of protected areas is larger in the developing countries than in developed countries. The smallest areas are in Europe. Figure 6 shows that the amount of nature conservation areas exceeded 10% of the land area in 12 developed countries in 1993. It is often difficult to know from the available data whether a conservation area network is operating effectively and how well the ecological criteria have been applied in their selection[10]. Figure 7 shows that the amount of threatened mammals and birds of the number of known species is higher in the temperate zone than in the boreal zone.

Table 9. Distribution of protected areas for areas larger than 1 000 ha[10].

Region	All million ha	All Number	Totally protected (Category I & II) million ha	Totally protected (Category I & II) Number	Partially protected (Category III, IV, & V) million ha	Partially protected (Category III, IV, & V) Number
World	8 619	792	2 546	465	6 073	328
Developing	3 281	486	1 164	296	2 117	190
Africa	704	139	260	91	444	48
Latin America	1 071	231	498	170	573	61
Asia	1 506	116	406	35	1 100	81
Developed countries	5 338	306	1 382	169	3 956	137
Europe	2 177	46	298	9	1 879	37
Former USSR	218	24	164	24	54	0
North America	1 348	147	422	66	926	81
Oceania/Asia	1 595	89	498	70	1 097	19

Category I: Scientific reserves and strict nature reserves possess outstanding, representative ecosystems. Public access is generally limited with only scientific research and educational use permitted.

Category II: National parks and provincial parks. Visitors may use them for recreation and study.

Category III: Natural monuments and natural landmarks containing unique geological formations, special animals or plants, or unusual habitats. May be managed for specific uses such as recreation or tourism or areas that provide optimum conditions for certain species or communities of wildlife.

Category IV: Managed nature reserves and wildlife sanctuaries are protected for specific purposes such as conservation of a significant plant or animal species.

Category V: Protected landscapes and sea-scapes may be entirely natural or may include cultural landscapes (e.g. scenically attractive agricultural areas).

Global forest resources

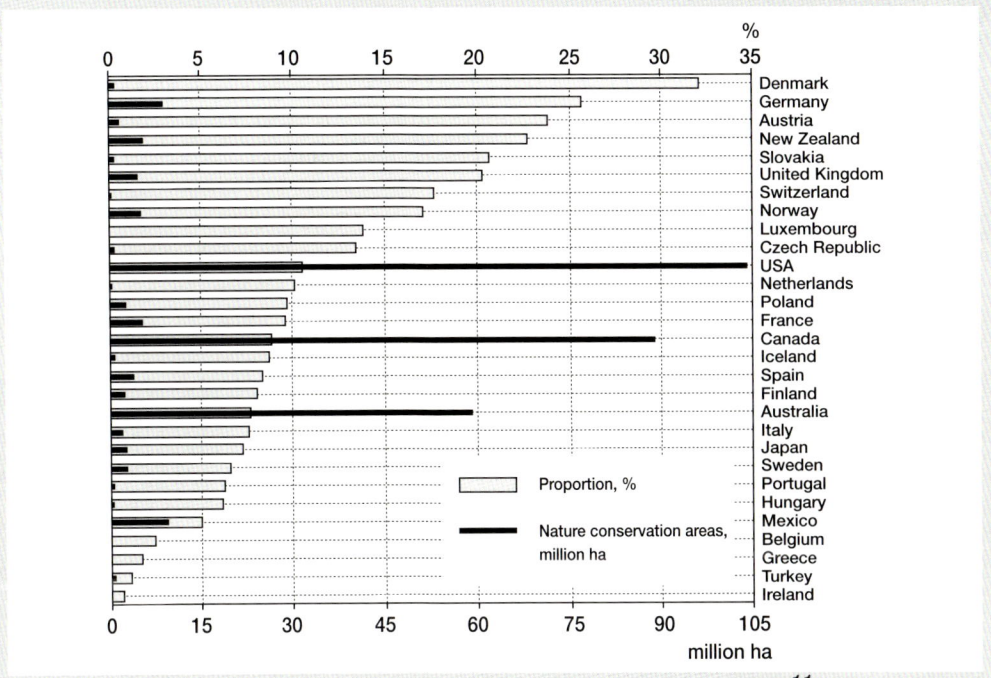

Figure 6. Proportion of nature conservation areas of the land area in some countries[11].

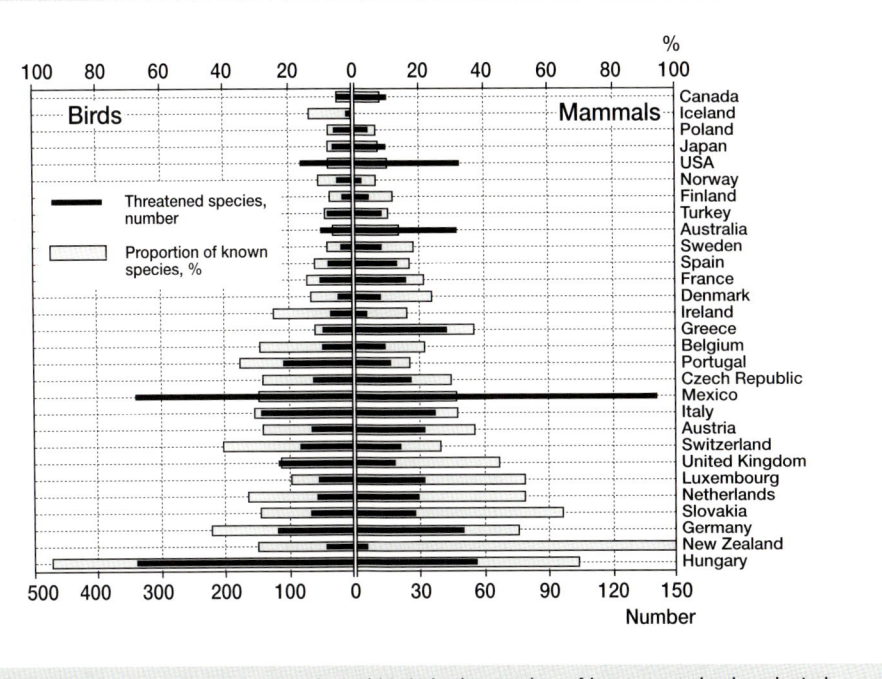

Figure 7. Amounts of threatened mammals and birds in the number of known species in selected countries[11].

CHAPTER 3

2.5 Forest health

Widespread forest decline has occurred at many locations in Europe and North America. Some of the causes of the decline include gaseous air pollutants, acidic precipitation, extreme weather conditions, etc. Transnational and national surveys show that forest health is still a serious matter of concern in Europe, as illustrated in Table 10[12]. Some reports indicate the situation has improved in certain locations, but the overall forest damage seems to be increasing on a regional level in Europe[13]. Drought with subsequent pest infestation and disease and air pollution are the prime factors contributing to the increase in damage. Developing countries also show evidence of forest decline due to air pollution[13].

Table 10. Percentages of defoliation and discoloration for nonconiferous and coniferous species in Europe from the 1995 survey of the United Nations Economic Commission for Europe (UN/ECE) and the European Commission (EC)[12] with defoliation being the fraction of needles or leaves lost compared to a reference tree of full foliage – defoliation larger than 25% is usually considered as damage – and with discoloration being the fraction of needles or leaves discolored compared to a reference tree.

Species type	Class						Number of trees
Defoliation	0–10%	>10–25%	>25–60%	>60%	Dead	>25%	
Nonconiferous	40.1	34.9	21.6	2.4	1.0	25.0	47 120
Coniferous	39.9	34.6	22.7	2.0	0.8	25.5	70 185
All species	40.0	34.7	22.2	2.2	0.9	25.3	117 305
Discoloration	0–10%	>10–25%	>25–60%	>60%	Dead	>10%	
Nonconiferous	88.7	8.2	2.1	0.4	0.6	11.3	46 169
Coniferous	90.3	7.3	1.8	0.3	0.3	9.7	65 636
All species	89.8	7.6	1.9	0.3	0.4	10.2	111 805

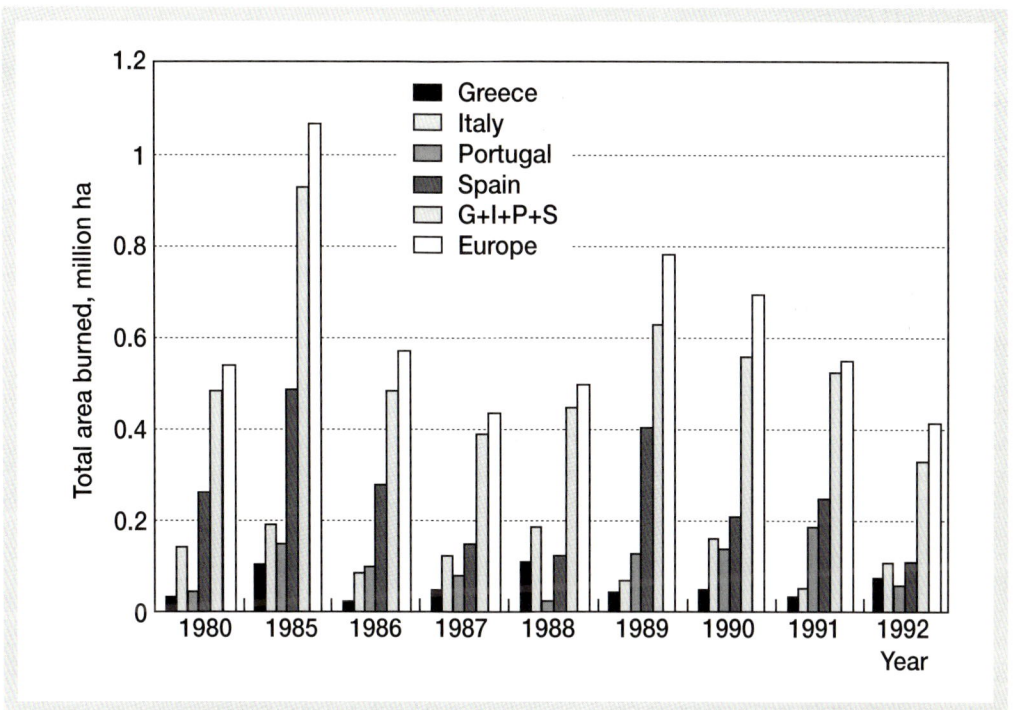

Figure 8. Area of forest fires in southern European countries and in Europe itself in 1980–1992 (G+I+P+S=Greece, Italy, Portugal and Spain)[16].

2.6 Forest fires

Fire significantly influences the structure and functioning of forests. Stand replacing fires burn almost all vegetation. Nonstand replacing fire burns only part of the vegetation. Fire frequency depends on the site conditions, structure of forests, the amount of flammable material, climate, and human impact. In a tropical rainforest, fire is usually rare and will not usually spread over wide areas because of an insufficient amount of dry and flammable plant material. In dry tropical forests, fires are frequent. In the temperate zone, forest fires play an important role in the ecosystem dynamics and occur despite control measures such as prescribed burning or slash burning in parts of North America, Australia, China, and Mediterranean Europe. In large areas of the boreal zone including Canada and the former USSR natural fires are also common. In Canada, the annually burned area increased in the 1980s to over 2 million ha. Earlier annually burned areas were 0.5–1.5 million ha[14]. In Russia, approximately 1.4–10 million ha of forested area burned between 1971 and 1991[15]. Figure 8 shows that the total burned area in Europe was 0.4–1.1 million ha in 1985–1992. Most burned area in Europe was in the Mediterranean region. In the Nordic countries, the frequency of fire is much lower today than in the past.

CHAPTER 3

2.7 Climate change

The productivity of forests and the number of species increase with increasing temperature, precipitation, and nutrient availability. Forests are sensitive to changing climatic conditions and may decline rapidly under extreme changes in water availability through drought or excessive rain. The Second Assessment Report of the Intergovernmental Panel on Climate Change[17] predicts that climate will change rapidly compared to the rate at which forest species grow, reproduce, and reestablish themselves. For the middle latitude regions, an average global warming of 1°C–3.5°C over the next 100 years would be equivalent to shifting isotherms toward the poles approximately 150–550 km or an altitude shift of 150–550 m. Temperatures would increase in areas of low altitudes. Simulations with vegetation models show that the area of tropical rain forests could decrease by 15%, tropical dry forests by 10%, and boreal forests by 17%. The area of temperate forests may increase by 5%. Greater losses due to more frequent fire and insect damage could offset the possible higher growth and yield due to higher temperatures and increased CO_2 concentration.

Terrestrial ecosystems exchange approximately 60 Pg carbon with the atmosphere annually. This is almost ten times more than the carbon emissions from land-use change (1.6 ± 1.0 Pg C yr.$^{-1}$) primarily due to deforestation and fossil fuel combustion (5.5 ± 0.5 Pg C yr.$^{-1}$)[18]. Even small changes in the functioning of the terrestrial carbon reservoir – forests – can therefore substantially influence global carbon balance. High and middle latitude forests have been a carbon sink of 0.7 ± 0.2 Pg C yr.$^{-1}$ in the 1980s[19]. This resulted in a net contribution of 0.9 ± 0.4 Pg C yr.$^{-1}$ from global forests to the atmosphere.

Brown *et al.* noted that the promotion of sustainable management of forests can help to mitigate climate change[20]. They list three management options in forestry to conserve and sequester carbon:

(i) Maintain existing carbon storage in forests through slowing deforestation, changing harvesting regimes, and protecting forests from other anthropogenic disturbances.

(ii) Increase carbon storage by increasing the area, carbon density, or both in native forests, plantations, agroforestry, and wood products.

(iii) Substitute fossil fuels with fuelwood from sustainably managed forests and substitute energy-intensive products such as steel, aluminium, and concrete with wood products.

Over the long term, the management of forests for substitution of fossil fuels is more effective than management to increase carbon storage in forests or products.

2.8 Plantation forests

The area of plantation forests increased in the developing countries by 3.2 million ha per year in the 1980s[4]. Table 11 shows that the area of plantation forests nearly doubled during the 1980s to approximately 68 million ha in 1990. The largest plantations in 1990 were in China at 31.8 million ha with an increase of plantations by 1.1 million ha per year. In China, the area of plantations increased even more than the area of natural forests decreased. The second largest area of plantations was in India. In Africa, the largest plantations were in South Africa and Algeria. The largest in Latin America were in Brazil and Chile. In tropical countries, eucalypts accounted for 23% of the plantations with pines at 11%, acacia at 8%, and teak at 5%.

The latest FAO report[13] indicates that the plantation area increased by 13 million ha between 1990 and 1995 to 81 million ha. The area of plantations increased most in China – 2 million ha in 5 years to 33.8 million ha in 1995. The area of plantations also increased substantially in India to 14.6 million ha. The forest plantations provide fuelwood, protection against erosion, and wood for industrial uses. Plantations can also serve the important function of reducing the pressure to clear natural forests. This helps to maintain biodiversity.

Table 11. Area of plantation forests in the developing regions in 1990 and change in the 1980s, 1 000 ha[4].

Region	Plantations Area in 1990	Plantations Annual change 1980–90	Plantations % of FOWL[1]	Natural forests Area in 1990	Natural forests Annual change 1980–90	Natural forests % of FOWL[1]	FOWL
Africa	4 416	157.7	0.4	540 669	-4 234	47.6	1 136 676
South Africa	965	15.5	2.3	7 243	-63	17.4	41 543
Algeria	485	18.3	12.3	1 554	-38	39.4	3 945
Asia/Pacific	56 264	2 718.4	8.5	441 095	-4 367	66.8	660 270
China	31 831	1 139.8	19.6	101 968	-400	62.9	162 029
India	13 230	1 009.0	16.0	51 729	-339	62.6	82 648
Indonesia	6 125	331.8	4.2	109 549	-1 212	75.5	145 108
Latin America/Caribbean	7 765	322.2	0.6	959 704	-7 682	76.2	1 259 717
Brazil	4 900	195.4	0.7	561 107	-3671	83.5	671 921
Chile	1 015	54.5	6.1	7 018	-60	42.3	16 583
Total developing countries	68 445	3198.3	2.2	1 941 468	-16 282	63.5	3 056 663

[1] FOWL = Forest and other wooded land (also includes other formations of woody vegetation such as open woodland, scrub and brushland, forests under shifting cultivation, etc.)

CHAPTER 3

Area of forest plantations is expected to increase substantially during the next decades[13]. In some countries plantations are already a substantial source of raw material for the forest industries. For example, in Chile plantations provide 95%, in New Zealand 93%, in Argentina and Brazil 60%, and in Zambia and Zimbabwe 50% of the industrial wood production[13].

Pine plantations in the southern hemisphere covered approximately 8.7million ha in 1990[21]. Latin America had 4.4 million ha of these plantations, Oceania 1.7 million ha, Asia 1.5 million ha and Africa 1.3 million ha. The proportion of young plantations is large, and therefore drain from these plantations will increase in the future. Assuming average annual increment of 20 m^3 ha^{-1}, the total net annual increment would be approximately 175 million m^3. Such amount of raw material is quite large if compared to the industrial roundwood production, i.e. 923 million m^3 in 1994. The annual available cut has been estimated to be over 100 million m^3 by 2000 and 110–130 million m^3 by 2010[21].

CHAPTER 3

References

1. Walter, H., *Vegetation of the earth and ecological systems of the geo-biosphere*, 2nd edn., Springer-Verlag, New York, 1979, 274 p.

2. Potter, C.S., Randerson, J. T., Field, C. B., et al., *Global Biogeochemical Cycles* 7(4):811(1993).

3. Whittaker, R. H., *Communities and Ecosystems*, 2nd edn., Macmillan Publishing Co., New York, 1975, 387 p.

4. Anon., *Forest resources assessment 1990, Global synthesis*, FAO Forestry Paper, United Nations, New York, 1995, 46 p.

5. Aber, J. D. and Melillo J. M., *Terrestrial Ecosystems*, Saunders College Publishing, Philadelphia, 1991, 429 p.

6. Larcher, W., *Physiological plant ecology*, Springer-Verlag, Berlin, 1975, 252 p.

7. Anon., *The forest resources of the temperate zones, The UN-ECE/FAO 1990 Forest resource assessment*, Vol. 1, New York, 1992, pp. 26 and 58.

8. Spiecker, H., Mielikäinen, K., Köhe, M., and Skoovsgaard, J. P. (Ed.), *Growth trends in European forests*, European Forest Institute Research Report No. 5., Springer-Verlag, Berlin, 1996, p. 369.

9. Ravindranath, N. H., Steward, R. B., Weber, M., and Nilsson, S., in *Climate Change 1995: Impacts, adaptations and mitigation of climate change: Scientific-technical analyses* (R. T. Watson, M. C. Zinyowere, and R. H. Moss, Ed.), Cambridge University Press, Cambridge, 1996, pp. 487–510.

10. Anon., *Protected area database produced in World resources 1994–1995 Report*, Oxford University Press, New York, 1993.

11. Anon., *OECD Environmental Data Compendium 1995*, OECD, Paris, 1996, 305 p.

12. Anon., *Forest Conditions in Europe, Results of the 1995 survey, 1996 Executive Report*, C-UN/ECE, Brussels, 1996, pp. 23–24.

13. Anon., *State of the world's forests 1997*, Words and Publications, Oxford, UK, 1997, pp. 13–16, 145–146.

14. Kurz, W. A. and Apps, M. J., *Water, Air and Soil Pollution* 82(1/2):321(1995).

15. Dixon, R. K. and Krankina, O. N., *Canadian Journal of Forest Research* 23(4):700(1993).

16. Anon., *Europe's Environment: Statistical Compendium for the Dobříš assessment*, ECSC-EC-EAEC, Brussels, Luxembourg, 1995, pp. 380–381.

17. Cannell, M. G. R., Cruz, R.V. O., and Galinski W., in *Climate Change 1995: Impacts, adaptations and mitigation of climate change: Scientific-technical analyses* (R. T. Watson, M. C. Zinyowere, and R. H. Moss, Ed.), Cambridge University Press, Cambridge, 1996, pp. 95–129.

18. Houghton, J. T., Meira Filho, L. G., Brucc, J., et al., *Climate change 1994, radiative forcing of climate change and an evaluation of the IPCC IS92 emission scenarios*, Cambridge University Press, Cambridge, 1995, 339 p.

19. Dixon, R. K., Brown, S., Houghton, R. A., et al., *Science* 263(2):185(1994).

20. Brown, S., Sathaye, J., Cannell, M., et al., *Commonwealth Forestry Review* 75(1):80(1996).

21. Hakkila, P., *Pine plantations of the southern hemisphere and tropics as a source of timber*, The Finnish Forest Research Institute, Research Papers 532:10(1994).

Global forest resources

CHAPTER 4

Structure and properties of wood and woody biomass

1	**Biomass components of the tree**	**117**
1.1	Introduction	117
1.2	Crown mass	119
1.3	Stump and root wood	121
2	**Wood as a raw material for pulp and paper**	**122**
2.1	Macroscopic characteristics of wood	122
2.2	Formation of wood	125
2.3	Chemical composition of wood	128
	2.3.1 Introduction	128
	2.3.2 Primary cell wall constituents	129
	2.3.3 Extractives	131
	2.3.4 Inorganic components	134
2.4	Structure of the cell wall	137
2.5	Woody cells	141
	2.5.1 Introduction	141
	2.5.2 Softwood cells	144
	2.5.3 Hardwood cells	147
2.6	Irregularities in wood	150
	2.6.1 Juvenile wood	150
	2.6.2 Reaction wood	152
	2.6.3 Heartwood	156
	2.6.4 Knots	158
2.7	Moisture content of wood	162
2.8	Basic density of wood	164
2.9	Heating value of wood and bark	173
3	**Nonwood fibers as raw material for pulp and paper**	**177**
	References	**182**

CHAPTER 4

Pentti Hakkila

Structure and properties of wood and woody biomass

1 Biomass components of the tree

1.1 Introduction

Forest biomass is the accumulated mass above and below the ground of the wood, bark, and foliage of woody shrub and tree species living and dead. Forest industries use this renewable source of raw material selectively. All tree species are not acceptable. Even among the preferred species, industrial use takes only the best part of the stem. Other parts of the tree remain as residue in the logging sites.

Sometimes yesterday's residue may be tomorrow's raw material for the production of fiber, chemicals, energy, etc. Before discussing the properties of stemwood in detail, this section will examine the other biomass components of the tree. Figure 1 shows the biomass components of the tree using the following nomenclature:

- Wood and bark compose the *full stem* of the tree. Depending on the minimum diameter requirement of the application, the stem can be *usable stem* or *unusable top*. The stem is the supportive, conducting, and storage organ of the tree.

- The crown contains the live and dead branches plus all foliage and reproductive organs of the tree. The crown is an assimilating and storage organ.

- The stump and roots comprise the *stump-root system. Stump* consists of the unused above-ground biomass below the bottom of the useable stem and its underground projection excluding the lateral roots. *Roots* include all side or lateral roots but not the taproot. This is a natural below-ground elongation of the stem and is therefore part of the stump. The stump-root system fixes the stem to the ground, takes up water and minerals, and functions as a food storage organ.

- The *whole tree* or *full tree* includes all the components of the tree biomass above the stump-root system – stem and crown.

- The *complete tree* refers to the entire tree including all biomass components above and below the ground.

CHAPTER 4

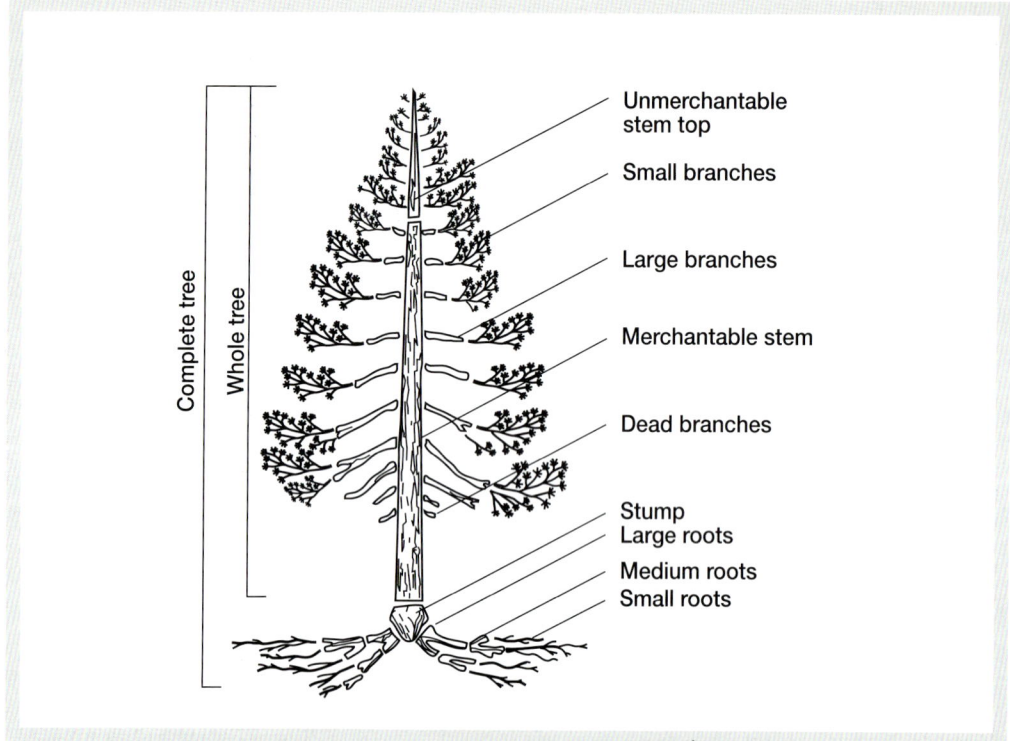

Figure 1. The biomass components of a tree (redrawn from Young et al.[1]).

Standing stems and conventional timber assortments are measured traditionally by volume with or without bark. For branches, foliage, stumps, and roots, the measurement of volume is impractical due to their peculiar shape and large differences in density. Mass rather than volume is therefore the most feasible unit for measurement and comparisons of tree components, the whole tree, and the complete tree.

In fresh condition, approximately half the total mass of a living tree is water. Moisture content varies between species, between trees within a species, and between the biomass components of a tree. The moisture content is not constant. It varies in live trees seasonally and even diurnally and during storage of wood and biomass depending on weather conditions. Dry mass rather than fresh mass is therefore the most precise and stable unit of measurement for the biomass components of a tree.

Hakkila has earlier published a comprehensive review on the quantity, properties, harvesting, and uses of residual forest biomass[2]. The ensuing text examines the crown and stump-root system as potential sources of fiber and energy compared with the stem mass using the following relations:

- dry mass of the component as a percentage of the dry mass of the full stem with bark
- dry mass of the component as a percentage of the dry mass of the whole tree.

Structure and properties of wood and woody biomass

1.2 Crown mass

The primary function of crown is to produce photosynthates for the tree. As the tree grows in height, lower parts of the crown remain in shade. The light conditions become less favorable for assimilation resulting in defoliation. The lower branches die and gradually fall. As the tree loses lower branches of the live crown and adds new growth to the upper crown, the entire canopy of a stand moves upward. The process is slowest in shade tolerant trees such as spruces that typically have a longer crown than the light-demanding trees such as pines.

The height of the live crown as a percentage of the total height of the tree is the crown ratio. It is characteric of a species, but the vitality of trees and the density of the stand – competition between individual trees – also affect it. A rapid reduction in the crown ratio usually indicates excessive stand density and the need for thinning.

Good management practice recommends thinning stands from below. Dominant trees remain to grow, and suppressed trees are removed. Consequently, trees harvested during late thinnings have a smaller crown ratio than trees in average stocking or trees removed in other cuttings. Table 1 shows results from extensive research of the Finnish Forest Research Institute.

Table 1. Crown ratio of Scots pine and Norway spruce in Finland.

Phase of development	Scots pine	Norway spruce
	Crown ratio, %	
Removal from early thinnings	57	78
Removal from late thinnings	43	67
Removal from regeneration cuttings	45	75
Trees of average stocking	54	79

The quantity of crown mass results from many factors such as tree age and size, stand density, the dominance of the tree in a stand, and tree species. Trees in dense unmanaged forests have less crown mass than trees of the same species in managed, repeatedly thinned stands. Variation between and within species is very wide. The data in Table 2 shows average crown mass in proportion to dry stem mass for a large number of trees with a breast height diameter of approximately 20 cm.

Table 2. Dry mass of branches, foliage, and crown in proportion to dry stem mass of several tree species in British Columbia and Maine[3, 4].

Species and region	Branches	Foliage	Crown
	Mass percent of stem mass		
17 softwood species in British Columbia	20	13	33
7 softwood species in Maine	20	14	34
9 hardwood species in Maine	14	5	18

CHAPTER 4

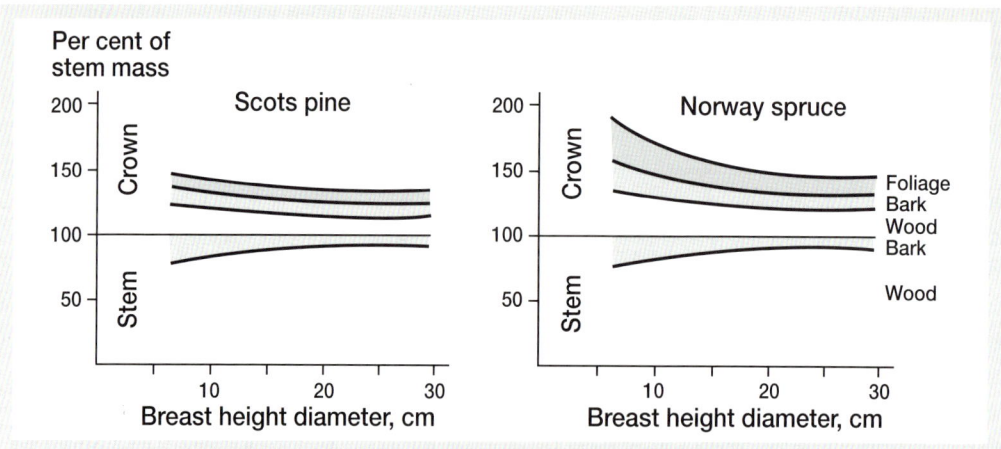

Figure 2. Biomass composition of dry whole-tree mass of Scots pine and Norway spruce in Finland as a function of dbh with relative quantity of stem mass at 100 as redrawn from Hakkila et al.[5].

The crown mass forms a substantial reserve of raw material. Recovery can use an integrated operation with the stem using the various techniques of whole-tree harvesting or a separate salvage operation of logging slash after the harvesting of conventional stemwood. Theoretically, three major use options exists – fiber, chemicals, and energy.

Figure 2 indicates that the potential gain of additional biomass is large in intensive whole-tree logging. Smaller trees have a greater relative gain. Less than half the crown mass actually contains wood.

When excluding foliage and recovering only branches, the percentage of bark in the additional raw material is still very high. On the average, in the branches of sawtimber-sized trees the percentage of bark is 20 to 45%. The proportion of bark is greater with thinner branches. For thin twigs or branch tips of less than 1-cm diameter, the mass of bark exceeds that of the wood. The high content of bark limits the use of the crown mass to purposes other than production of energy or chemicals. Besides the high contents of bark and foliage, the crown mass also has other disadvantages for pulp and paper making:

- The proportion of bark and foliage is typically more than 50%.

- In foliage, the proportion of fibrous cells is very low, and the amount of extractives is high. In sulfate pulping, the yield of pulp is only approximately 20%. The quality of pulp is inferior because of the high percentage of nonfibrous tissue and the short length of the cells.

- Due to the presence of reactive materials, the chemical composition of branch wood differs considerably from stemwood. In softwood branches, the amounts of lignin and extractives are high, and the amount of cellulose is low. In hardwood branches, the amounts of cellulose and extractives are high, and the amount of lignin is low.

- The dimensions of normal branch wood cells are smaller than those of stemwood. In addition, branch wood contains a considerable amount of reaction wood with inferior fiber properties. Table 3 shows differences in the average data for the cell length of several tree species in the eastern United States for stemwood and branch wood[6].

Table 3. Cell length of stemwood and branch wood of several tree species in the eastern United States[6].

Cell type	Stemwood	Branchwood
	Cell length, mm	
Tracheids in 8 softwoods	3.44	1.81
Fibers in 8 diffuse-porous hardwoods	1.16	0.83
Fibers in 4 ring-porous hardwoods	1.24	0.94
Vessel elements in 8 diffuse-porous hardwoods	0.54	0.45
Vessel elements in 4 ring-porous hardwoods	0.31	0.25

For these reasons, the crown mass is not an attractive raw material for the production of pulp. It does have considerable potential as a reserve of clean, renewable energy for replacing fossil fuels to reduce emissions of carbon dioxide into the atmosphere. For example, each cubic meter of stemwood harvested in regeneration cuttings in Finland has an energy potential in the crown mass of 0.4 MWh for Scots pine stands and 1.0 MWh for Norway spruce stands.

1.3 Stump and root wood

The root system has three functions. It absorbs and conducts water and minerals, it transfers and stores food, and it fixes the stem firmly to the ground. The three distinct sections in the stump-root system are the root crown at the base of the stem, the rapid-taper section comprising the root parts that provide lateral support and anchorage, and the wide network of thinner roots. Only the stump and sections of the roots measuring over 5–10 cm could have any potential use.

Little information is available on the amount of stump and root mass compared to the mass of the stem. The soil and site cause variation, but the variation among trees is considerably less than in the crown mass. The rule of thumb for boreal forests is that the dry mass of a stump-root system is approximately 30% of the stem mass, 25% of the whole-tree mass, and 20% of the complete-tree mass. The technically harvestable portion of the stump-root system is somewhat smaller, since thinner roots cannot be recovered. Research results for pines in the southern part of the United States and Scots pine in Finland and western Russia show that the harvestable quantity of stump and root wood is 20%–24% of the stemwood mass. In drained peatlands where trees develop a large butt swell, the proportion may be over 30% in tree species with a superficial root system such as Norway spruce.

CHAPTER 4

Many properties of stump and root wood of mature saw timber trees do not differ from those of stemwood. The following characteristics are important for chemical pulping:

- The percentage of bark does not differ significantly from that in the stem.
- In the transition zone between the stem base and roots, the cells are shorter than normal, but their length increases along the root. The cells of root wood have large diameter, wide lumen, and thin walls. These properties become more distinct at greater distances from the root base. In all stump-root systems, pine has longer tracheids, but some other softwoods have shorter tracheids than the stemwood. In hardwood stump-root systems, the differences in average fiber length are small. Vessel elements are fewer but wider in diameter.
- Differences in the chemical composition are small, but stump wood contains more extractives. Only root wood sections thinner than 2 cm in diameter have more lignin and less cellulose.

Stump and root wood are suitable for sulfate pulping because of their fiber properties. The cost of uprooting, splitting, transporting, cleaning, and reducing to chips is very high. If a serious shortage of fiber should occur, stump and root wood could be sources of marginal raw material for pulp. The Swedish and Finnish pulp industries did use stump and root wood from regeneration cuttings of Scots pine and Norway spruce in the 1980s, but the high cost of chips forced discontinuation of the operations.

2 Wood as a raw material for pulp and paper

2.1 Macroscopic characteristics of wood

Wood is *anisotropic in nature.* Its appearance and physical properties vary according to its sectioned plane. Figure 3 shows the following three planes:

- *transverse or cross-section* surface perpendicular to the long axis of the stem
- *radial surface* across the growth rings through the pith and parallel to the stem axis
- *tangential surface* tangent to the growth rings and parallel to the stem axis.

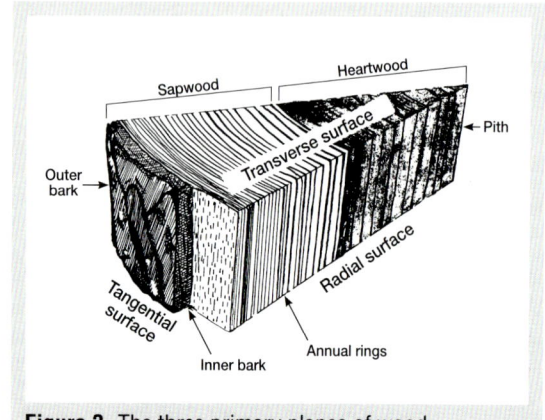

Figure 3. The three primary planes of wood.

Visible in Fig. 4 at the cross-section surface of the stem are *the pith* at the center, the *secondary xylem or wood* surrounding the pith, and a mantel of bark in which living, light-colored *inner bark or phloem* and dead, dark-colored *outer bark or rhytidome* occur. Between the wood and inner bark is the vascular cambial zone not visible with the naked eye or a lens.

In the boreal and temperate climates, the tree stem has regular concentric annual increments or *growth rings* that result from differences in the structure, size, and proportions of various types of cells produced in the beginning and at the end of the growing season. The volume growth of a tree is usually rapid during the early season. Then it starts to slow and finally ceases before the end of the growing season producing a band of lighter and darker wood each year.

Figure 4. Cross-section of an 85 year old *Larix sibirica* stem.

In softwoods, *earlywood cells* form in the beginning of the season when a growth hormon called auxin is abundantly available. Their primary function is to transport water. They have large radial diameter, wide lumen, and thin walls. *Latewood cells* form at the end of the season when the supply of photosynthate is plentiful. Latewood gives mechanical strength to the stem. Figure 5 shows that the latewood cells have smaller radial diameter and have a small lumen and thick walls. Due to the differences in the diameter, wall thickness, and coarseness, earlywood and latewood cells differ considerably in their paper making properties.

According to Mork's[7] definition, latewood in softwoods includes tracheids in which the common wall in the tangential direction between two cells is exactly half or over half the radial width of the lumen. If the joint width of two walls is less, the cells belong to earlywood. Although the definition was originally for spruce, it has use for other softwoods also. The transition from *light-colored earlywood to dark-colored latewood* in a softwood may be gradual or abrupt depending on the tree species, age from the pith, and to a smaller extent on environmental conditions.

The proportion of latewood volume to the total volume of wood denoted as *percentage of latewood* is an important indicator of wood quality when using wood for lumber, pulp, or other purposes. The percentage of latewood denotes only the proportion of wood volume that meets the minimum cell-wall thickness as defined above. It does not provide any indication of the variation in wall thickness within the earlywood and latewood zones. Its unsuitability for hardwoods limits its use.

CHAPTER 4

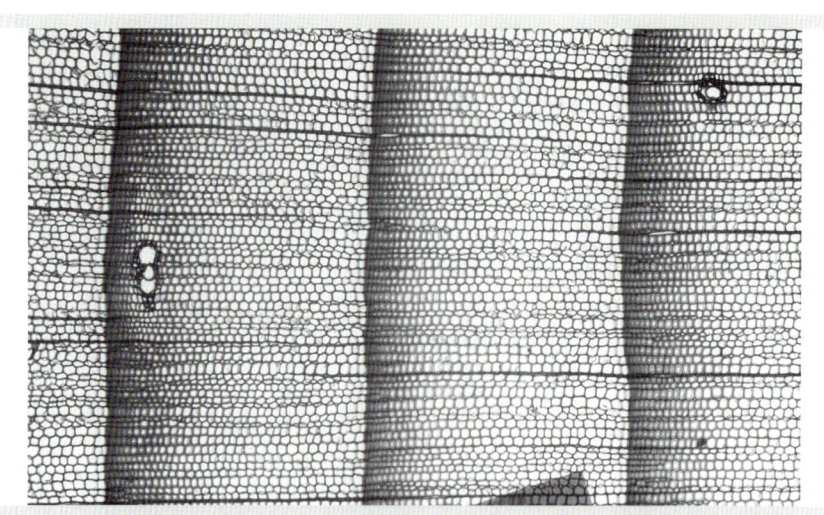

Figure 5. Cross-section surface of a softwood showing three annual rings and two resin canals for *Picea abies* (Photo by Pekka Saranpää).

In hardwoods, the radial cell arrangement and the alternating bands of earlywood and latewood are less distinct. In some hardwood species such as oaks and beeches, large-diameter pores or vessel elements whose main function is water transport may be particularly plentiful and large during the early season. These species are *ring-porous hardwoods* since the growth rings have the large-diameter earlywood pores arranged in a ringlike fashion indicated in Fig. 6. In *diffuse-porous hardwoods* such as poplars and birches the pores have the uniform size and even distribution shown in Fig. 7. Some species do not clearly fall in either category and exhibit a gradual transition of vessel size from earlywood to latewood. They are *semi-diffuse-porous* or *semi-ring-porous hardwoods*.

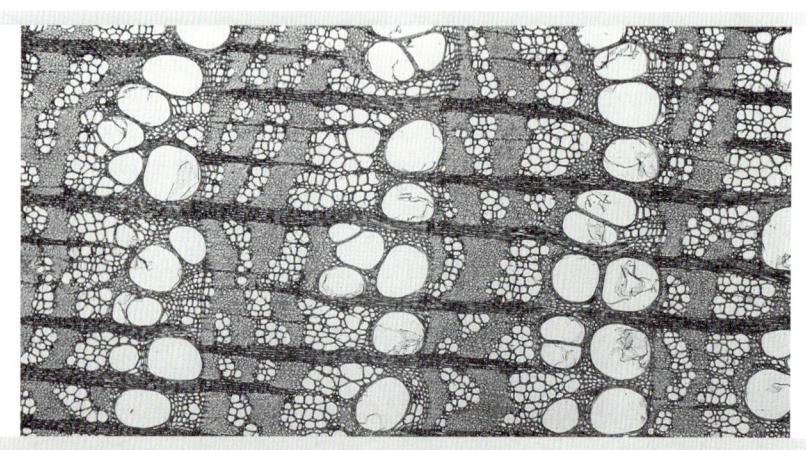

Figure 6. Cross-section of a ring-porous hardwood showing three annual rings with the diameter of vessels wide in the beginning of the season and small at the end of the season for *Ulmus glabra* (Photo by Pekka Saranpää).

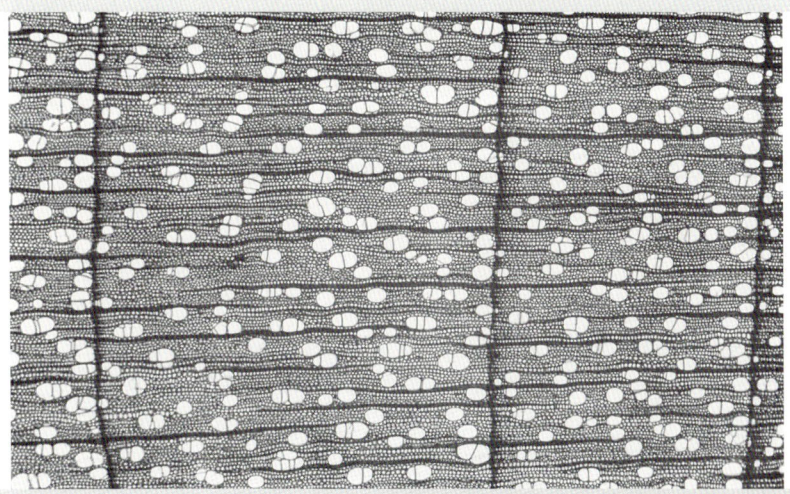

Figure 7. Cross-section of a diffuse-porous hardwood showing three annual rings with the diameter of vessels unchanged through the ring for *Betula pendula* (Photo by Pekka Saranpää).

Prosenchymatous cells function as conducting and supporting tissues. They loose their protoplasm and die immediately after the thickening and lignification of the walls. Only some parenchymatous cells continue to function as longitudinal or radial storage tissues and retain the protoplasm. As long as the parenchyma cells remain physiologically active, the xylem is called *sapwood*. After many years, the protoplasm also dies in the parenchyma cells depending on the tree species, growth conditions, and vitality of the tree. Followed by secondary changes, this process leads to formation of physiologically dead *heartwood* in the center of the stem. The naked eye can distinguish the heartwood from sapwood in most tree species due to differences in color, moisture content, or both (cf. Figs. 3 and 4).

2.2 Formation of wood

Plants have many types of *cells* with varying form and functions. Aggregations of cells of similar type and functions form *tissues*. An essential characteristic of woody plants is vascular tissue – specialized tissue for conducting water.

Division of cells produces woody biomass. The process of growth proceeds simultaneously at two levels – growth in length and growth in diameter. *Elongation* or *longitudinal growth* originates from the buds of the trees at or near the apical growing points called *apical meristems* or *primary meristems* at the tips of the stem, branches, and roots. The primary meristem tissue consists of thin-walled cells rich in protoplasm and capable of repeated division. Although this *primary growth* controls the height of the stem and the entire architecture of the tree, it is responsible for only a tiny fraction of the volume growth of the stem. Behind the thin apical zone of cell formation, the new cells form *permanent tissues* through changes in shape, size, and function. The visible result of the primary growth in timber is the *pith* in the center of the stem.

CHAPTER 4

Behind the points of primary growth and elongation, groups of primary cells outside the apical growing zones remain meristematic and form a sheath of plasma-rich cells called *vascular cambium* instead of changing into permanent tissues. This one-cell-wide cylindrical mantle extends from the tip of the stem to the growing tips of the branches and roots. During the growing season, its cells continuously divide to produce *secondary xylem or wood* toward the inside and *secondary phloem or inner bark* toward the outside. This causes the *secondary* or *radial growth* of the tree.

The meristematic cells of the vascular cambium are *cambial initials* consisting of fusiform initials and cambial ray initials. The longitudinal or axial cells of the xylem and phloem originate from the *fusiform initials*. These are thin, elongated tapering cells with variable length. Depending on tree species and the age of the cambium, the length of the fusiform initials ranges in softwoods from less than 1 mm to more than 9 mm. The diameter is usually 30 micrometers or more. In less specialized hardwoods such as birch (*Betula* spp.) or yellow-poplar (*Liriodendron tulipifera*), the length of the fusiform initials ranges from 1 to 2 mm. In highly developed hardwoods such as black locust (*Robinia pseudoacacia*), the range is from 0.3 to 0.6 mm[8]. The length of the fusiform initials has considerable importance because it determines to a large extent the length of the longitudinal cells in wood.

The cambial *ray initials* give rise to parenchymatous xylem and phloem cells that form radial rays. When the stem grows in thickness, only a thin layer of vertical cells in the vicinity of the cambium remains alive. Other vertical tissues die after the lignification process. Narrow strips of radially oriented short parenchymatous cells remain alive forming *primary rays* that connect the pith with the cambium and phloem. When new growth rings form and the diameter of the stem increases, radial *secondary rays* appear. The physiological function of the ray tissues is to store and distribute horizontally the assimilation products in the tree. The longitudinal or axial cells primarily form *conducting tissues* for water transport and *mechanical tissues* to support the stem, branch, and root mass.

In softwoods, Fig. 8 shows that the rays are usually only one cell wide in the tangential direction and typically 1 to 20 cells high occupying 5%–11% of the wood volume. In hardwoods, Fig. 9 shows that the rays vary more in size and volume. For example, the rays in aspen are uniseriate or only one cell wide in the tangential direction, in birch they are 1–3 seriate, and in oak they are 1–30 seriate. Ray height may vary in hardwoods from 1 to several hundred cells. In oak, the height may be 50 mm making the ray clearly visible with the naked eye. Ray volume ranges in hardwoods from 7% to 30%[9] and strongly influences the figure of wood for carpentry and the fiber properties for pulp.

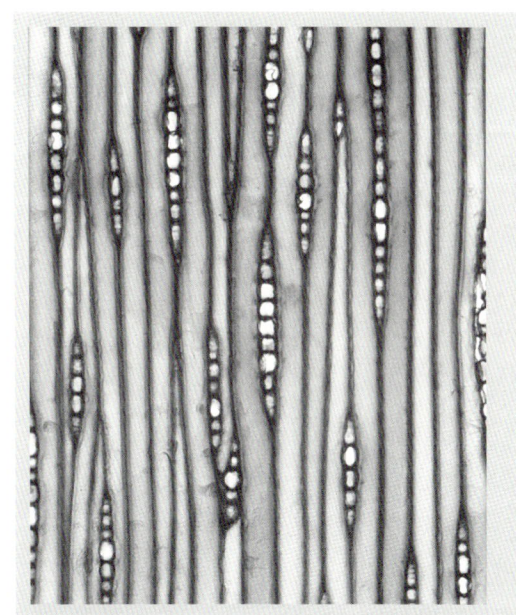

Figure 8. Rays on the tangential surface of a softwood, *Pinus sylvestris* (Photo by Pekka Saranpää).

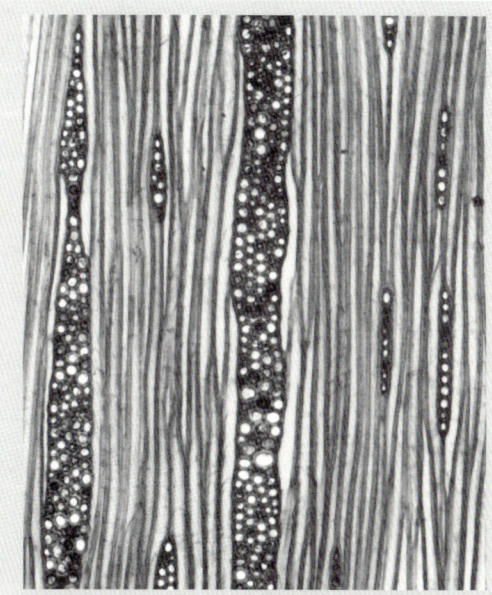

Figure 9. Rays on the tangential surface of a hardwood, *Acer platanoides* (Photo by Pekka Saranpää).

The cells of the vascular cambium – the cambial initials – can divide repeatedly. They divide in half periclinically in the tangential plane parallel to the mantle of the stem to increase the stem diameter or anticlinically in the radial plane to increase the circumference of the cambium itself as the diameter of the tree grows. The cambial circumference also increases through a gradual increase in the length of the fusiform initials. For example, the length of the fusiform initials in eastern white pine (*Pinus strobus*) was only 0.9 mm in the first growth ring but 4 mm in the 60th ring. The cell diameter simultaneously increased from 16 m to 42 m[10]. The average length of the fusiform initials increases in softwoods 100%–400% and in hardwoods 50%–100% during the decades as the cambium moves outward with the increasing girth of the stem[8]. The consequences of this gradual elongation of the cambial fusiform initials are of great practical significance in the pulp and paper industries. It is the ultimate reason why raw material harvested from early thinnings of young, fast-growing plantations has short fibers and chips reduced from sawmill slabs yield pulp with especially long fibers.

When a fusiform initial divides in the tangential-longitudinal direction to form two identical cells, one will continue to function as a cambial fusiform initial and retain its ability to divide bidirectionally indefinitely. The other will become a *mother cell* with a limited capability to divide and then only periclinically. The mother cell divides further to form two *daughter cells* that usually divide once. The new cells differentiate to form secondary wood inside the cambium and secondary inner bark or phloem outside the cambium. Since fewer mother cells form outside the cambium and their ability to divide further is weak, the growth of bark is much slower than that of wood.

CHAPTER 4

The one-cell-wide cambium and the surrounding mother and daughter cells form a thin sheet of undifferentiated tissue called *cambial zone*. Because of the division pattern of the fusiform initials in the cambial zone, the cells of wood and bark align in radial rows. Figure 5 shows that softwoods retain this *radial cell arrangement,* but the post-cambial enlargement of vessel elements disturbs it in hardwoods as Figs. 6 and 7 indicate. Any disturbance or injury in the cambial zone will result in a defect in the stem.

After division stops in the cambial zone, cells may still enlarge in size. In softwoods, the *post-cambial diameter increase* is most conspicuous in the radial direction of earlywood tracheids, and in hardwoods it is most conspicuous in the earlywood vessel elements. The *post-cambial elongation* is less than 10%–15% in softwood tracheids. In hardwoods, the fibrous elements more than double their lengths during maturation from the cambial initials.

With the initiation of daughter cells through cell division at the cambial zone, new pectic-rich *cell walls* form. These are a primary wall that encloses the new unit and the middle lamella or intercellular layer that separates adjoining cells. At the completion of enlargement of the new cells, a rigid secondary cell wall begins to develop on the inside surface of the primary wall. During the phase of cell wall thickening that follows, cellulose and hemicelluloses form within the cell and deposit on the primary wall and a new secondary wall. The formation of lignin begins before this phase is complete. It starts at the cell corners and spreads along the primary wall and middle lamella.

With the enlargement and formation of the secondary wall, the maturing cells simultaneously undergo *further modifications* such as formation of pits, spiral thickening, and perforations on the cell wall. In prosenchymatous cells, the death and disappearance of the living cell contents follow this maturing process. The entire process of cell maturation normally requires a few weeks[8]. The parenchymatous cells remain alive and physiologically active for many years until the formation of heartwood begins.

2.3 Chemical composition of wood

2.3.1 Introduction

The woody biomass has three *principal elements*: carbon (C), oxygen (O), and hydrogen (H). Approximately 50% of the dry mass of wood is carbon, more than 40% is oxygen, and 6% is hydrogen. Small, variable amounts of nitrogen (N) and *mineral elements or ash* are also present. The content of ash in temperate-zone trees is below 0.5% in the wood but much larger in the bark and foliage.

Combinations of carbon, oxygen, and hydrogen form carbohydrates and lignin – the *primary cell wall constituents* of wood. The carbohydrate portion of wood comprises cellulose and hemicelluloses jointly known as *holocellulose*. Cellulose, hemicelluloses, and lignin intermix in the cell wall in a complex way.

The proportions and composition of the different chemical constituents vary among species. In addition, prosenchymatous tissues differ from parenchyma tissues, earlywood differs from latewood, juvenile wood differs from mature wood, sapwood differs from heartwood, normal wood differs from reaction wood, etc. Even greater differences occur between the wood, bark, and foliage components of a tree. Fengel and

Grosser compiled data on the chemical composition of wood in 153 species of the temperate zones[11]. Table 4 shows the average amount of primary cell wall constituents in temperate softwoods and hardwoods. Trees also contain smaller amounts of nonstructural carbohydrates such as starch and sucrose. Instead of being cell wall material, these carbohydrates function primarily as energy reserves and are classified as *extractives*. Both the primary cell wall constituents and extractives are polymeric. They contain large molecules formed by the combination of small repeating structural units.

Table 4. The average amount of primary cell wall constituents in temperate softwoods and hardwoods[11].

Cell wall constituent	Softwoods	Hardwoods
	Proportion in dry matter, %	
Cellulose	40–45	40–45
Hemicelluloses	25–30	25–35
Lignin	25–35	20–25

2.3.2 Primary cell wall constituents

Cellulose is the most abundant organic material on earth. It is present in all the higher plants. Occupying 40%–50% of the wall mass, it is the main constituent of wood. It occurs predominantly in the secondary cell wall usually in association with hemicelluloses and lignin. It is a polysaccharide with high resistance to chemicals. This property is very important in the production of chemical pulp. Its insolubility in aqueous alkaline solutions distinguishes it analytically from the hemicelluloses. It is different from lignin by its resistance to oxidizing agents and susceptibility to hydrolysis by acids and from the extractives by its insolubility in water and organic solvents. Cellulose dissolves in strong mineral acids by hydrolysis to dextrins and oligosaccharides and ultimately to D-glucose[12].

Cellulose consists of glucose molecules ($C_6H_{12}O_6$) produced by the tree through photosynthesis. These monomer units are first transformed into glucose anhydrides ($C_6H_{10}O_5$) and then joined through oxygen atoms end-to-end to form long-chain molecules or polymers $(C_6H_{10}O_5)_n$. The number of monomers per macromolecule indicates the *degree of polymerization* in cellulose. According to Timell[13], the value of n is 8 000–10 000 for wood cellulose in the natural state. In the cell wall, cellulose macromolecules are parallel to each other with lateral linkage through hydrogen bonding to form long strands called *microfibrils*.

The yield of chemical pulp and the strength of fiber correlate positively with the content of cellulose in wood. Besides the strength of the single fiber, another major physical factor influencing the properties of a pulp sheet is the degree of polymerization of the cellulose. In chemical pulping, the degree of polymerization of cellulose is normally 1 000–2 000[14].

Hemicelluloses constitute 25%–35% of the cell wall mass. The proportion is somewhat less in softwoods than in hardwoods. They are heterogeneous low molecular weight polysaccharides formed from glucose and other six-carbon and five-carbon sugar molecules in the apical and cambial growing tissues of the tree. They are insoluble in water but soluble in aqueous alkali. This property allows separating them from the total carbohydrate fraction or holocellulose to yield essentially pure cellulose or α-*cellulose*. Hemicelluloses also readily hydrolyze to their monomeric components in dilute acid forming sugars and sugar acids in a process called *hydrolysis*. Upon complete hydrolysis, the hemicelluloses yield monosaccharides – hexoses (D-*glucose,* D-*mannose,* D-*galactose*) and pentoses (D-*xylose,* L-*arabinose*) – and monomethyl *glucuronic acid* or oligosaccharides containing this acid. Similar to the respective monosaccharides, homopolymers that contain only identical sugar units are called *glucan, mannan, galactan, xylan, and arabinan*. The unsystematic terms *hexosans* and *pentosans* often apply to those polymers that yield hexose and pentose sugars, respectively, on hydrolysis[12].

Hemicellulose molecules have only 150–200 monomer units. They are therefore shorter than cellulose and have a low degree of polymerization. A cellulose chain molecule is linear and composed exclusively of glucose molecules. Hemicellulose molecules contain branching and include a variety of monosaccharides. Two main classes of substances are *xylans* and *galactoglucomannans*. *Xylans* form by polymerization of the anhydro forms of five-carbon sugars and 4-O-methyl-D-glucuronic acid. *Galactoglucomannans* form by polymerization of six-carbon sugars. Timell discusses the properties of hemicelluloses in detail[15, 16].

The galactoglucomannans predominate in *softwood hemicelluloses* and account for 15%– 20% of the dry mass of wood. The glucose to mannose ratio is approximately 1:3. The ratio of galactose to glucose can vary from 1:1 to 1:10. Softwoods also contain approximately 10% of a xylan. The major *hardwood hemicellulose* is an acidic xylan present at 25%±5% by dry mass of extractive-free wood. In a few species such as some birches, the xylan content can reach 35%. Hardwoods also contain approximately 5% of a glucomannan. In papermaking, hemicelluloses serve as bonding, plasticizing, and swelling agents.

The term hemicellulose is neither systematic nor definitive. The distinction between the hemicelluloses and certain water-soluble extractives is not always clear. For example, some noncellulosic polysaccharides such as arabinogalactans are largely soluble in cold water and are therefore usually classififed among the extractives. Arabinogalactan is characteristic to the genus *Larix* where it occurs in the lumina of heartwood tracheids at 5%–30%. It greatly increases the durability of wood.

Lignin characteristically differentiates wood from other cellulosic materials found in the plant kingdom. In its natural state as it occurs in woody cells, it is an amorphous, indefinitely large polymer known as *native lignin or protolignin*. Its most important property is the rigidity and increased stiffness it imparts to cell walls. It is present in the fine cavities within the cell wall where it acts as a bulking agent. Because of its low hygroscopicity, it improves the dimensional stability of the cell wall.

Lignin is quite insoluble. Its basic structural unit consists of phenylpropane with a phenol ring that is substituted by zero, one, or two methoxyl groups. The addition of one methoxyl group to the phenol ring produces a *guaiacyl unit*, and the addition of two methoxyl groups results in a *syringyl unit*. Softwoods contain a guaiacyl lignin, and hardwoods have a guaiacyl-syringyl lignin that is a copolymer of guaiacyl and syringyl residues[17].

The amount of lignin in normal wood is 20%–35% depending on the tree species. The figures in Table 5 show average Klason lignin content for a large number of softwoods and hardwoods in various parts of the world. The proportion of lignin is considerably less in temperate hardwoods than in softwoods and tropical hardwoods.

Table 5. The average Klason lignin content and the standard deviation between species for a large number of softwoods and hardwoods in various parts of the world[18].

Country	Average content of Klason lignin, %	
	Softwoods	Hardwoods
USA	28.8±2.6	23.0±3.0
Former USSR	29.0±1.6	21.9±3.2
Japan	29.6±2.6	22.1±3.0
Taiwan		25.0±3.8
The Philippines		29.4±5.6
Mozambique		27.3±3.4

Chemical pulping removes lignin from the cell wall. Higher lignin contents yield lower amounts of chemical pulp. In practice, some lignin remains in the pulp. Decreasing the lignin content of a pulp results in improved strength because any remaining lignin impairs the bonding ability of the cells. The two groups of lignin – guaiacyl and syringyl lignin – behave differently in pulping. This complicates the mixed pulping of softwoods and hardwoods.

In softwoods, latewood has a lower content of lignin caused indirectly by a difference in cell wall thickness. The decrease in lignin with increase in the percentage of latewood appears to be primarily a dilution effect of the middle-lamella lignin with an actual decline across the secondary wall. Another difference is that the latewood hydrolyzate yields more mannan but less arabinan and xylan. These changes are quantitative in the sense that they directly relate to wall thickening. They are also qualitative, since definite compositional changes occur during latewood wall development[19].

2.3.3 Extractives

Cellulose, hemicelluloses, and lignin primarily determine the chemical and physical properties of wood. In addition, a variety of *secondary or extraneous components* occur in the cell walls and cell lumina. Extraneous components that are at least somewhat readily soluble in neutral organic solvents or water are *extractives*. Some other

CHAPTER 4

extraneous components such as proteins, salts of organic acids, and inorganic materials are partly or wholly insoluble in solvents used to remove the extractives[20].

Extraction with neutral solvents such as hot and cold water, petroleum ether, ethyl ether, ethylene dichloride, dichloromethane, ethanol, or acetone will remove the extractives from wood. No single solvent can remove all extractives. For reasonably complete removal of extractives, it is necessary to employ a sequence of two or more solvents. For example, successive treatments with ether, alcohol, and water may remove the extractives.

Extractives include a large number of compounds involved in the metabolic processes of a tree. The *primary metabolites* are the bioorganic intermediates common to essentially all organisms. It is often possible to interconvert them to primary metabolites. They typically include simple sugars, aminoacids, simple fats, and various carboxylic acids. These always occur in the extractives of living trees. The amounts will vary depending on the time in the growth cycle, the nutritional state, and the tissue.

Secondary metabolites are more complex compounds, and their formation within the organism is essentially irreversible. Glucose – the primary product of photosynthesis – is the starting material for producing not only cell wall components but also most secondary metabolites including starch, sitosterol, simple terpenoids, chlorophyll, phenylpropanoids, the common flavonoids, and simple tannins. In some species or tissues, these intermediates accumulate in pools that can be isolated in large quantities. In others, they metabolize further as fast as they form[21].

Most extractives are intermediate compounds from the metabolic processes of a tree. The qualitative and quantitative variations are considerable and depend on the producing tissue and the environment influencing the organism. Since photosynthesis occurs in the leaves, the foliage synthesizes and stores the carbohydrates and other low molecular weight compounds. Foliage therefore contains unusually large amounts of extractives with a complex composition. Whole-tree chips reduced from trees containing limbs deteriorate rapidly when stored.

Sugars, starch, fats, and fatty acids form a reserve food supply for a tree. Terpenes, resin acids, and phenols are partly protective elements and partly toxic waste from the metabolic processes. Sterols probably function as growth hormones, and components of protein function as enzymes in the synthetic processes[22]. Since the needs of the tree change during the year, the concentrations of constituents vary accordingly.

The rays and other parenchyma tissues in *sapwood* and inner bark are rich in simple monomers and nutrients such as fats, starch, sucrose, simple sugars, inositols, simple glycosides, free and esterified sterols, phenylpropanoids, and other simple phenolics. *Heartwood* and outer bark are deficient in nutrients, glycosides, and metabolic intermediates but rich in compounds such as hydrolyzable and condensed tannins and many other phenolics, alkaloids, resins, essential oils, and a wide variety of specialized compounds. Heartwood and outer bark may also contain the various gums, kinos, and balsams that have evolved as part of the wound-response mechanisms, as well as the compounds capable of protecting these metabolically inactive tissues against biological attack[21, 23].

In stemwood, the concentration of extractives is highest in the heartwood. To protect the inactive center of the stem against attacking micro-organisms, the content of many extractives such as pinosylvin is higher at the periphery of the heartwood zone than in the center. Species growing in temperate zones usually have only a small percentage of extractives in the heartwod. The levels can be much larger in tropical and subtropical species.

With the increase in the amount of extractives when heartwood forms, new compounds also appear. Hardwoods usually have more extractives than softwoods, although the amounts vary from species to species. In some tropical hardwoods, the proportion of extractives may be up to 20%–30% of the dry mass. In temperate tree species, it is generally less than 5%–10%. The average amount of extractives in 22 hardwood species in the southern United States was 4.9% in stemwood and 14.4% in stem bark, when the samples were extracted first with alcohol-benzene and subsequently with ethanol[21]. The values in Table 6 show the amount of acetone extractives in debarked Scots pine (*Pinus sylvestris*) and Norway spruce (*Picea abies*) pulpwood in Finland.

Table 6. The amount of acetone extractives in pulpwood in Finland.

Pulpwood assortment	Sapwood	Heartwood	All wood
	Acetone extractives, % of dry mass		
Scots pine in southern Finland	3.0	5.0	3.3
Scots pine in northern Finland	3.2	5.7	4.0
Norway spruce in southern Finland	1.5	1.6	1.5
Norway spruce in northern Finland	1.9	1.9	1.9

Although the extractives are not a structural part of the woody tissue, they contribute to many properties of wood such as odor, color, durability, permeability, and basic density. They are of considerable importance in understanding the biochemistry and taxonomy of trees, and they also contribute to the raw material properties of the biomass especially in the pulping and biochemical industries. Hillis has published comprehensive analyses of the significance of wood extractives[23, 24].

Most extractives usually dissolve rapidly in the alkaline cooking liquor during delignification by the pulping process. The pulp produced retains them only slightly. The *yield of pulp decreases* with increasing concentration of extractives when calculated per unit mass of raw material. For example, larch wood gives low yields of kraft pulp on a dry mass basis of wood due to its high content of water-soluble arabinogalactan.

Extractives are often responsible for *increased chemical consumption,* since they may decompose the cooking chemicals. They may also *inhibit pulping reactions* and reduce delignification by preventing diffusion and penetration of the cooking chemicals into the raw material. Extractives may also be detrimental to the color, bleachability, and wettability of chemical pulps. During the preparation of mechanical pulps and newsprint and during storage of the latter, polyphenols and other extractives can cause *color prob-*

lems. Changes in color may result from the oxidation of chemical compounds when exposed to air and sunlight, an enzymatic catalytic effect of cell sap in rays and other parenchyma tissues, and a combination of polyphenols or other compounds with metals[25].

Free resin coagulates and adheres to fibers and metal surfaces causing *pitch troubles* in pulp and paper manufacture. Pitch deposits may interfere with the paper machine operation and cause spots or holes in the finished paper. Resin deposited on fiber surfaces prevents bonding between the fibers and between a glue and the fiber material. The ability to form hydrogen bonds is a fundamental property of the fibers and a necessary requirement for paper strength. Pitch comes from the resin acids incompletely removed in acidic and mechanical pulping processes or precipitated from the pulping liquor due to sudden changes in temperature, pH, or other conditions. Alkaline processes can tolerate resin. Pitch usually presents no problems in sulfate pulping of softwoods. In the mechanical pulping of tree species containing high resin levels such as yellow pines in the southern United States, chemicals, the selection of juvenile raw material with a lower resin content, or both are necessary to control pitch troubles.

After cutting of timber, the extractives content decreases, and the composition changes. Lingering life processes of ray cells are initially involved through respiration. This gradually changes to degradation by microorganisms as they invade the biomass. The *resin reactions* involve simultaneous oxidation by air and enzymatic hydrolysis. Fats and waxes mainly hydrolyze enzymatically. Because of hydrolysis and oxidation, the hydrophilicity of the resin constituents increases. When storing raw material as chips, the resin reactions are rapid due to the high temperature in the chip pile. The presence of materials such as foliage and inner bark with a high content of live parenchyma cells accelerates the reactions.

In acid sulfite pulping, changes in resin are beneficial. Storing wood for several months to minimize pitch problems and lower the resin content of the pulp is possible. In kraft pulping, the storage of wood has detrimental consequences, since yields of both turpentine and tall oil decrease[26]. As a consequence of microbiological activity, polysaccharides undergo attack during long storage resulting in reduced yield and quality of pulp.

The extractives present a great opportunity and challenge to the chemical industries. *The resins* and related materials comprising oleoresins from pines are examples of extractives of considerable importance. Commercial resin, fatty acids, and turpentine products are tapped from living pines, extracted from chips made from mature pine stumps, and to a much larger extent recovered as by-products in sulfate pulping.

2.3.4 Inorganic components

Woody biomass contains small amounts of extraneous inorganic components that the tree takes from the soil through its roots and sap stream. When the concentration of a chemical element in the soil liquid outside the root is greater than in the liquid inside the root, that element will flow through semipermeable membranes into the cells of the root hairs and root tips and further to the stem and crown. The amount of chemical elements

Structure and properties of wood and woody biomass

in a tree biomass therefore reflects not only the requirements of the tree but also the amount in the ground. Wood, bark, and foliage may also contain elements that they obviously do not need such as heavy metals.

In temperate-zone trees, the quantity of inorganic elements in stemwood rarely exceeds 0.5%–0.7%. The inorganic elements occur in trees as components of the extractives or as crystals. The alkaline earths such as calcium, potassium, and magnesium usually are the main inorganic components. Certain tree species may contain large amounts of silica particularly at the base of the stem. This can cause chain saws, chipper knives, and other tools to become dull. Inorganic components also promote wear and tear and chemical problems in pulping especially in the chemical recovery systems. In composite board, they are a serious defect.

Young trees have a higher concentration of ash than mature trees. Hardwoods have more ash than softwoods, and tropical hardwoods have more ash than temperate hardwoods. In Table 7, each figure represents the average ash content of stem wood for a large number of softwoods and hardwoods in several countries.

Table 7. The average ash content of stemwood and the standard deviation between species for a large number of softwoods and hardwoods in several countries[18].

Country	Softwoods	Hardwoods
	Ash in stem wood, % of dry mass	
USA	0.3±0.1	0.5±0.3
Former USSR	0.5±0.4	0.6±0.4
Japan	0.4±0.4	0.5±0.2
Taiwan		0.9±0.4
The Philippines		1.2±0.7
Mozambique		1.6±1.1

When using biomass for the production of energy, all tree parts may be involved. Compared to stemwood, other tree components contain a much higher proportion of ash. In the southern hardwoods in the United States, the proportion of ash is tenfold in the stem bark compared to the stemwood and sevenfold in branch bark compared to branch wood as Table 8 showsUnder boreal conditions, tree biomass contains less ash. The average ash content of the five major tree species of Finland is 0.46% in stemwood, 2.97% in stem bark, 1.52% in unbarked branches, and 4.97% in foliage.

Table 8. The proportion of ash in the wood and bark of stem and branches in the southern hardwoods in the United States[22].

Tree component	Wood	Bark
	Ash, % of dry mass	
Stem	0.75	7.87
Branches	0.94	6.76

CHAPTER 4

Table 9 shows the average concentration of important minerals in the dry mass of wood and bark in 22 hardwood species in the southern United States and in the major softwood and hardwood species in Finland. Without exception, the concentrations of all primary elements are greater in bark and especially in the inner bark.

Table 9. Concentration of some mineral elements in the stemwood and stem bark of 5 major Finnish tree species[27] and 22 hardwood species in the southern United States[21].

Tree component	Concentration of primary elements, %				Concentration of trace elements, ppm				
	P	K	Ca	Mg	Mn	Fe	Zn	S	B	Cu
2 softwoods in Finland
Stemwood	0.01	0.06	0.12	0.02	147	41	13	116	3	2
Stem bark	0.08	0.29	0.85	0.08	507	60	75	343	12	4
3 hardwoods in Finland
Stemwood	0.02	0.08	0.08	0.02	34	20	16	90	2	2
Stem bark	0.09	0.37	0.85	0.07	190	191	131	341	17	13
22 hardwoods in southern United States
Stemwood	0.02	0.16	0.19	0.04	89	67	11			5
Stem bark	0.03	0.20	3.05	0.11	568	135	35			13

When burning woody biomass, the mineral elements remain in the ash residue as oxides. The ash does not contain any nitrogen and chlorine because they are volatile and escape in the combustion process.

The prevailing element in wood ash is calcium. With magnesium, this makes ash an effective neutralizing agent for acid soils. Bark is especially rich in calcium. Potassium and phosphorus are common fertilizers, and many trace elements are also necessary for trees. Since these elements are vital for the growth of the trees, recycling of wood and bark ash from wood and chip fired heating and power plants will help to maintain the nutrient balance and sustainability of forest soil.

2.4 Structure of the cell wall

Woody cells have a hollow center called *lumen* and layered *cell walls* that are typically perforated with small openings. The thickness, structure, and chemical composition of the wall greatly influence the behavior of wood in various uses. In chemical pulping the cooking liquor either penetrates the wood through the cell lumens along the grain or more importantly diffuses into the wood through the cell walls with equal speed in all directions.

Figure 10 shows that the *primary cell wall constituents* are cellulose, hemicellulose, and lignin. These polymer substances form from the combination of smaller structural units in a highly ordered manner. In addition, woody cells may contain minor components or *extraneous materials* such as organic extractives and inorganic materials or ash. The secondary components are partly in the lumen, and they do not substantially affect the structure of the cell wall.

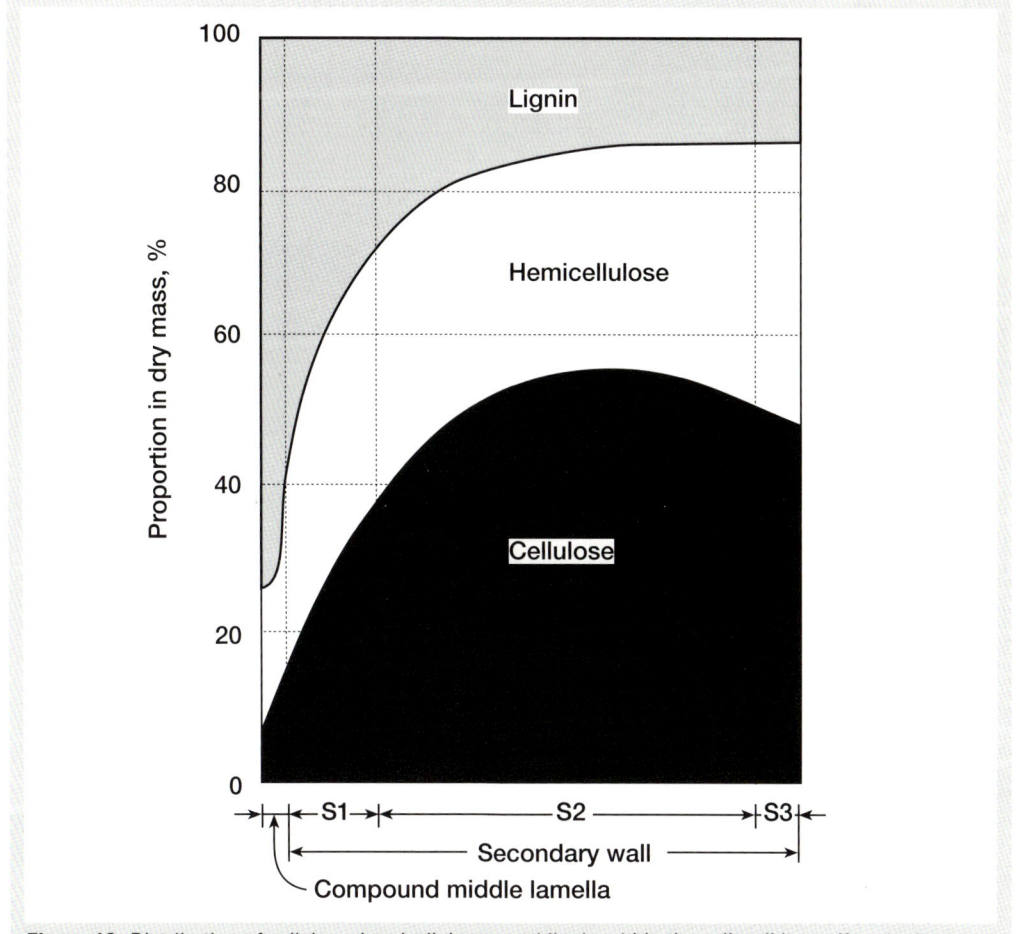

Figure 10. Distribution of cellulose, hemicelluloses, and lignin within the cell wall in conifers (redrawn from Panshin and de Zeeuw[8]).

CHAPTER 4

Cellulose is the major component of the mass of the cell wall and has the most pronounced effect on the properties of wood. The basic unit of cellulose is the glucose molecule that forms through assimilation in the foliage and then goes through the vascular phloem tissues to the apical and cambial growing points of the tree. After transformation into glucose anhydrides there, they link end-to-end to form a long chain polymer, $(C_6H_{10}O_5)_n$, or cellulose. The polymeric cellulose molecules are less than 5 μm long and approximately 1 nm in diameter – too small for observation in an electron microscope.

In a woody cell, the polysaccharide molecules have an arrangement in a precise pattern as long strands called *microfibrils* shown in Fig. 11. The core of a microfibril contains a bundle of cellulose molecules encased in a shell of short chain hemicellulose molecules. Microfibrils are 10–12 nm wide and 5–6 nm thick[28] – visible with an electron microscope. In most parts of a microfibril, cellulose molecules have a parallel arrangement like a crystal that gives rise to the name *crystallites* or *micelles*. Amorphous areas where the cellulose molecules do not have a precise arrangement interrupt the crystallite areas. Approximately two-thirds of the cellulose is in crystalline form. The hemicelluloses occur primarily in the amorphous areas in the spaces between microfibrils to bond them together[29].

The mechanism of incorporation of lignin in the cell wall is not well known. It is probably present in the amorphous areas of microfibrils and between microfibrils to form a rigid and stable wall structure. Tiny voids occur between the microfibrils allowing water to penetrate into the cell wall. This causes swelling and shrinkage in wood when it absorbs or loses water in a moisture content below the fiber saturation point.

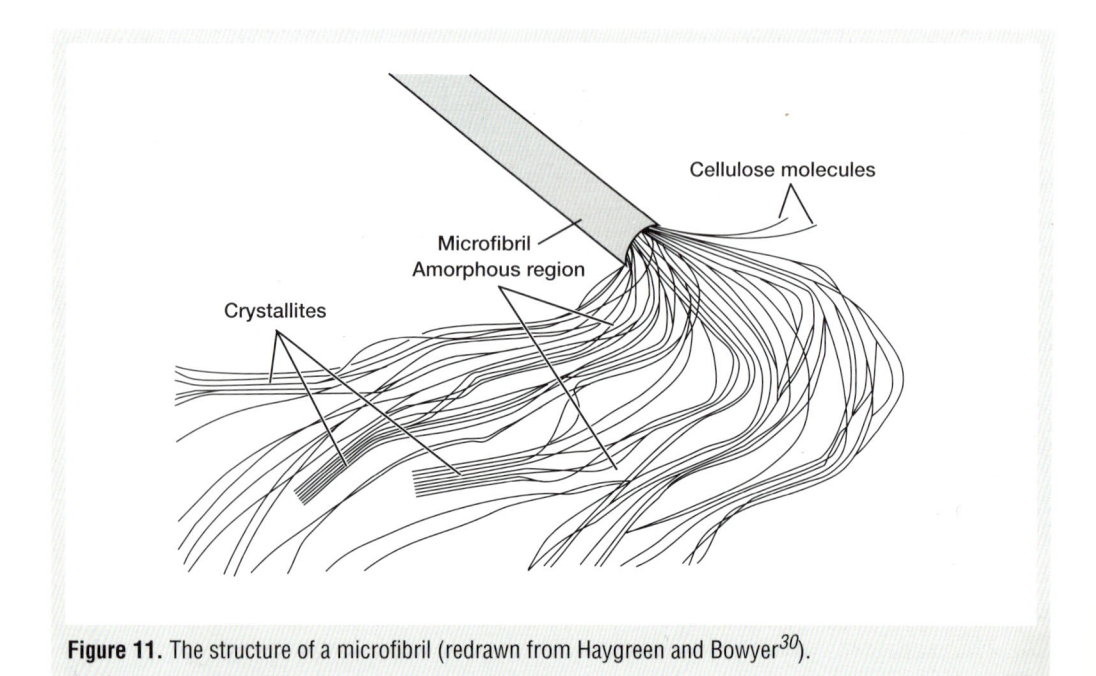

Figure 11. The structure of a microfibril (redrawn from Haygreen and Bowyer[30]).

Immediately after the final division of a cambial daughter cell, a newly formed wood cell is filled with fluids and covered only by a thin membrane-like *primary wall* (P) that still allows the cell to enlarge. The primary wall is poor in cellulose but rich in lignin and pectins that are complex colloidal compounds with high molecular weight. The pectin-rich material between the primary walls of two adjacent cells is the *middle lamella*. Because the thickness of the primary wall is only approximately 0.1 μm and that of the middle lamella even less, it is difficult to separate the region between two primary walls. For certain practical purposes, the combined true middle lamella and the two adjacent primary walls have the name *compound middle lamella*.

After completion of the cell enlargement, the *secondary wall* (S) starts developing through deposition of materials on the inside of the primary wall. In the secondary wall, the microfibrils have a parallel arrangement. They spiral around the long axis of the cell forming thin *lamellae* shown in Fig. 12. When a piece of dry wood absorbs water and swells, the angle of the microfibrils determines how the swelling will distribute in the longitudinal, radial, and tangential directions of the piece.

In the beginning of the cell wall thickening process, the first 4–6 lamellae of the secondary wall form a 0.1–0.2 μm thick lignin-rich S_1-*layer*. The angle of microfibrils in this layer is 50–70°.

As the thickening of the cell wall proceeds, the *orientation of microfibrils* changes, and the angle decreases to 10–30°. In temperate softwoods, this S_2-*layer* of the secondary wall varies widely in thickness. In latewood walls, it consists of approximately 30–40 lamellae and contains more cellulose and less lignin than the P- and S_1-layers. In earlywood walls, the S_2-layer is considerably thinner. This is why the content of lignin is higher in thin-walled earlywood cells than in thick-walled latewood cells. Figure 10 shows the distribution of the cellulose, hemicelluloses, and lignin in the cell wall in conifers.

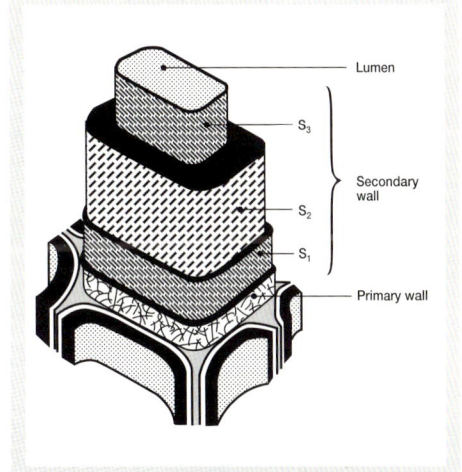

Figure 12. Cell wall layers and organization of microfibrils in a softwood tracheid (redrawn from Côte[31]).

Before the thickening of the cell wall ceases, a change in the orientation of microfibrils occurs again. An S_3-*layer* forms on the inside of the wall. In this layer, the angle of microfibrils to the cell axis is 60–90°. The S_3-layer is even thinner than the S_1-layer. It is poor in lignin but rich in hemicelluloses[29]. On the inside of the S_3-layer, many hardwoods and few softwoods have helical ridges formed by bundles of microfibrils. These *spiral thickenings* do not affect the properties of pulp. Since their frequency, size, form, and location in the cell are specific to species, they are useful in microscopic identification of wood and separate fibers.

CHAPTER 4

The wall thickness varies depending on tree species, cell type, season of formation, and other factors. The following averages for *Abies balsamea, Picea abies* and *Pinus sylvestris* demonstrate the relative volumetric proportions of different cell wall layers in northern softwoods: middle lamella plus primary wall 2%, S_1-layer 16%, S_2-layer 74%, and S_3-layer 8%[8]. Due to its high proportion, the S_2-layer largely determines the properties of wood as a raw material for pulp and lumber.

In all types of cells, tiny recesses through which fluids and gases can pass from cell to cell interrupt the secondary walls. These regions lacking the secondary wall are called *pits*. The primary parts of a pit are *pit membrane* and *pit cavity*. The membrane consists of the middle lamella and primary wall that control the flow of fluids. The cavity is the space within the recess of the secondary wall. The lumen end of the cavity is the *aperture*.

Two main types of pits are the *simple pits* and the *bordered pits*. The *simple pits* have a straight-walled cavity and a single circular outline in the surface view. They occur in all parenchyma cells and also in some hardwood prosenchyma cells. In the *bordered pits,* the cavity forms a conical chamber that is concave toward the membrane. In the frontal view, a border surrounds the opening or aperture. In softwood tracheids, all pits are bordered.

A complementary pit in an adjacent cell wall usually matches a pit to form a pit pair. If two adjacent pits are similar, they form a *simple pit pair* or *bordered pit pair*. If the pits are dissimilar such as between a parenchyma cell and a tracheid in softwoods or a parenchyma cell and vessel element in hardwoods, they form a *half-bordered pit pair* as Fig. 13 shows.

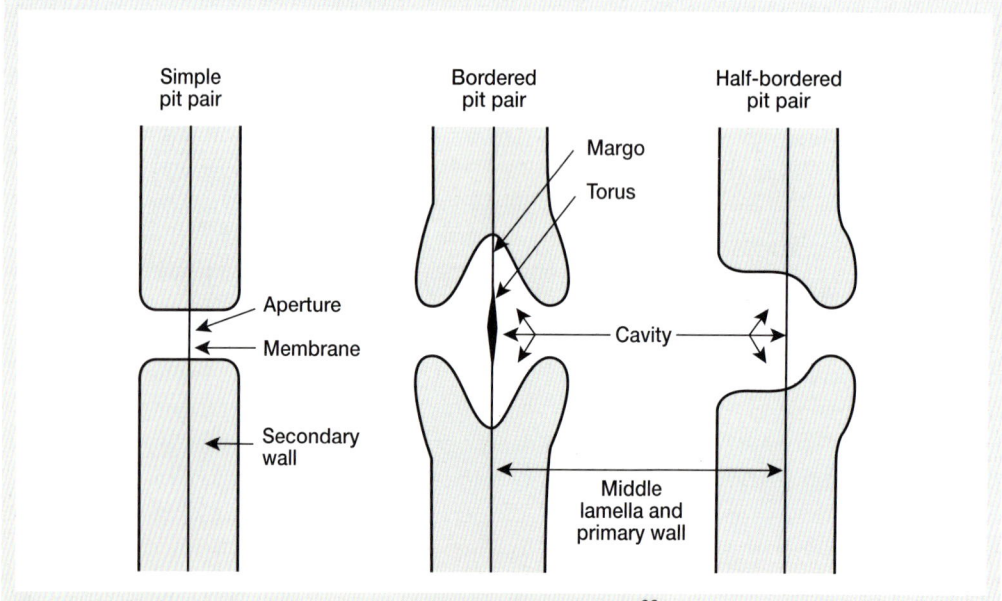

Figure 13. Profiles of pit pairs (redrawn from Haygreen and Bowyer[30]).

In softwoods, the membrane of a bordered pit has two parts. A disk-shaped center called *torus* is slightly wider in diameter than the pit aperture. A *margo* is formed by the surrounding membrane region extending from the pit border to torus. Openings between the microfibril bundles of the marco allow liquids and gases to pass the pit membrane. The torus may close the pit aperture and form an *aspirated pit*. This occurs with heartwood formation in some softwoods to prevent the passage of liquids. In such heartwood, penetration of preservatives and pulping chemicals is slower and the cooking time is longer than in sapwood with open bordered pits.

2.5 Woody cells

2.5.1 Introduction

Wood contains many types of cells that all have specific functions. A cubic centimeter of wood contains hundreds of thousands of cells.

The function determines the size, form, wall thickness, and perforation of a cell. Three major categories are conducting cells, supporting cells, and storage cells. The first two groups are longitudinal or axial and have an elongated form. The storage cells are primarily radial but may also be longitudinal. They are typically short and thin-walled. The conducting and supporting cells are *prosenchymatous,* and the storage cells are *parenchymatous*.

Softwoods have a simple structure. They posses only few types of cells, and the cells have a regular alignment. Figures 14–16 show that hardwoods have a more complex structure with a larger number of cell types specialized for different functions. All tree species do not have all types of softwood or hardwood cells. Table 10 gives examples of the amounts of different cell types in selected tree species. Since cell types differ greatly in dimensions, wall thickness, and other properties, their proportional distribution affects the behavior of wood in pulping and the properties of pulp.

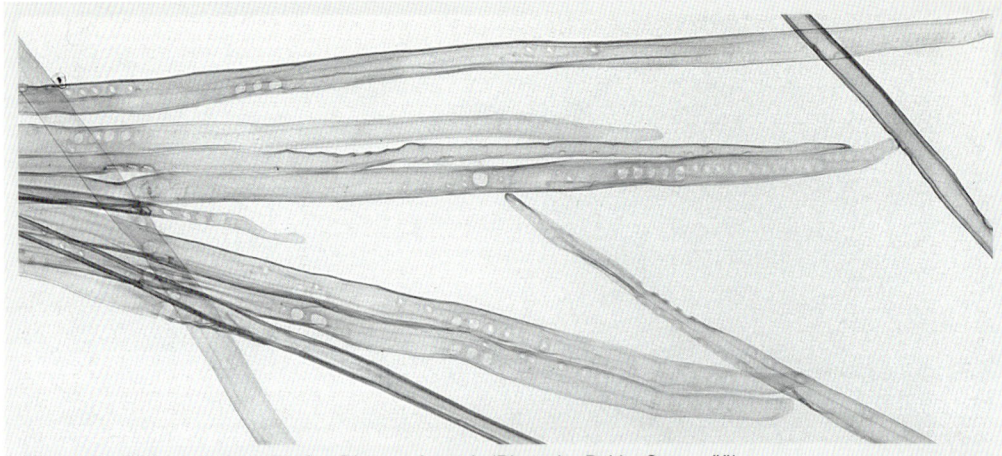

Figure 14. Softwood tracheids for *Pinus sylvestris* (Photo by Pekka Saranpää).

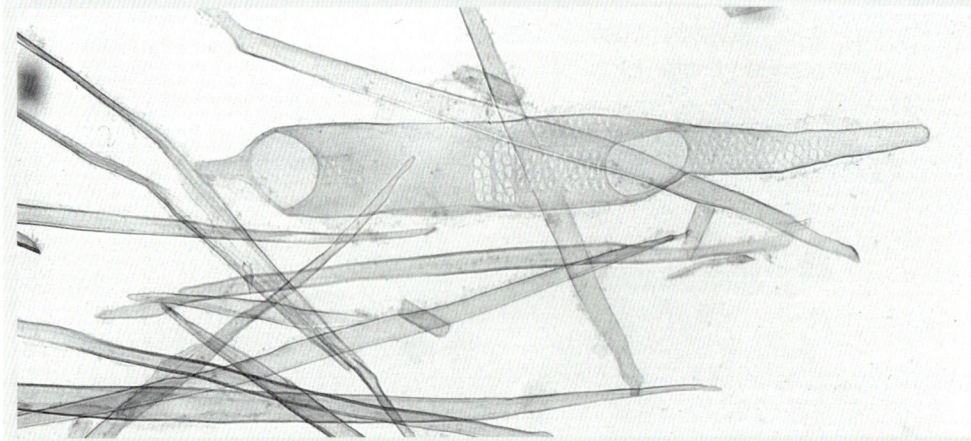

Figure 15. Fibers and a vessel segment of a diffuse-porous hardwood for *Populus tremula* (Photo by Pekka Saranpää).

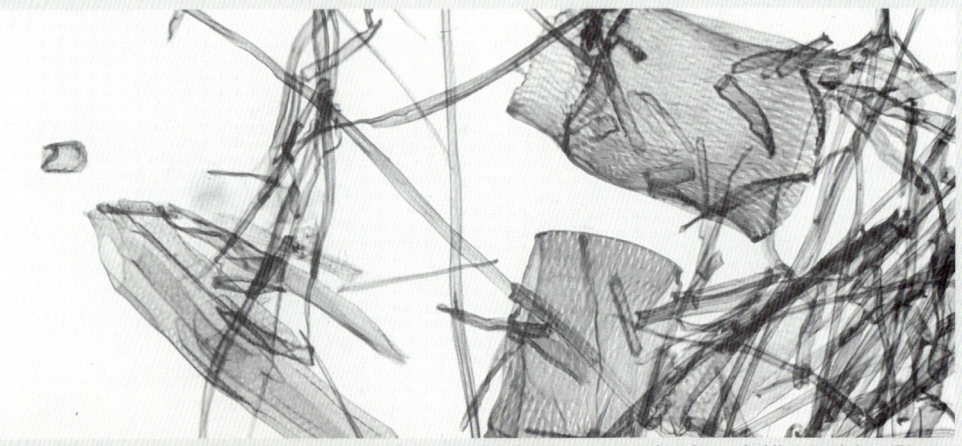

Figure 16. Fibers and wide and narrow vessel segments of a ring-porous hardwood, *Ulmus glabra* (Photo by Pekka Saranpää).

Table 10. Proportion of different types of cells in the wood of selected tree species according to Ilvessalo-Pfäffli[9] from original sources Isenberg[32], Wagenführ and Schreiber[33], and Panshin and de Zeeuw[8] (* Including epithetial cells).

Tree species	Tracheids	Rays	Longitudinal parenchyma	Resin canals
Softwoods	\multicolumn{4}{c}{Percent of wood volume}			
Abies alba	90.4	9.6	Trace	90.4
Abies balsamea	94.3	5.7		
Araucaria angustifolia	94.5	5.5	Trace	
Larix decidua	91.2	8.8	0.9	
Larix laricina	89.0	10.0	0.9	
Picea abies	95.3	4.7	1.4*	
Picea mariana	94.8	5.0		0.2
Picea sitchensis	92.5	7.2		0.3
Pinus palustris	90.8	8.4		0.8
Pinus ponderosa	93.0	6.7		0.3
Pinus radiata	88.6	11.2		
Pinus strobus	94.0	5.3		0.7
Pinus sylvestris	93.1	5.5		
Pseudotsuga menziesii	89.0	9.0	2.0	

	Fibers	Vessels	Rays	Longitudinal parenchyma
Hardwoods	\multicolumn{4}{c}{Percent of wood volume}			
Acer rubrum	68.0	18.0	13.3	0.1
Acer saccharum	61.0	21.0	17.9	0.1
Betula papyrifera	5.7	10.6	11.7	2.0
Betula pendula	64.8	24.7	8.5	2.0
Eucalyptus globulus	49.0	21.0	14.0	16.0
Fagus grandifolia	56.7	21.4	20.4	-
Fagus sylvatica	37.4	31.0	27.0	4.6
Liquidambar styraciflua	26.6	54.9	18.3	0.2
Liriodendron tulipifera	49.0	36.6	14.2	0.2
Nyssa sylvatica	45.0	38.4	17.6	-
Populus deltoides	53.1	33.0	13.7	0.2
Populus tremula	60.9	26.4	12.7	-
Quercus alba	47.8	16.1	28.0	8.0
Quercus robur (narrow-ringed)	44.3	39.5	16.2	-
Quercus robur (wide-ringed)	58.1	7.7	29.3	4.9

2.5.2 Softwood cells

As much as 90%–95% of the volume of softwood consists of vertical or *longitudinal tracheids*. They are elongated, tubelike cells with a square cross-section having blunt and closed ends. Softwood tracheids are the primary source of fiber in the pulp and paper industries. They produce strong paper and make it possible to run paper machines at high speeds. The length, diameter, wall thickness, and their relations in tracheids are of paramount importance for the users of long-fiber pulp.

The dimensions of tracheids vary between tree species. Environmental factors and the age of the cambium – the distance of the annual ring from the pith in years – affect them. Pulpwood from young trees or tops of older trees has shorter tracheids than pulpwood from mature trees or sawmill slabs. The average length of mature tracheids for the most common pulpwood species is usually 3–4 mm in mature wood. Spruces have longer tracheids than pines. Table 11 shows that *Araucaria angustifolia* and *Sequoia sempervirens* are tree species with unusually long tracheids. Note that a load of pulpwood logs despite the age of the trees always includes wood that is not fully mature and therefore has shorter tracheids.

Table 11. Dimensions of longitudinal tracheids in different softwoods compiled by Ilvessalo-Pfäffli[9].

Tree species	Tracheid length, mm		Tracheid width, μm	
Abies alba	3.7	1.6–5.7	38	18–58
Abies balsamea	3.5	1.9–5.6	30–40	
Araucaria angustifolia	7.2	5.6–9.0	47	19–60
Larix decidua	3.5	1.4–6.2	38	24–52
Picea abies	3.4	1.1–6.0	31	21–40
Picea glauca	3.5		25–30	
Picea sitchensis	5.6	3.6–7.3	35–45	
Pinus banksiana	3.5	1.5–5.7	28–40	
Pinus caribea	4.6		41–52	
Pinus contorta	3.1		35–45	
Pinus elliottii	4.0		43	
Pinus nigra	3.2	0.5–4.9	39	16–60
Pinus palustris	4.0		41	
Pinus radiata	3.0		44	
Pinus resinosa	3.4	1.2–5.2	30–40	
Pinus sylvestris	3.1	1.8–4.5	35	14–46
Pinus taeda	4.0		45	
Pseudotsuga menziesii	3.9	1.7–7.0	35–45	
Sequoia sempervirens	7.0	2.9–9.3	50–65	
Tsuga canadensis	3.0		28–40	
Tsuga heterophylla	4.2	1.8–6.0	30–40	

The tangential width of a tracheid is approximately 30–50 μm. The length-to-diameter ratio of softwood tracheids is of the magnitude 100:1. This ratio does vary between species, from tree to tree, and within a tree.

Structure and properties of wood and woody biomass

For earlywood tracheids, the main function is water conduction. For latewood tracheids, it is to support the structure. The cell wall thickness reflects this difference. Earlywood tracheids have thin walls, pronounced lumen, and more pits for easier water conduction. Figure 17 shows that latewood tracheids have thick walls, smaller radial diameter, smaller lumen, and fewer and smaller pits. In papermaking, they behave quite differently, since stiff and thick-walled latewood tracheids retain their original cross-section form and thin-walled earlywood tracheids collapse and become ribbonlike.

Some softwoods have short, rectangular, longitudinal cells placed endwise in strands. They are prosenchymatous *strand tracheids* or *longitudinal parenchyma cells*. Due to their small level, they have little importance in paper making. In pine wood, they do not occur.

Rays composed primarily of *radial parenchyma cells* occupy approximately 5%–10% of softwood volume. They are typically full of material, thin-walled, and brick-like. Since the length is only 0.1–0.2 mm, they are not desirable in paper making. Radial parenchyma cells remain living until the formation of heartwood. In some softwood species, rays may contain some prosenchymatous *ray tracheids* that are approximately the same size as the ray parenchyma cells but are dead and have bordered pits instead of simple pits.

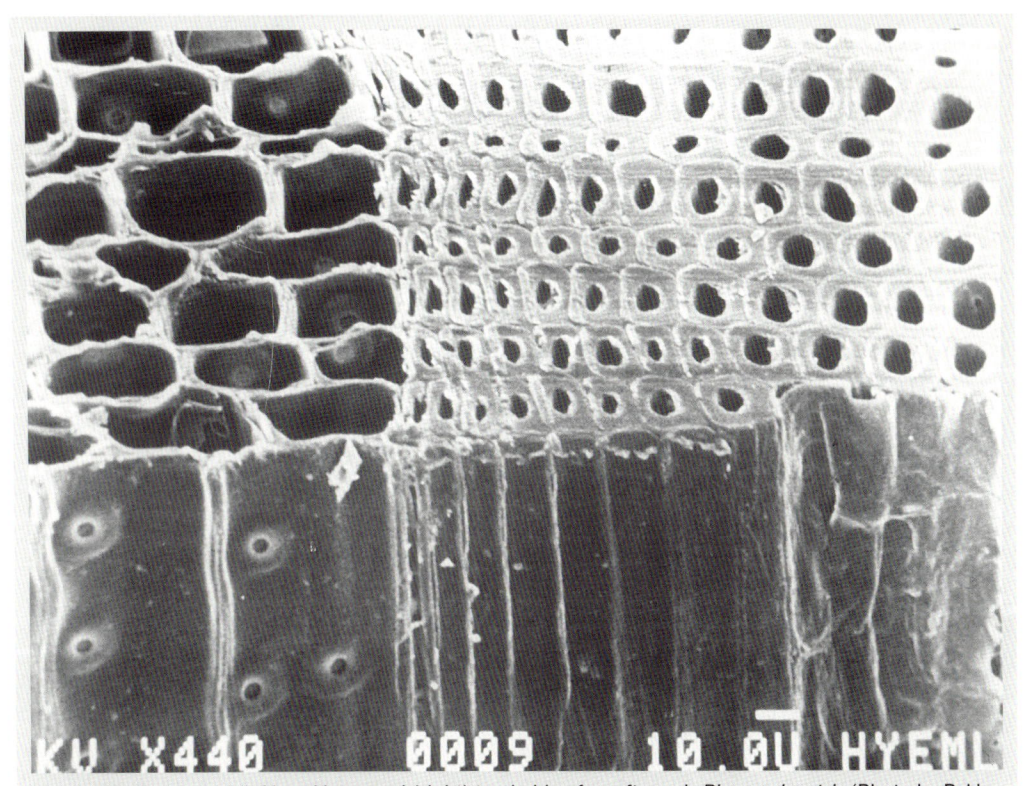

Figure 17. Earlywood (left) and latewood (right) tracheids of a softwood, *Pinus sylvestris* (Photo by Pekka Saranpää).

CHAPTER 4

Larix, Picea, Pinus, and *Pseudotsuga* have radial and vertical *resin canals* that form a network for resin transport in the tree. The canals are intercellular spaces extending vertically among longitudinal tracheids and radially within rays. The resin canals have a lining of resin producing parenchyma cells called *epithelial cells* shown in Fig. 18. Wounding of the cambium causes formation of *traumatic resin canals*. Even in pines where the resin canals are the most frequent, their volume is less than 1%. They are nevertheless important in pulping and papermaking because of the resin they contain.

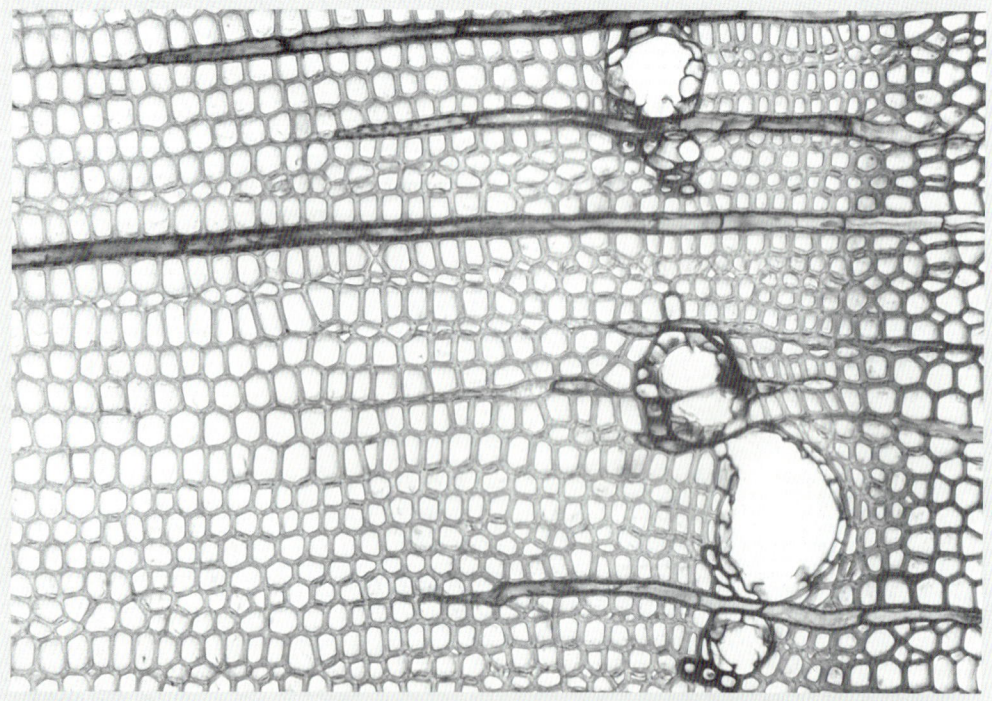

Figure 18. Resin canals on the cross section surface of a softwood, *Picea abies* (Photo by Pekka Saranpää).

2.5.3 Hardwood cells

Hardwoods have a more complex cell structure than softwoods. They contain several types of specialized cells in widely varying proportions. Specific characteristics of the hardwoods are a lack of radial alignment of cells, variable size and composition of cells, abundance of rays, and the presence of pores or vessel elements. The four major cell types shown in Fig. 19 occur in hardwoods: fibers, vessel elements, tracheids, and parenchyma cells. Wood *fiber* is also a generic term for all woody cells used by the pulp and paper industries.

In most hardwood species, *fibers* occupy 40%–75% of wood volume. The primary function of fibers is to support the structures of the tree, although they can also conduct water. The fibers are therefore long, tapered narrow cells with closed ends and very thick walls. A higher amount of fibers gives a greater hardwood strength for a given density. Two types of fibers exist: *fiber tracheids* and *libriform fibers*. The former have bordered pits, and the latter have simple pits.

Fibers are the most important component of hardwood pulps. In most hardwoods, the length of the fibers is 0.7–1.2 mm, the width is 10–30 µm, and the length-to-diameter ratio is approximtely 1:50 as Table 12 shows. The narrow form that results in the smooth surface and even texture of a paper sheet compensates somewhat for the short length. Although they have lower strength properties than softwood fibers, a certain proportion of hardwood fiber in the raw material of high-grade printing papers is necessary for these reasons.

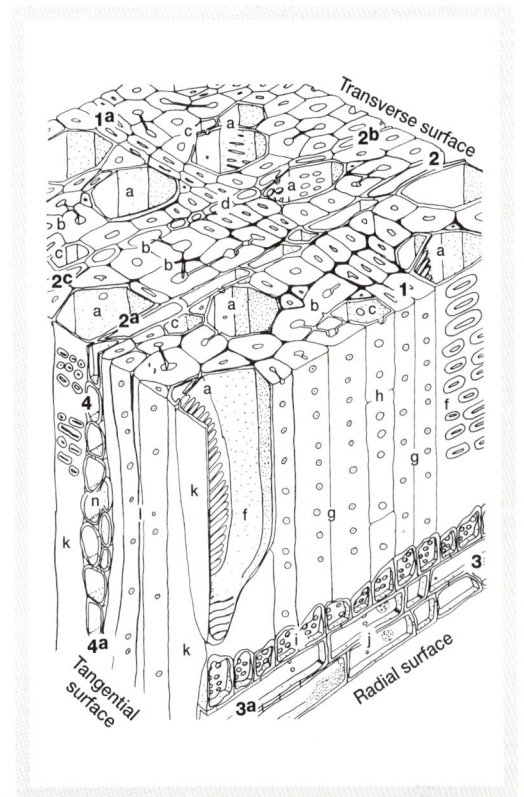

Figure 19. Schematic three-plane drawing of diffuse-porous hardwood for *Liquidambar styraciflua* (redrawn from Panshin and de Zeeuw[8]) showing Transverse surface: 1–1a, boundary between two annual rings (growth proceeded from right to left); 2–2a, wood ray consisting of procumbent cells; 2b–2c, wood ray consisting of upright cells; a, vessels; b, fiber tracheids; c cells of longitudinal parenchyma and Radial surface: f, portions of vessel elements; g, portions of fiber tracheids; h, a strand of longitudinal parenchyma; 3–3a, upper portion of a heterocellular ray comprised of a marginal row of upright cells and two rows of procumbent cells.
Tangential surface: k, portions of vessel elements; l, fiber tracheids; 4–4a, portion of wood ray; n, procumbent cells in the body of the ray.

CHAPTER 4

Table 12. Examples of the dimensions of hardwood fibers compiled by Ilvessalo-Pfäffli[9].

Tree species	Fiber length, mm Average	Fiber length, mm Range	Fiber width, µm Average	Fiber width, µm Range
Acacia auriculiformis	0.8		14	
Acer rubrum	0.8	0.3–1.1	16–30	
Acer saccharum	0.8	0.3–1.3	16–30	
Albizia falcataria	1.0–1.1		24–42	
Alnus rubra	1.2		16–40	
Betula papyrifera	1.3		25	
Betula pendula	1.3	0.8–1.8	25	18–36
Eucalyptus globulus	1.1	0.3–1.5	20	10–28
Eucalyptus saligna	0.8–0.9		16–2	
Fagus grandifolia	1.2	0.6–1.9	16–22	
Fagus sylvatica	1.2	0.5–1.7	21	14–30
Gmelina arborea	1.0		28–38	
Liquidambar styraciflua	1.7	1.0–2.5	20–40	
Liriorendron tulipifera	1.9	0.8–2.7	24–4	
Populus deltoides	1.0		25–40	
Populus tremula	0.9	0.2–1.6	19	13–30
Populus tremuloides	1.0	0.4–1.9	10–27	
Quercus alba	1.4		14–22	
Quercus robur	1.1	0.5–1.6	23	14–30
Salix alba	1.1		22	

An essential anatomical characteristic of hardwoods is the presence of *vessel elements* or *vessel members* that do not occur in softwoods. They are short, large-diameter, and thin-walled cells with perforated ends known as *perforation plates*. They link end-to-end along the longitudinal axis of the stem to form tubelike structures of indefinite length called *vessels*. In the cross-section phase of wood, they appear as pores. Hardwoods sometimes have the designation of *porous woods*. The vessels function as avenues of water distribution.

The proportion of vessels in most hardwoods is 10%–40% of the volume, but less of the mass. As a rule the vessel elements are shorter than hardwood fibers as Table 13 shows.

Table 13. The length of fibers and vessel elements in seven hardwood species in the United States[18].

Tree species	Vessel elements	Fibers
	Average length, mm	
Acer rubrum	0.42	0.92
Alnus rubra	0.85	1.19
Betula papyriferia	1.00	1.35
Fagus grandifolia	0.61	1.28
Liquidambar styraciflua	1.32	1.82
Populus tremuloides	0.67	1.32
Quercus alba	0.40	1.39

As noted above, the diameter of vessels varies greatly from species to species. In certain hardwoods, it can even vary within a growth ring. Depending on the range and pattern of the annual variation, hardwoods are diffuse-porous with little variation in the vessel diameter, ring-porous with large vessels in earlywood and small vessels in latewood, or semi-ring-porous.

Small-diameter vessels are approximately 20 µm wide. Large vessels in the earlywood of oak may be 300 µm wide. In lumber sawn from ring-porous wood such as oak, large earlywood vessels may form striking species-specific scratches interrupted by zones of latewood with small vessels.

Pulping separates the single vessel elements, and they break. Their papermaking properties are inferior, since they do not bond well and contribute little to the strength of paper. They can occasionally lift from the surface of a sheet during printing. Wood with a high proportion of vessels gives a low yield of pulp. Identification of hardwood pulps uses primarily the dimensions, pitting, and perforation pattern of vessel elements.

In some hardwood species, *tyloses* may block the vessels. They are saclike outgrowths of protoplasm from an adjacent parenchyma cell through a pit cavity into vessel lumen before the formation of heartwood. Wood with well developed tyloses is difficult to dry or impregnate with preservatives or chemicals. A common example of the effect of tyloses on the properties of wood is the difference between the heartwoods of white oak and red oak. The former is an excellent material for barrels because of its plugged vessels, and the latter is not suitable for this purpose due to its open vessels.

Hardwood tracheids occur in small amounts in a few species. They are small, longitudinal conducting cells usually densely covered with borded pits. They act as transition elements between major cell types. Two types of hardwood tracheids exist. *Vascular tracheids* are very similar to small vessel elements of latewood except their ends are imperforate. *Vasicentric tracheids* are short cells with closed ends and irregular shape resembling axial parenchyma cells in cross-section. Because of their small amount, their role in paper making is not significant.

Longitudinal parenchyma cells are more common in hardwoods than in softwoods. They are present in most hardwoods but seldom occupy more than 2% of wood volume. Longitudinal parenchyma cells occupy 16% of volume in *Eucalyptus globulus*[34] and even more than half the wood volume in some tropical hardwoods. Parenchyma cells are short, brick-shaped, and thin-walled. The function of parenchyma cells is storage and conduction of carbohydrates. They therefore live until the formation of heartwood. The most common type of longitudinal parenchyma is *strand parenchyma*.

Rays are the only transverse elements in hardwoods. They consist entirely of parenchyma cells and normally occupy 10%–20% of wood volume. In oaks, the ray volume may exceed 30%. Fast-grown wood has more rays than slow-grown wood. In a *homocellular ray,* the cells are approxiamtely the same size and shape. If a ray has more than one type of parenchyma cell, it is a *heterocellular ray*. The cells in a hardwood ray are normally radially elongated or *procumbent,* but they may be also vertically oriented or *upright*.

Wood identification uses the configuration of various types of cells in the ray. In pulp, distinguishing between longitudinal and ray parenchyma cells is difficult. Both are short and thin-walled. They frequently contain extraneous materials such as crystals, resins, tannins, oils, gums, latex, starch, etc., that may accelerate the deterioration of chips during storage. Silica is common in the ray cells of tropical woods. It causes wear and tear when cutting or machining wood.

2.6 Irregularities in wood

2.6.1 Juvenile wood

Three periods occur in the formation of wood during the life cycle of a tree. At a young age, the cambial zone forms immature juvenile wood. Mature or adult wood forms later. At an older age, overly mature wood forms. The transition from one phase to another is gradual without any distinct demarcation.

Juvenile or crown-formed wood differs in structure from normal mature wood. Its presence affects the technical properties of wood in the mechanical and chemical forest industries[35]. The following characteristics are typical of the juvenile wood of softwoods compared with mature wood:

- shorter tracheids
- thinner cell walls
- smaller mass-to-length ratio of tracheid
- lower proportion of latewood
- gradual transition from earlywood to latewood
- larger fibril angle
- higher proportion of lignin
- lower proportion of cellulose
- higher proportion of reaction wood
- higher proportion of knot wood
- lower basic density.

The following are the important pulp and paper characteristics of juvenile wood:
- lower yield of pulp
- lower yield of crude tall oil
- faster beating time
- higher proportion of rejects
- lower tear factor
- higher burst factor
- higher tensile factor
- higher fold endurance
- higher light scattering coefficient
- higher opacity
- higher sheet density and lower bulk
- smoother sheet surface
- lower consumption of energy in mechanical pulping.

Numerous studies on within-stem variation in cell length consistently show a significant increase from pith to bark. In the ring nearest the pith, the tracheid length in softwoods is 0.5–1.5 mm, and the fiber length in hardwoods 0.1–1.0 mm. It first increases rapidly outward. After a few rings, the rate of increase declines until reaching a maximum length that is 3–5 times the initial length.

The age of the cambium at the tree height under consideration measured in the number of annual rings from the pith rather than by the biological age of the tree affects the structure of the cells formed. A young tree produces only juvenile wood, but a mature tree produces mature wood at the butt end and juvenile wood at the top end of the stem. All wood above a certain height in the stem is juvenile.

Juvenile wood occurs in a cylindrical core along the length of a stem consisting usually of 10–20 growth rings. It forms in the region of the active crown as the result of prolonged influence of apical meristem on the cambium. As the tree crown moves further upward in the growing tree, the influence of the apical meristem on a given cambial zone decreases and mature wood forms[8]. The differences between juvenile and mature wood in spruces compared with pines may be a result of relative differences in the length of the living crown.

For commercial use, the juvenile wood and more mature wood from the same stem may behave like two diffeent materials for some tree species. Although changes in the anatomical structure and chemical composition from the pith outward are gradual and no distinct boundary occurs, juvenile wood is actually a core of uniform diameter along the length of the stem.

The diameter of the juvenile cylinder depends on the growth rate. Faster growth gives a larger volume of juvenile wood with a greater effect on the properties of timber. Its occurrence is therefore particularly problematic in fast-growing tropical and sub-tropical pine plantations with a short rotation and wide spacing of trees.

The number of years during which juvenile wood forms at a given stem height is specific to species. It is usually smaller in short-lived trees. It is 15–25 years in slow-

CHAPTER 4

growing Scots pine and Norway spruce but only 10–15 years in fast-growing radiata and yellow pines. At the age of 20 years, approximately 60% of the total stem volume of loblolly pine consists of juvenile wood. At the age of 40 years, the volume is approximately 20%. Each wood characteristic does not necessarily require the same number of years until it reaches the status of normal mature wood. For example, softwood tracheids reach their full length earlier than their full wall thickness.

There is no region in softwood stems where compression wood occurs more frequently than in the first juvenile growth rings near the pith. The reason is that the stem of a very young tree and the upper top portion of a mature tree are lithe and slender. Competing vegetation, wind, snow, or ice can therefore easily displace them from their vertical position. Since righting in young trees and tree tops is generally quite rapid, the compression wood close to the pith is usually limited in amount and restricted to a few growth rings. In stems that have remained inclined for several years, the compression wood at the center occupies the entire increment and is a more severe type. In addition to the high content of compression wood, juvenile wood has larger amounts of knots than mature wood[14].

Pulpwood from early thinnings and tops of saw timber trees are assortments characterized by a high content of juvenile wood. Since juvenile wood differs in its pulping behavior from normal wood – particularly from chips reduced from sawmill residues, the best yield and quality result when pulping juvenile wood-rich raw material under its optimum conditions separately from normal wood. Separate pulping results in a special product called *juvenile fiber* or juvenile pulp. Products such as printing and writing papers should use this juvenile fiber rather than something requiring high tear strength.

In the future, the raw material for the pulp industries will increasingly come more from fast-growing plantations of a short rotation cycle. The proportion of juvenile wood in the raw material flow will therefore increase. With a larger supply of small-sized wood, the separate debarking, chipping, pulping, and bleaching of juvenile wood from early thinnings will become more feasible.

2.6.2 Reaction wood

Introduction

When wind, snow, careless planting, or other factors force a stem to lean or bend, the tree tries to regain its correct orientation. It starts growing specific wood tissue to exert pressure along the grain and restore the vertical growth pattern. This only afffects the supporting cells – softwood tracheids and hardwood fibers. Since the development of this abnormal tissue is a result of the tree's reaction to disturbance, the term for it is *reaction wood*. It is present in small amounts in virtually all stems. It is especially common in the stemwood of young trees. It is always present in branches and knots.

In conifers, reaction wood forms on the lower side, and in hardwoods it forms on the upper side of a leaning or bended stem and branches. In conifers, reaction wood is *compression wood* and in hardwoods is *tension wood*. The tissue on the opposite side of the stem is *opposite wood*. The abnormal properties of reaction wood are usually undesirable for commercial use especially when it is very severe.

Compression wood in softwoods

Compression wood has several distinct anatomical features. Although the wood is hard and dense, it is weaker than normal wood of the same mass.

Eccentric growth rings and dark, reddish color usually indicate the presence of compression wood. It is easily visible to the naked eye on a cross-section surface of a log or in a piece of lumber. In a cross-section of a softwood branch, the pith is nearer the upper side, and compression wood appears on the lower side of the branch as a dark arch or crescent-shaped patch. Compression wood contains an exceptionally large proportion of latewood in the region of fastest growth. Figure 20 shows severe compression wood in a leaning pine stem.

Figure 20. Severe compression wood on a cross-section surface of a leaning softwood stem in *Pinus sylvestris* (Photo by Hannu Kalaja).

Compression-wood tracheids are shorter than those of normal wood. Their tips have frequent distortions. Normal tracheids usually have a very simple outline. They have a round cross-section compared with the angular outline of normal tracheids. As a result of the rounded form, Fig. 21 shows that the tracheids are not close-fitting but exhibit intercellular spaces.

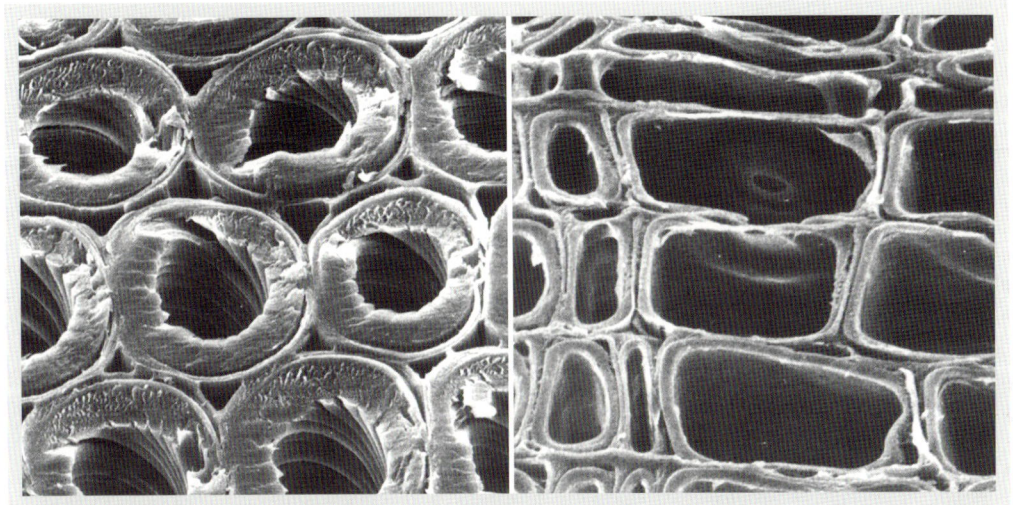

Figure 21. Compression wood from the middle of a growth ring (left) and opposite wood from the boundary of two growth rings (right) in branch wood of a softwood, *Pinus sylvestris* (Photo by Pekka Saranpää).

CHAPTER 4

Compression wood tracheids have a much thicker wall and a narrower lumen than the tracheids in normal earlywood. The secondary cell wall lacks the S_3-layer. The S_2-layer has a 30–50° microfibrillar angle to the cell axis compared with 10–30° in normal wood. Narrow helical cavities approximately 0.1 μm wide spiral around the tracheid parallel to the cellulose microfibrils. Between the cavities are 1–2 μm wide *helical ribs*. Warts often cover the helical cavities and ribs. When subjected to stress, the compression tracheid wall will crack at its weakest point – usually along the helical cavities – to form oblique 0.5–1.5 μm wide *helical checks*[36].

The abnormally large fibril angle and helical cavities contribute to the high longitudinal and low transverse shrinkage of coniferous compression wood. A tenfold increase in longitudinal shrinkage compared with normal wood is not unusual. Where compression wood occurs adjacent to normal wood, drying distortions and unequal longitudinal shrinkage cause internal stresses resulting in bending, twisting, and splitting of lumber.

Significant abnormalities occur in the chemical composition of compression wood. The differences are quantitative rather than qualitative. Characteristic features are the high content of lignin and galactan and low content of cellulose and galactoglucomannan as Table 14 shows for the average total composition of 27 conifer species. The chemical differences between normal and compression woods are less pronounced qualitatively, although the lignin in compression wood differs in its structure from that in normal softwood[36].

Table 14. The average chemical composition of normal and compression woods in 27 conifer species[36].

Constituent	Normal wood of softwood	Compression wood of softwood
	Proportion in extractive-free wood, %	
Lignin	30	30
Cellolose	42	30
Laricinan	Trace	2
Galactoglucomannan	18	9
Galactan	Trace	10
Xylan	8	8
Other polysaccharides	2	2

Tension wood in hardwoods

Compared with compression wood, tension wood is harder to detect by the naked eye. The most characteristic macroscopic feature is the appearance of a *woolly surface* on boards. Due to differences in chemical composition, one can easily detect tension wood under microscope examination by treatment with dyes.

The characteristic anatomical modifications of tension wood occur primarily with the fibers. The *percentage of fibers is greater* than in normal wood. Tension-wood fibers have *smaller diameters* and *greater lengths*, and they exhibit unusually *thick walls* with rounded outlines. Figure 22 shows that the *frequency and size of vessels may be* less.

The innermost, rather thick, unlignified part of the fiber wall has a gelatinous appearance. This is the *gelatinous or G-layer*, and the fiber itself is *gelatinous fiber*. The G-layer formed almost exclusively of pure cellulose microfibrils may sometimes nearly fill the fiber lumen. In longitudinal sections, gelatinous fibers almost always show oblique, incipient slip planes and minute compression failures.

Because of the presence of the G-layer, the S_2-layer is generally much smaller in width. This makes it possible for the S_1-layer to exert a dominating influence on the longitudinal shrinkage. Despite almost no axial orientation of microfibrils, the longitudinal shrinkage of tension wood is therefore approximately 5 times that of normal wood.

Tension wood also has less strength compared with normal wood except for tensile strength. The lack of strong bonding between the G-layer and the S_2-layer essentially eliminates a large volume of high-strength cellulose from effective influence on the strength of the wood.

In the tension wood, the earlywood fibers are chemically and anatomically modified, and the latewood fibers, vessels, and rays remain unaffected. The G-layer largely consists of highly crystalline cellulose microfibrils oriented parallel to the fiber axis. Hardly any lignin occurs in this additional cell-wall layer. Table 15 shows the average

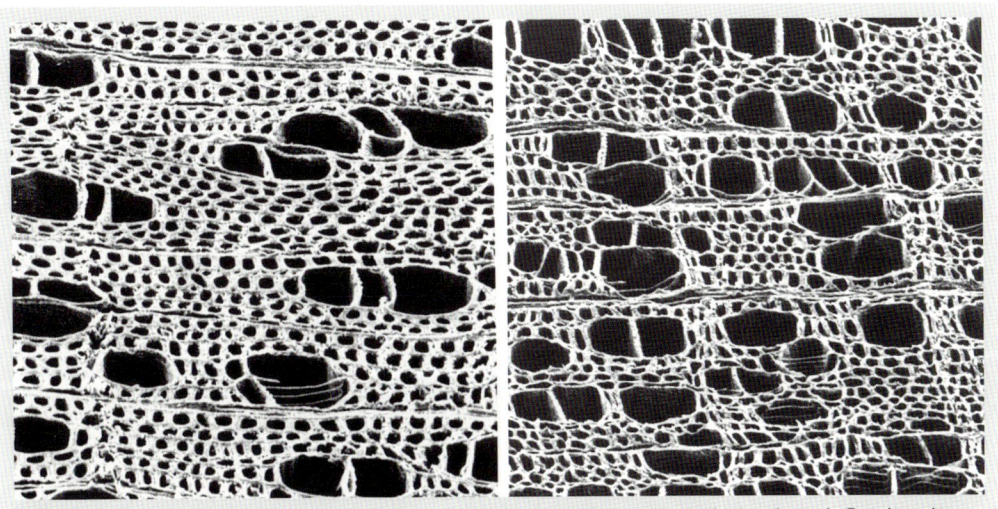

Figure 22. Tension wood (left) and opposite wood (right) in branch wood of a hardwood, *Betula pubescens* (Photo by Pekka Saranpää).

CHAPTER 4

chemical composition of normal and tension woods in ten hardwood species. While the compression wood of softwoods is exceptionally poor in cellulose, the tension wood of hardwoods has a high cellulose content often at least ten percentage points higher than normal stem wood.

Table 15. The average chemical composition of normal and tension woods in ten hardwood species in the northeastern United States[37].

Constituent	Normal wood of hardwood	Tension wood of hardwood
	Proportion in extractive-free wood, %	
Lignin	23	19
Cellulose	42	53
Glucomannan	4	2
Galactan	2	5
Xylan	28	21

Normal and tension wood fibers contain approximately the same absolute amount of lignin, but the relative lignin content is lower in tension wood fibers due to the presence of the additional unlignified G-layer. The xylan content in tension wood is also low. This is a result of the incorporation of additional cellulose into tension wood fibers.

Modifications of tissues occur in the opposite wood but only affect the fibers. The narrow zone of opposite wood on the lower side of the pith has shorter than normal fibers compared with the longer fibers in tension wood.

2.6.3 Heartwood

Wood is susceptible to deterioration by foreign organisms. To increase the natural durability of wood and the longevity of the trees, sapwood changes to physiologically dead heartwood when not needed for water conduction and food storage any more. A number of biochemical changes occur:

- The nuclei of the live parenchyma cells disappear.
- Storage of reserve foods such as starch and sugars decreases and disappears in the parenchyma cells.
- Polyphenolic extractives toxic to wood-deteriorating organisms form through decomposition of sugars at the boundary zone of sapwood and heartwood. They accumulate in the cell lumina or cell walls. In many species, these extractives give heartwood a darker, characteristic color.
- In most species – especially softwoods, the water content decreases radically. Table 17 shows that the moisture content may be higher in some hardwoods.
- To further decrease the permeability of wood, in some species the bordered pits of tracheids are closed or aspirated as Fig. 23 shows. In some hardwoods such as white oaks, ashes, and hickories, the vessels become plugged with tyloses. The cell structure remains otherwise unchanged.

Structure and properties of wood and woody biomass

In the yellow pines in the southern United States, the formation of heartwood starts at the age of 15–20 years[38]. In the slow-growing and long-lived Scots pine trees in Finland, heartwood formation starts considerably later at the age of 30–45 years.

Heartwood usually starts to form at 1–3 m above ground and tapers from the level of first initiation toward both the butt and crown. The extent of taper of the heartwood zone toward the stem apex depends on the tree species[23]. The most vigorous trees of a stand maintain the widest zone of sapwood. In a wood quality survey of lodgepole pine, the highest proportion of heartwood occurred in large, old trees with low crown ratios, large branches, and large within-crown taper[39].

Heartwood forms a conical core inside the sapwood that spreads year after year upward and outward in the stem. In Scots pine, the percentage of heartwood reaches its highest value at heights corresponding to 20%–30% of the tree height. The percentage of heartwood depends on the species, age, and physiological condition of the tree. Pulpwood from old mature trees is rich in heartwood, but pulpwood from early thinnings, stem tops, and sawmill slabs has very little heartwood.

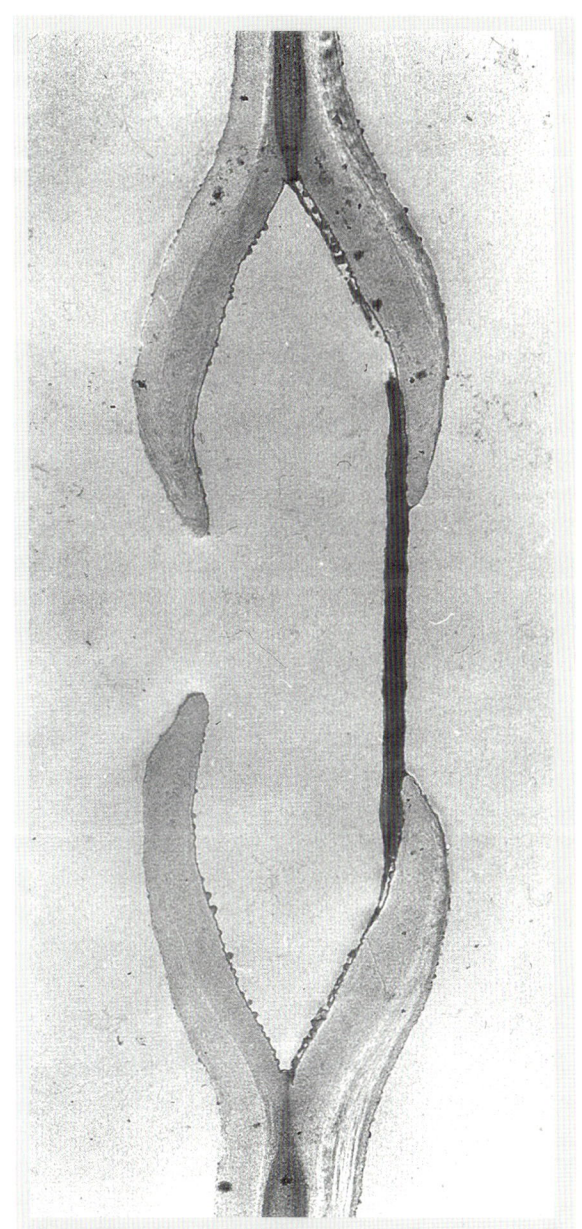

Figure 23. An aspirated bordered pit in the heartwood of a softwood, *Pinus sylvestris* (Photo by Pekka Saranpää).

Koch[39] found the percentage of heartwood in lodgepole pine correlated positively with latitude. In *Pinus contorta* var. *latifolia,* the proportion of heartwood increased from 17% at latitude 45° to 42% at latitude 60°. In Finland, the percentage of heartwood in Scots pine pulpwood increases similarly from south to north. Table 16 shows that this is a direct consequence of the decrease in growth rate and increase in age at the time of harvesting[40].

Table 16. The percentage of heartwood in Scots pine pulpwood in various parts of Finland[40].

Latitude in Finland	Percentage of heartwood in Scots pine pulpwood
60°–62°	16.4
62°–64°	21.2
64°–66°	27.2
66°–68°	35.5

Heartwood is advantageous for transportation, because it reduces the moisture content and weight of the wood. It allows a larger load size in trucking and improves the buoyancy of timber in floating.

The increase in extractives, closure of bordered pits, and formation of tyloses makes it difficult to penetrate heartwood with liquids such as preservatives or pulping chemicals. When pulping, a longer cooking time is necessary. The extractives increase the yield of tall oil in sulfate pulping. In mechanical pulping, the low moisture content of heartwood in softwoods may be a negative factor. The extractives interfere with bleaching and require additional treatments. Acid extractives may cause corrosion in metals.

Sapwood and heartwood may therefore behave like two different tree species in pulping. Pulping pure pine or pure spruce wood is a mixed process. When the harvesting of pulpwood in many countries shifts from slow-growing natural stands to fast-growing plantations of shorter rotation, the proportion of heartwood and its variation in the raw material decreases.

2.6.4 Knots

Branches normally originate at the pith of the stem and are organized according to a species-specific pattern. The cylindrical sheath of cambium mantles the tree and interconnects the tissue systems formed through cell division and differentiation in the stem and branches.

Each year a continuous growth increment forms over the stem and branches. When the stem grows in diameter, branch bases become embedded and are gradually occluded in the wood of the stem. Since the embedded base of a branch cannot grow in thickness, it develops a conical tapering toward the pith. This section of the branch inside the stem is the *knot* shown in Fig. 24. On a sawed surface of lumber, a knot appears round when sawed at right angle to its length. If the cut is diagonal, the knot appears oval. When sawing is lengthwise, the result is a spike knot.

In the stemwood, the grain orientation becomes distorted around the knot. A thicker branch has a greater distortion. As long as the branch remains alive, the continuous sheath of cambium incorporates the embedded branch base into the stem causing an *intergrown or tight knot* that does not loosen with drying of the wood.

When the tree grows in height, the oldest branches at the crown base become shaded. They die and are shed. The branch cambium ceases to function. The stub of the dead branch imbedded in the stemwood does not become tightly interconnected in the surrounding stemwood. It is therefore an *encased or loose knot* shown in Fig. 25. The cells of the bole are not continuous with the cells of the knot. The

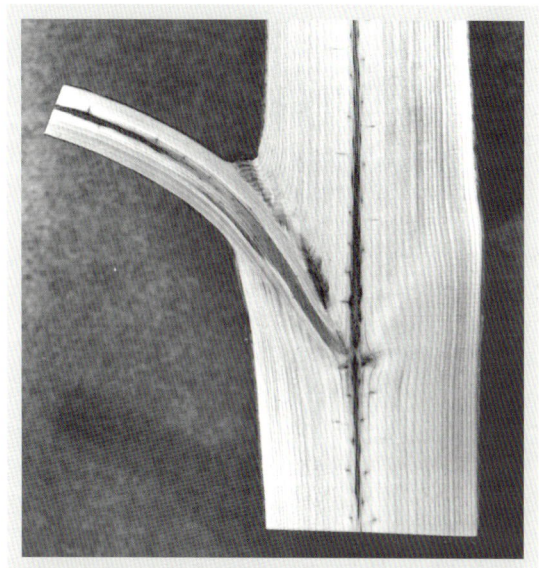

Figure 24. Embedded base of a branch inside the stem, *Pinus sylvestris* (Photo by Hannu Kalaja).

grain in the surrounding stemwood therefore has less distortion around an encased knot, and it lowers the mechanical strength of lumber less than an intergrown knot. It is an especially harmful defect in lumber because of its tendency to fall out and leave a knot hole when wood dries and shrinks.

Dead branches deteriorate and gradually break, but short branch stubs remain on the tree for several years or even decades depending on the tree species, branch thickness, stand density, and climatic factors such as temperature, humidity, snow, and wind. Spruces retain dead branches longer than pines, and northern pines retain dead branches longer than the yellow pines in the southern United States. This is probably due to the slower deterioration of dead branch wood in a cold climate.

When the branch stub is completely lost from the stem, its end is

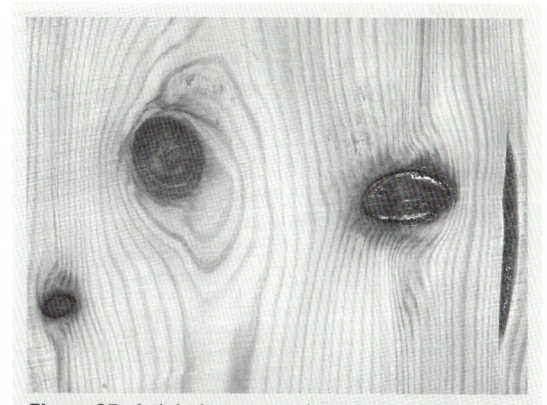

Figure 25. A tight intergrown knot (upper left) and two loose encased knots in a piece of lumber, *Pinus sylvestris* (Photo by Pekka Saranpää).

overgrown by wound tissue. The stem then begins to produce *knot-free, clear wood*. To speed up this development and increase the amount of valuable clear wood, the butt end of the tree may be pruned artificially by hand saw or other means. The use of prun-

CHAPTER 4

ing as a silvicultural measure is especially feasible in the fast growing radiata pine plantations in New Zealand and some other southern hemisphere countries where self-pruning of exotic pines is poor but production of clear wood after pruning is rapid.

The proportion of knots in the stem volume is a result of many factors including the number, size, and angle of the branches and especially by the self-pruning behavior of the tree. It therefore depends on the tree species, density of the stand, position of the tree in the stand, tree age, etc. The knot volume is lower in the outer parts and butt section of the stem compared with the inner parts and top. In lodgepole pine plantations in Finland, the proportion of knotwood was 0.4% at the butt but increased to 2.4% at the 90% height of the stem[41].

In the Nordic countries, the knots consist of approximately 1% of the stem volume of Scots pine and Norway spruce. The amount is slightly less in saw logs and more in pulpwood logs. Unusually small amounts of knots occur in chips made from sawmill slabs for use as raw material for the pulp industries. Small-sized pulpwood from early thinnings is rich in knotwood particularly when the spacing of trees is wide and the site is fertile. Pulpwood from the tops of saw timber trees also has a higher proportion of knotwood than pulpwood from suppressed trees from later thinnings. When measured on the dry mass basis instead of volume, the proportion of knotwood is at least 1.5-fold in Scots pine and 2-fold in Norway spruce due to differences in basic density.

Figures 21 and 22 show that branchwood has many differences from normal stem wood. The differences are largely a result of the abundance of reaction wood in branches. Information on the technical properties of branch wood is plentiful[2], but few studies refer directly to the embedded part of the branches – knots. The following conclusions are possible:

- Considering the proportions of different types of cells, stemwood and knotwood do not differ greatly. Softwood knots have more but smaller resin canals, and hardwood knots have more vessels than stemwood.

- Softwood knots have a low content of cellulose and galactoglucomannan and a high content of lignin and galactan. Hardwood knots have a low content of lignin, xylan, and glucomannan and a high content of cellulose and galactan. Compared with stemwood, the content of cellulose in softwood knots is approximately 10 percentage points lower and in hardwood knots is 10 percentage points higher.

- Fiber dimensions and their variation are significantly smaller in knots. Some dimensional similarities between tracheids from softwood branches and fibers from hardwood stems exist, but their properties differ in papermaking. Many studies of a large number of tree species in Europe and North America show the following when comparing stem wood cells:

    - Branch wood tracheids of softwoods are 45%–50% shorter and 25% thinner.

    - Branch wood fibers of hardwoods are 25%–30% shorter and 15%–20% thinner.

    - Branch wood vessel elements of hardwoods are 15%–20% shorter.

Structure and properties of wood and woody biomass

- Due to the thick cell walls and narrow lumina of reaction wood and a greater content of extractives, the basic density of branch wood is unusually high. Figure 26 shows that the highest density values of each tree species usually occur in knots. For example, knot wood in Scots pine is 1.5–2 times and in Norway spruce 2–2.5 times as dense as stemwood. Differences are relatively smaller in hardwoods especially in tree species with dense stemwood.

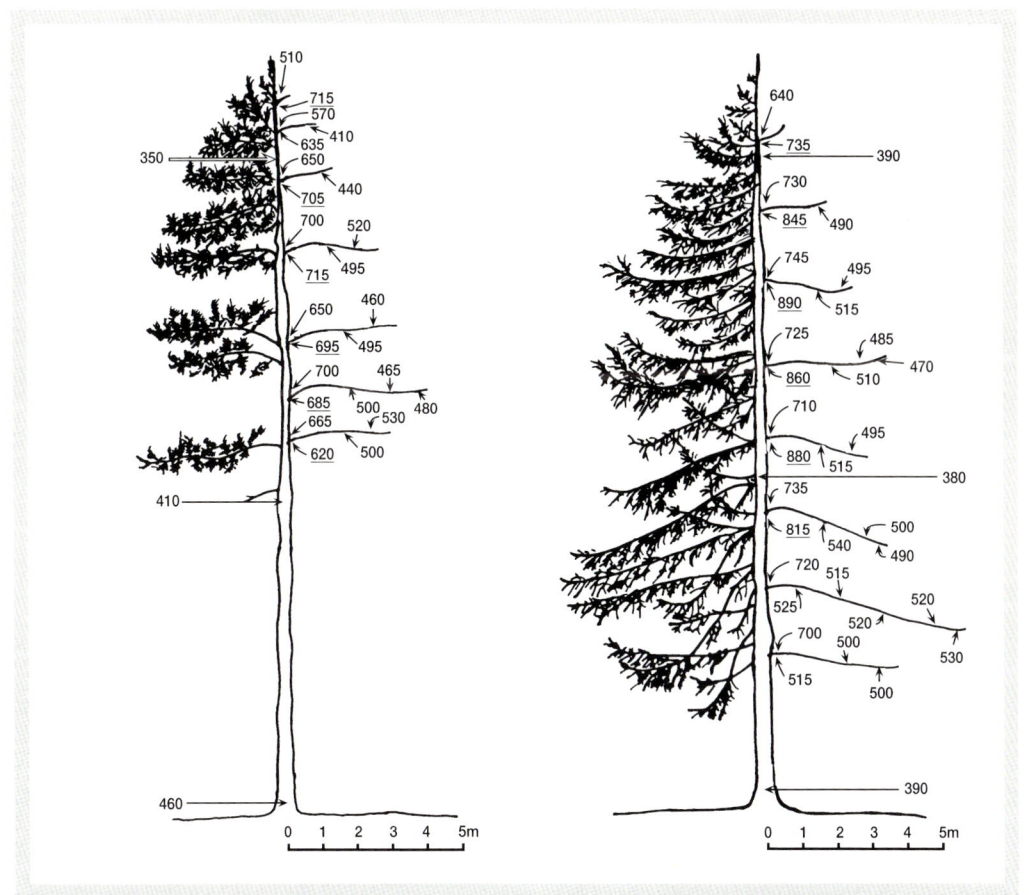

Figure 26. Variation of basic density in branch wood in a *Pinus sylvestris* (left) and a *Picea abies* tree. Underlined numbers refer to the embedded part of the branch (redrawn from Hakkila[42]).

The presence of knots in timber is often undesirable. Their effect on the value of wood is far greater than their volume. In lumber, the severity of a knot defect depends on the size, organization, location, angle, shape, soundness, and slope of grain. Due to the distortion of grain, knots reduce the bending strength of lumber, increase the forces required in cutting and chipping, result in uneven cutting surfaces, cause checking when drying lumber, and cause problems in painting.

Due to their high density, high content of extractives, low moisture content, and distorted grain, knots are also harmful in the pulp industries. When chipping pulpwood logs, the high density and distorted grain increase the amount of irregular, excessively thick particles. Knots and surrounding wood tear from the log rather than undergo chipping. In chemical pulping of softwood chips, these thick and dense particles of lignin-rich, resin-impregnated wood remain undercooked resulting in an increase in rejects and lower yield of pulp. They also increase the consumption of chemicals in cooking and bleaching, the consumption of energy in mechanical pulping, and the time required for beating. All these factors strain the capacity of the pulp mill. Crushing the excessively thick particles from chip screening between rotating steel rolls before pulping may reduce some processing problems. The fibers still have an inferior quality.

2.7 Moisture content of wood

Wood is a *hygroscopic* material – it absorbs and looses water. Determination of the moisture content (MC) can use either dry mass or wet mass as the following equations show:

$$MC \text{ (dry mass basis)}, \% = 100 \cdot \frac{\text{mass of water in wood}}{\text{dry mass of wood}} \tag{1}$$

$$MC \text{ (wet mass basis)}, \% = 100 \cdot \frac{\text{mass of water in wood}}{\text{total mass of wood with water}} \tag{2}$$

Measurement on dry mass basis is important in the sawmill and board industries, and measurement on wet mass basis is important in the pulp industries and when using wood for fuels.

In live trees, the mass of water is often greater than the mass of dry matter. In softwoods, the moisture content is many times greater in sapwood than in heartwood. In hardwoods, the differences are smaller, and Table 17 shows that the average moisture content is usually lower. The moisture content of pulpwood is especially important in mechanical pulping. To ensure high brightness of mechanical pulp, the wood should be fresh.

The moisture content of live trees varies seasonally and even diurnally depending on the weather. When the measurement of pulpwood at the mill uses weight scaling, one must consider the fluctuation of moisture content. In birch trees (*Betula pendula, Betula pubescens*) in Finland, the average moisture content is 45% when the trees are dormant and leafless in the winter, rises to 48% in the spring when the trees are still leafless but the root system is already actively conducting water to the stem, decreases to 39% in the summer when transpiration from leaves is at the maximum level, and increases again to 45% in the fall when the trees loose their foliage. In conifers, the seasonal changes are generally smaller.

Table 17. Moisture content of fresh sapwood and heartwood on the wet mass basis in some northern American softwoods and hardwoods[43].

Softwoods	Sapwood	Heartwood	Hardwoods	Sapwood	Heartwood
	Moisture content, %			Moisture content, %	
Picea rubens	56	25	Acer saccharum	42	39
P. glauca	56	25	Betula papyrifera	42	47
P. mariana	56	25	Fagus grandifolia	42	35
Pinus resinosa	57	24	Liriodendron tuliphera	51	45
Pinus contorta	55	29	Platanus occidentalis	57	53
Pinus taeda	52	25	Populus tremuloides	53	49
Pinus palustris	51	24	Quercus alba	44	39
Pseudotsuga menziesii	53	27	Quercus rubra	41	44
Tsuga canadensis	54	49			
Picea sitchensis	59	29			

Water occurs in wood in two forms – bound water in the cell walls and free water in the cell lumina. These two states of water affect wood properties differently.

Bound water diffuses into the disordered amorphous areas of the microfibrils and void spaces of the cell wall and is chemically united with the cellulose, hemicelluloses, and lignin through hydrogen bonding. As water enters the cell wall, the framework of microfibrils expands laterally. The expansion of wood in its longitudinal, radial, and tangential direction depends on the angle of microfibrils to the long axis of the cells. At the *fiber saturation point,* all sites available for hydrogen bonding in the cell wall are full of water, but no free water exists in the cell lumina. Then the wood has reached its maximum swollen volume. For most tree species, the moisture content of wood is approximately 23% on the wet mass basis or somewhat less at the fiber saturation point.

Water accumulating in wood above the fiber saturation point is not bound physico-chemically to the cell wall substance but remains free in cell lumina. The maximum amount of *free water* in wood therefore depends on the volume of void spaces. Consequently, in live trees dense wood typically has a lower moisture content than low-density wood.

The moisture content of wood products fluctuates with the relative humidity and temperature of the atmosphere. The state when wood reaches a steady moisture content in certain atmospheric conditions is *the equilibrium moisture content.* In normal outdoor conditions under cover, this is usually 11%–13%, but in heated buildings it is only 4%–8% on the wet mass basis due to the lower relative humidity of the air.

When wood tries to reach the equilibrium condition, water moves into or from it. As a result of the movement, *a moisture gradient* develops between the surface and center zones of the piece. The movement takes place 12–15 times faster along the grain than across the grain. In pulpwood logs with undamaged bark, drying occurs through the end surfaces.

In service, wood undergoes continuous exposure to changes in the moisture content. In conditions below the fiber saturation point, the changes occur in the water bound between the microfibrils causing shrinkage or swelling of wood. The orientation of microfibrils determines the distribution of volumetric change into the three phases of wood. When wood of northern American tree species dries from the fresh to oven-dry condition, the range of shrinkage is that of Table 18. High-density woods approach the upper limit and low-density woods the lower limit of the range.

Table 18. The range of shrinkage of northern American softwoods and hardwoods when wood dries from the fresh to oven-dry condition[43].

Dimension	Softwoods	Hardwoods
	Shrinkage, % of fresh dimensions	
Longitudinal	0.1–0.3	0.1–0.3
Radial	2.5–5	3.5–6
Tangential	5–8	7–11
Volumetric	8–13	11–17

The flow rate of gases and liquids is a result of the *permeability of wood* determined by the size of the passages in and between the cells. In the softwoods, the openings in the margo of the pit membranes or the closure of bordered pits with the formation of heartwood are usually the limiting factors of the longitudinal permeability. In hardwoods, the size of the vessels and the possible presence of tyloses primarily control the longitudinal permeability. In all tree species, the permeability is very small across the grain.

Permeability is very important when using wood as a raw material in the forest industries. It affects the drying of lumber, treating of wood with preservatives, and chemical pulping of wood.

2.8 Basic density of wood

Density is a useful indicator of pulpwood quality because of its relationship to certain wood and fiber properties such as the thickness of cell wall. It closely correlates with the mechanical strength of lumber, pulpwood, and poles. It reflects the behavior of raw material in pulping, consumption of wood volume per Mg of pulp, and the properties of pulp and paper. It is also the conversion factor for comparison of volume and mass measurements of pulpwood denoting the mass per unit volume of a substance in kg m^{-3} solid.

For *mass density*, both the mass and volume of wood are measured at the same moisture content as received. Mass density therefore has a direct relation to the moisture content of the substance. The minimum values occur when wood is absolutely dry, and the maximum occur when fully saturated. In the boreal and temperate regions, the mass density of freshly felled pulpwood for most tree species is 800–1 000 kg m^{-3}, but in many tropical hardwoods it exceeds 1 000 kg m^{-3}. When the moisture content of wood falls below the fiber saturation point, the loss of water and the volumetric shrinkage of wood and bark affect the mass density.

The *basic density* of wood is the oven dry mass per green volume in kg m^{-3}. Volume is measured in the unshrunken condition – the moisture content must be above the fiber saturation point. For a large majority of pulpwood species, the basic density varies between 330–600 kg m^{-3}.

Dividing the basic density of wood by the density of water at a temperature of +4°C (1 000 kg m^{-3}) gives the *specific gravity* or *relative density*. This is a quantity without units that is independent of the measuring system. The specific gravity of common pulpwood species is 0.330–0.600. The basic density and specific gravity of wood are a result of the following factors:

- *The density of the cell wall substance itself* that is distinct from the cell wall that may contain voids: Only little differences occur between tree species depending on the proportions of the cell wall substances. The density in wood cellulose is approximately 1 560 kg m^{-3}, in hemicelluloses it is 1 500 kg m^{-3}, and in lignin it is 1 300–1 350 kg m^{-3}. Due to the lower content of lignin in hardwoods, the density of the cell wall substance is approximately 2% higher than in softwoods[44, 45]. The density of the cell wall substance in normal wood averages 1 480–1 500 kg m^{-3}.

- *The volume of cell walls in relation to the cavities* occupied by cell lumina and intercellular spaces: For example, the basic density of latewood in softwoods with the thick-walled fibers is 2–3 times as high as earlywood that has thin-walled fibers by definition. According to Spurr and Hsiung[46], the basic density of latewood in softwoods is 600–900 kg m^{-3} and in earlywood is 250–320 kg m^{-3}. Figures 27 and 28 give examples of density variation in latewood and earlywood of softwoods.

- *The amount of extraneous substances* in wood: Extractives and inorganic components increase the basic density since they fill cavities. Juvenile wood around the pith in old trees is therefore higher in density than young trees due to the accumulation of extractives in the heartwood. The formation of extractives hinders the use of basic density as a pulpwood quality indicator such as calculating wood consumption in pulping. Removing the extractives and determining the basic density of extracted wood can increase the use of density as a tool.

Large differences in basic density occur between tree species. Due to the wide variation among trees, accurate average values by species are not available. Hardwoods usually have higher density than softwoods. Ring-porous hardwoods produce denser wood than diffuse-porous hardwoods, and tropical hardwoods are normally denser than temperate and boreal hardwoods.

CHAPTER 4

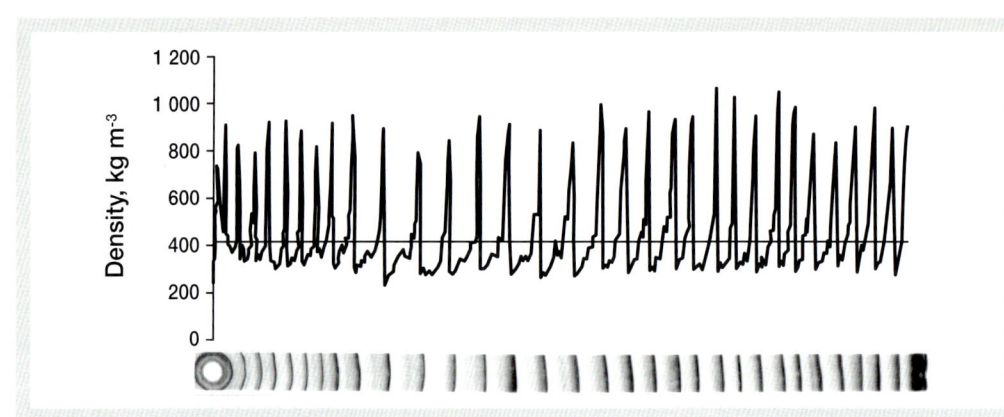

Figure 27. Variation of basic density of wood within and between growth rings of *Picea abies* (Photo by Kari Sauvala).

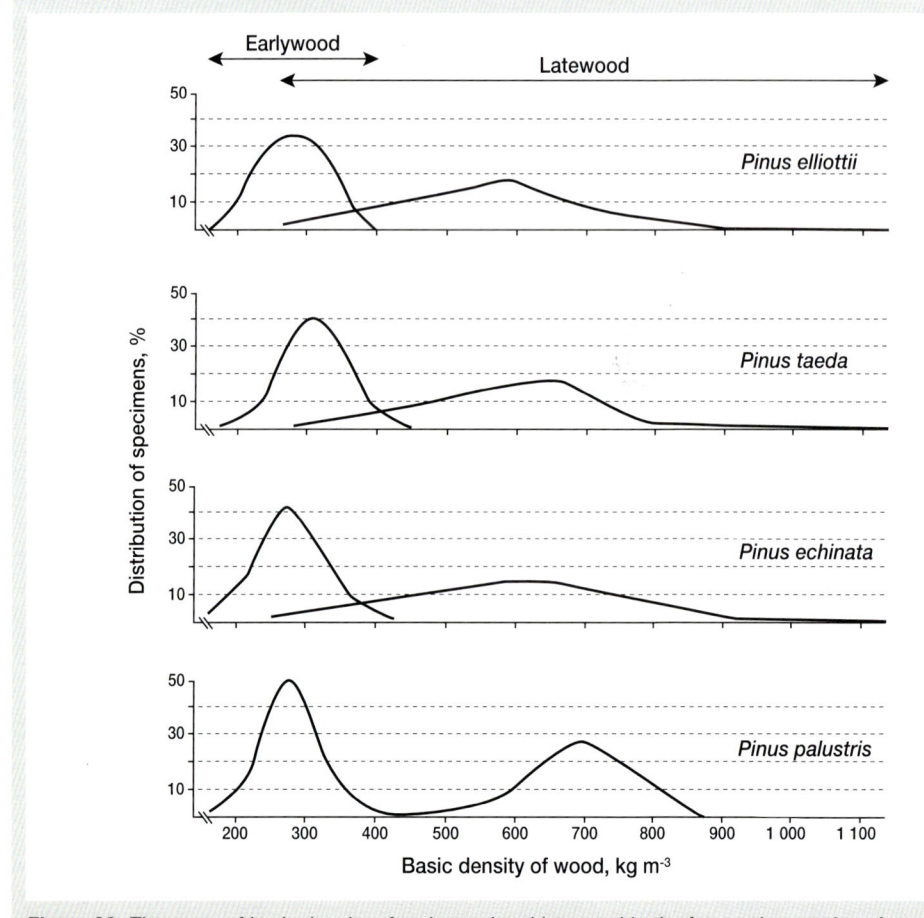

Figure 28. The range of basic density of earlywood and latewood in the four major species of southern pine in the United States (redrawn from Paul[47]).

Table 19 gives examples of the basic density of some major industrial tree species in Canada and Finland. The Canadian figures are whole-stem averages calculated from many studies compiled by Conzales[48] from naturally regenerated stands in Canada and northern portions of the United States. The Finnish figures represent managed stands and younger trees. Note that the basic density of pulpwood is lower than the whole-stem averages in the table, since the high-density butt logs from the largest stems are usually cut for saw logs.

Table 19. Examples of average basic density in major Canadian and Finnish softwoods and hardwoods.

Softwoods	Basic density, kg m^{-3}	Hardwoods	Basic density, kg m^{-3}
Canadian species			
Larix laricina, tamarack	474	*Carya ovata*, shagbark hickory	674
Pseudotsuga menziesii, douglas fir	438	*Quercus alba*, white oak	623
Picea mariana, black spruce	423	*Acer saccharum*, sugar maple	591
Tsuga heterophylla, western hemlock	410	*Fagus grandifolia*, American beech	586
Pinus banksiana, jack pine	399	*Quercus rubra*, red oak	575
Pinus contorta var. *latifolia*, lodgepole pine	399	*Betula alleghaniensis*, yellow birch	568
Pinus ponderosa, ponderosa pine	392	*Betula papyrifera*, white birch	507
Tsuga canadensis, eastern hemlock	387	*Populus tremuloides*, trembling aspen	386
Pinus resinosa, red pine	381	*Populus grandidentata*, large-tooth poplar	382
Picea rubens, red spruce	380	*Populus balsamiphera*, balsam poplar	363
Picea glauca, white spruce	359		
Pinus strobus, eastern white pine	344		
Abies balsamea, balsam fir	333		
Thuja plicata, western red cedar	319		
Thuja occidentalis, eastern red cedar	302		
Finnish species			
Larix sibirica, Siberian larch	490	*Betula pendula*, silver birch	490
Pinus sylvestris, Scots pine	420	*Populus tremula*, European aspen	400
Picea abies, Norway spruce	385	*Alnus incana*, grey alder	360

The within-species variation is generally wider in softwoods than in hardwoods. Several components of variation occur within a species – within-tree variation, tree-to-tree variation, stand-to-stand variation, and geographical variation.

Within-tree variation has two components – variation in the cross-section of the stem *from pith to cambium* and variation along the stem from butt to top. A density increase toward the cambium is characteristic of most softwoods especially pines. The tendency of the proportion of latewood in the annual ring and latewood's density to increase year by year up to a certain ring age causes this[49]. For example, the proportion of latewood in Finnish conditions in the first five annual rings of Scots pine and Norway spruce is approximately 15%, but it increases by the 30th annual ring to 29% in Scots pine and 22% in Norway spruce[50]. The basic density changes in the radial direction of the stem accordingly.

In many tree species, the percentage of latewood is unusually low in the juvenile wood near the pith of the stem resulting in a low initial wood density. In pines, the percentage of latewood and the wood density increase from the pith toward the cambium until they reach a maximum. An overmature pine may start producing less dense wood again. In spruces, the density first decreases but then increases from the pith to the cambium. Changes in the cell wall thickness are usually smaller in hardwoods. Infiltration of extractives with subsequent heartwood formation disturbs the radial variation of density.

The density variation from pith to cambium has great significance in wood use. For example, chips reduced from sawmill slabs usually contain mature wood from the outer growth rings of saw logs. They therefore yield pulp with long thick-walled fibers characterized by excellent tear strength.

The radial variation of wood density causes further *variation in the axial or longitudinal direction* of the stem. In most conifers, wood density decreases from butt to top. Changes in ring width, proportion of latewood, and mostly the presence of juvenile wood affect it. Some conifers – especially spruces – show only a slight longitudinal variation in wood density.

Differences among species are larger in hardwoods. Among diffuse-porous hardwoods, density first decreases notably and then increases toward the top or increases or decreases steadily from butt to top. In various ring-porous hardwoods such as oaks, the longitudinal density variation is often negligible.

In the northern European tree species, the average wood density of a stem is approximately the same as the density at 25% stem height[50–52]. Figure 29 illustrates patterns of longitudinal variation in wood density in five indigenous and an exotic tree species in Finland. *Pinus sylvestris* and *Larix sibirica* represent species with a distinct difference between the butt and the top. Differences are minor in *Picea abies* and other tree species with a long live crown.

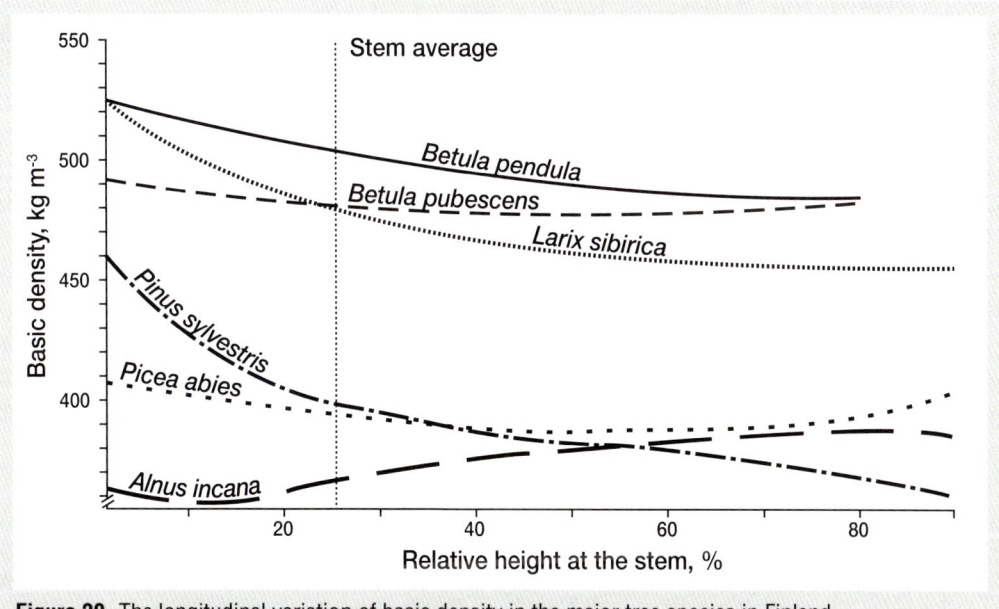

Structure and properties of wood and woody biomass

Figure 29. The longitudinal variation of basic density in the major tree species in Finland.

The longitudinal density variation from butt to top has considerable practical significance. For example, a 3-m butt log may have 15%–20% higher density than the top log from the same Scots pine stem. If using the butt logs for saw timber and only the top logs for pulpwood, the basic density and fiber characteristics of the raw material change radically. One can also control the quality of pulpwood by segregating butt logs and top logs from each other to form two or more pulpwood assortments. The sorting allows separate processing of low-density and high-density wood to produce special pulps of homogenous quality.

Figure 30 gives an example of the longitudinal variation of the basic density and percentage of bark by volume in young pine trees from the first commercial thinnings in Finland. These two factors largely determine the dry mass of wood in a cubic meter of unbarked pulpwood and the consumption of unbarked wood in pulping. Differences between the butt and the top of the stem are very distinct.

Considerable *variation occurs from tree to tree* in the same stand due to differences in tree age and size, growth rate, stem taper, and heredity. The relationships are complex, differences among species are great, and research results are sometimes contradictory.

In most conifers, age is probably a more decisive factor for wood density than growth rate[46], but the effects of various factors are species-specific. In pine trees, tree age best explains the tree-to-tree variation. In spruce trees, the growth rate is the dominant factor. The effect of ring width on wood density leads to the following conclusions[29]:

CHAPTER 4

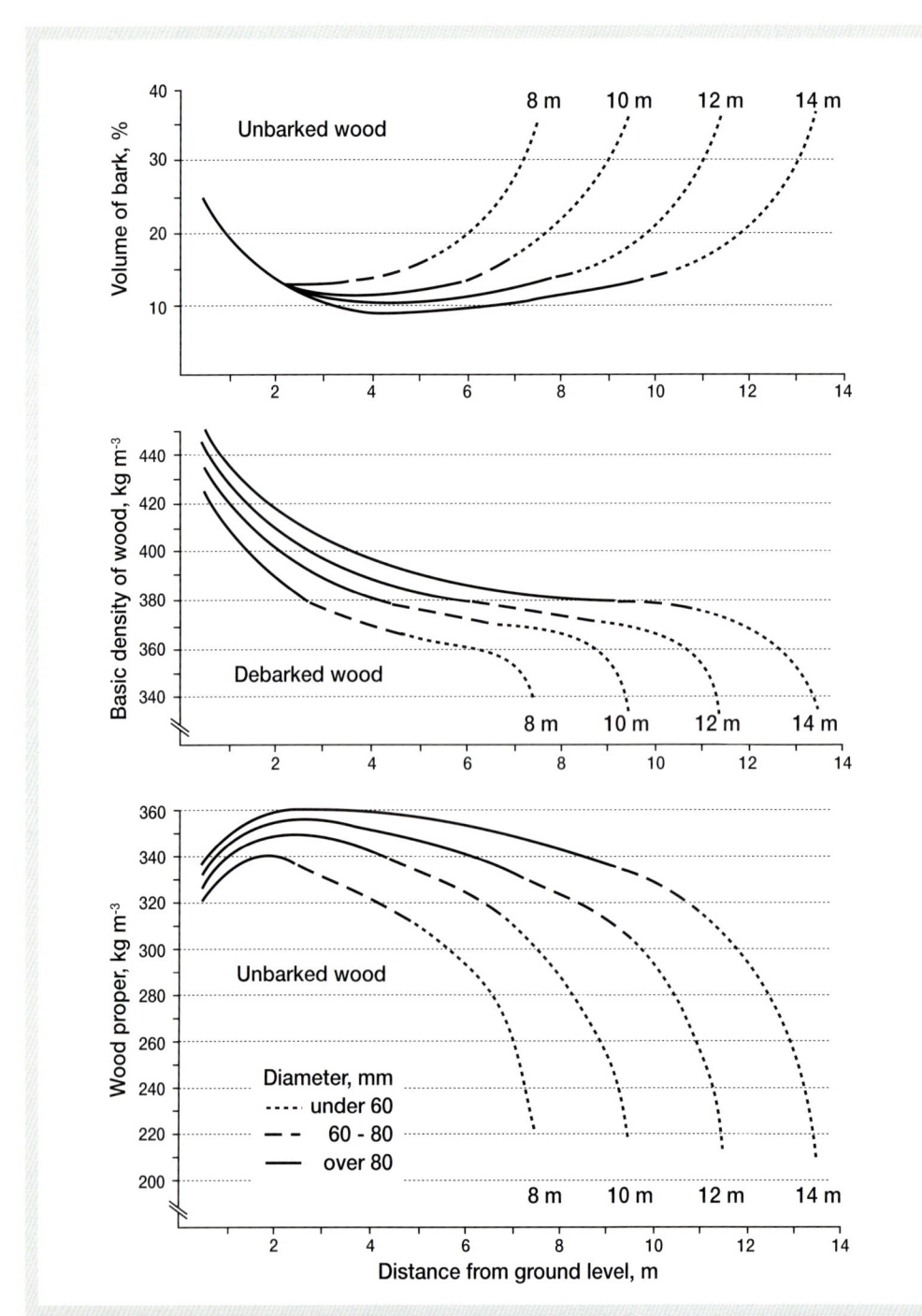

Figure 30. The longitudinal variation of basic density, percentage of bark by volume, and the dry mass of wood proper in a cubic meter of unbarked pulpwood in 8, 10, 12 and 14 m high stems from the first commercial thinning of *Pinus sylvestris* in Finland[53].

- Among the coniferous woods, the density of spruces increases with decreasing width of growth rings. The density of pines and larches first increases and then decreases with the maximum density at 1–1.5 mm ring widths. The pattern of firs and Douglas-fir apparently lies between the two.
- In diffuse-porous hardwoods, ring width is not a significant criterion of density.
- In ring-porous hardwoods, wide annual rings imply wood of high density.

The same environmental, heredity, and forest management factors that are responsible for variation among trees also cause *stand-to-stand variation*. For example, pine pulpwood has low density and thin-walled fibers when harvested from the first commercial thinning of a young plantation, but spruce pulpwood has lowest density when it originates from a fast-growing stand that is young or old. While the average basic density of Norway spruce of all ages in Sweden is 400 kg m^{-3}, the basic density of very wide-ringed spruce wood grown on fertile abandoned farm land in southern Sweden is only 250–275 kg m^{-3}. This means that 40% more wood volume is necessary in pulping[35].

The effect of ring width on the basic density of wood is the same whether the fast growth is a result of fertile site, fertilization, or wide spacing of trees. Figure 31 shows the basic density of Norway spruce from naturally regenerated managed stands and plantations as a function of stand age in southern Finland. At the same stand age, wood from plantations is 5% lower in density due to faster growth. Since plantation-grown wood is cut from younger stands, the difference in the basic density of the pulpwood crops is greater.

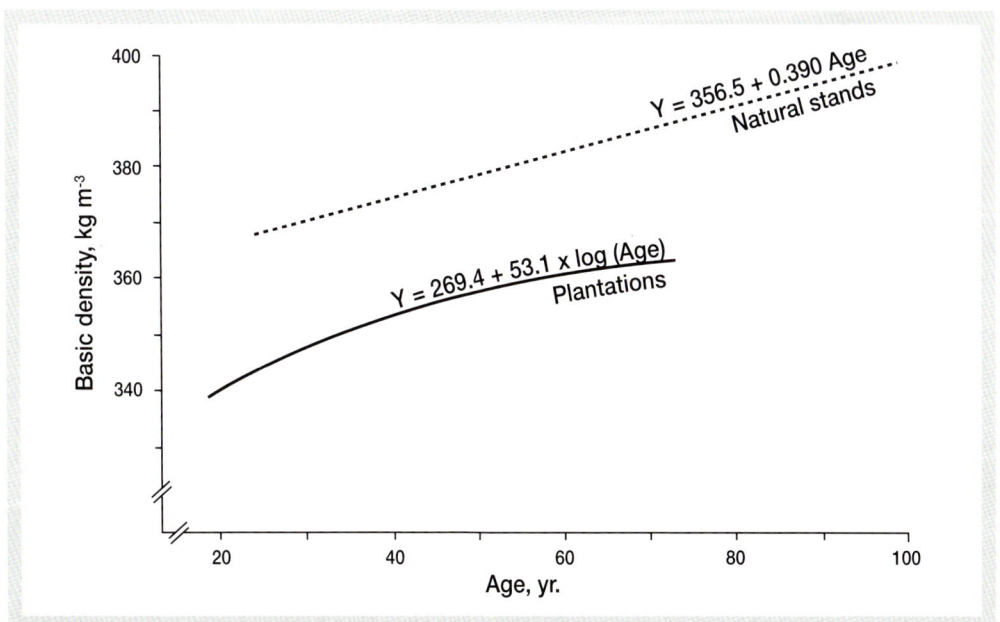

Figure 31. Basic density of wood as a function of tree age in naturally regenerated managed forests and plantations of *Picea abies* in Finland.

CHAPTER 4

The *geographical location* of the stand also affects the wood density. Differences in climate, soil fertility, forest management practice, age structure of forests, and hereditary factors may cause this variation. In northern forests, the basic density of a species frequently decreases toward the north, although this is not always the case. In Finland, Scots pine has the lowest basic density in northern Lapland because the cold climate shortens the growing season and reduces the proportion of thick-walled latewood fibers. Consequently, the consumption of pine wood per ton of pulp is higher and the tear strength of paper is lower in the most northern sulfate mills[40]. Similarly, the basic density of loblolly pine in the southeastern United States decreases toward the north due to a shorter growing period and lower proportion of latewood[54]. When cultivating loblolly pine in fast-growing plantations in Brazil, Argentina, Australia or other southern hemisphere countries, the average basic density of pulpwood remains lower than in the United States due to the fast growth, short rotation, and high content of juvenile wood.

Since variation of wood density within a tree species depends on variation in cell wall thickness in relation to the diameter of lumen, wood density can predict and control the properties of fibers and pulp. Thick-walled fibers from dense wood are stiff and retain their original cross-section form in paper making. Due to the strength of the single fibers, they produce paper with high tear resistance. A sheet of paper having a given weight per square meter made from such thick-walled fibers contains fewer fibers per unit area and less interfiber surfaces resulting in poor bonding.

Thin-walled fibers from low-density wood collapse and become ribbonlike. This increases the bonding of the fibers and formation of dense, nonporous, opaque sheets. Low-density wood increases the tensile and burst strength and folding endurance and produces a smoother and closer sheet of paper. The percentage of collapsed fibers depends on the cell wall thickness and also on the pulping process. It increases with decreasing yield due to higher delignification.

For high tear strength – an important property especially in kraft paper, thick-walled cells and dense wood are desirable. If paper having great tensile or burst strength is necessary, thin-walled cells and a low wood density are advantageous. If the smoothness and closeness of a sheet of paper are of prime consideration, a high proportion of cells with thin walls will help. In the manufacture of newsprint and thin translucent papers with high resistance to fold and burst and with good tensile strength, low-density wood is often preferable.

Low-density wood absorbs water and chemicals more readily than dense wood in pulping. Mixed cooking of chip particles with a wide density range therefore causes yield losses in chemical pulping, since low-density wood will overcook or high-density wood will undercook. Both conditions result in excessive amounts of rejects. High amounts of thick-walled, coarse fibers also lengthen the beating time requirement and make sheet formation more difficult. Because of their pale color, thin-walled, earlywood fibers require less bleaching to reach a desirable degree of whiteness. In mechanical pulping, the consumption of energy per ton of pulp increases with the basic density of wood.

No wood is ideal for all paper qualities. Thick-walled and thin-walled fibers have their advantages and disadvantages. Sorting of pulpwood according to basic density and coarseness of the fibers could help to control the pulping process and the quality of the end product. Knowing the variation of basic density makes it possible to sort the raw material according to given criteria to improve the process control and product quality. It is therefore important to know how the basic density of wood varies in the raw material and where to obtain wood of a certain density or fiber coarseness.

The effect of process conditions on the properties of pulp and paper may be minor compared with variations in the raw material. If controlling the earlywood-to-latewood ratio or cell wall thickness were possible, altering pulp quality would be possible to produce pulp with predetermined characteristics[55]. Three approaches are available:

- Splintering chips into springwood and summerwood fragments and then segregating them according to basic density. The treatment is impractical, since it damages fibers.
- Fractionating fibers after pulping but before beating. For example, it could be technically possible with multistage hydrocyclone fractionation to produce an accept pulp with a high content of thin-walled fibers for products requiring excellent printing properties and tensile strength and a reject pulp with a high content of thick-walled fibers for products such as tissue and board requiring high compressibility, stiffness, porosity, absorbency, and tear. At present the technology is not economic in a mill-scale operation.
- Selecting raw material using the fundamental knowledge of the variation of critical wood properties between trees and within a tree. Many pulp mills around the world use this approach particularly by blending chips from sawmill slabs in various proportions with normal pulpwood chips. In New Zealand, three radiata pine market kraft pulp categories exist: low, medium, and high coarseness. In a study by Kibblewhite and Bawden[56], pulp from young thinning wood had short and thin-walled fibers with coarseness 0.218 mg m^{-1} and slabwood furnish from saw logs contained long, thick-walled fibers with high 0.306 mg m^{-1} coarseness.

2.9 Heating value of wood and bark

Trees convert atmospheric carbon dioxide and water in photosynthesis to simple carbohydrates. These are the basic raw materials for further biochemical synthesis in the formation of wood and other biomass components. In the same process, the trees absorb solar energy through chlorophyll of the foliage.

In a natural cycle, the carbon dioxide, water, and energy release from the biomass through an oxidation process in decomposition or burning. For tree components such as hardwood leaves, this cycle may require only a year, but for stemwood it may last hundreds of years. Accumulating biomass therefore acts as a carbon sink to the carbon dioxide content of the atmosphere and slows the greenhouse effect. As a source of energy, biomass is almost carbon neutral due to its renewability.

CHAPTER 4

Forests are a huge reserve of renewable energy. Those parts of the tree that are not suitable for raw material of the forest industries can find use for the production of energy. In managed forests, over 40% of biomass of removed trees remains unused after logging operations as crown mass, stumps, roots, and low-quality stem parts. In natural, unmanaged forests, the amount of logging residue is even larger. Recovery of the biomass of the felled trees is therefore less than 40%–60%. When using this biomass in the mechanical or chemical forest industries, 20%–50% of the harvested material becomes solid or liquid process residue that has no use except for fuel. Forest and mill residue are therefore extremely important sources of green energy for the modern forest industries. This is particularly true for the sulfate pulp industries.

The dry mass of wood and bark averages 48%–52% carbon, 38%–42% oxygen, 6.0%–6.5% hydrogen, 0.2%–0.5% nitrogen, and 0.3%–5% mineral elements. Softwoods typically have a slightly higher content of carbon than hardwoods, and bark always has a considerably higher content of mineral elements or ash than wood.

The thermal energy of a fuel depends on the chemical composition of the substance. The combustable elements are carbon and hydrogen. In the complete combustion of biomass, oxygen combines with carbon to produce carbon dioxide and with hydrogen to produce water. Thermal energy once absorbed by trees from solar radiation simultaneously releases according to the following formulas:

$$C + O_2 \rightarrow CO_2 + 32.8 \text{ MJ kg}^{-1} \text{ of } C \tag{3}$$

$$2H_2 + O_2 \rightarrow 2H_2O + 142.2 \text{ MJ kg}^{-1} \text{ of } H_2 \tag{4}$$

The total heat released from the fuel is the *calorimetric or higher heating value*. When biomass burns in a furnace, some released heat must be consumed for vaporizing water that originates from two sources. Water is present in wood and bark at 150–550 kg m^{-3}. When hydrogen and oxygen combine, they generate 140–350 kg m^{-3} water depending on the density of the wood and bark.

In most combustion systems, the heat consumed in vaporization is lost to the atmosphere. Even in complete combustion, it is therefore impossible to exploit fully the calorimetric heating value of the biomass. In practical calculations, the energy spent on vaporization is therefore deducted from the calorimetric heating value assuming that the water cools to its initial temperature but retains the vapor form. This concept is *the effective or lower heating value*, and it depends on the moisture content of the fuel.

If the biomass is absolutely dry, the following equation gives the conversion between the calorimetric and effective heating values. In the equation, 2.45 MJ kg^{-1} is the heat energy necessary to vaporize water at 20 °C, and the factor 0.09 is due to one part of releasing hydrogen and eight parts of oxygen combining to form nine parts of water. If the hydrogen content of the dry forest biomass is 6%, then the effective heating value is 1.3 MJ kg^{-1} less than the calorimetric heating value as Eq. 5 shows:

$$W_{ea} = W_c - (2.45 \cdot 0.09 \cdot H_2 = W_c - 0.22 H_2) \tag{5}$$

where W_{ea} is the effective heating value of absolutely dry biomass [MJ kg^{-1}], W_c is the calorimetric heating value of absolutely dry biomass [MJ kg^{-1}], and H_2 is the hydrogen content of absolutely dry biomass [%].

Biomass always contains water. Due to the latent heat of condensation, a further deduction from the calorimetric heating value is necessary depending on the moisture content of fuel. The equation is then as follows:

$$W_{em} = W_{ea} - 2.45 \frac{MC}{100 - MC} \tag{6}$$

where W_{em} is the effective heating value of biomass with moisture [MJ kg^{-1} dry mass] and MC is the moisture content in the biomass on a fresh mass basis [%].

Differences in heating values between tree species are narrow. The calorimetric heating value is approximately 20.5 MJ kg^{-1}, and the effective heating value of dry wood is approximately 19.2 MJ kg^{-1}. Lignin, resins, terpenes, and waxes increase the heating values, and inorganic materials decrease them. According to Kollmann[57], the effective heating value is 17.4–18.2 MJ kg^{-1} in cellulose, 25.5 MJ kg^{-1} in lignin, and 35.6–38.1 MJ kg^{-1} in resins.

Due to differences in the content of lignin and extractives, heating values are slightly higher in softwoods than in hardwoods. Differences are greater between tree components than between tree species. For example, birch has a calorimetric heating value of 19.9 MJ kg^{-1} in wood, 20.2 MJ kg^{-1} in inner bark, and an unusually high 33.3 MJ kg^{-1} in outer bark because of its high content of suberine[58]. The values in Table 20 give an example of the heating values of different tree components of the pines in the southern United States. Wood properties such as basic density, percentage of latewood, and ring width do not explain the differences. Extractive content accounts for most of the variation.

Table 20. Heating values of different three components of the pines in the southern United States[59].

Material	Calorimetric heating value of southern pines, MJ kg $^{-1}$
Liquid resin	34.0–37.8
Commercial charcoal from wood and bark	26.1–29.6
Resinous wood from mature stump	23.8–25.2
Needles	20.8–21.2
Stem bark at tree butt	20.6–21.3
Stem bark at tree top	19.9–20.5
Stem wood	19.3–21.7
Early wood	19.7–20.4
Late wood	19.5–20.4
Root wood	19.9–20.2
Old cones	18.8–19.0
Dried kraft black liquor after removal of tall oil	13.5–14.3

Wood fuels are frequently measured by volume rather than by weight. When determining energy content on the volume basis, the basic density of biomass becomes an important factor, since differences in the density between species are greater than those in the heating value of dry mass. High basic density means high volumetric heating value. Table 21 shows the effective heating value of Scots pine and birch wood as a function of moisture content per kg of dry mass and per m^3 solid.

Table 21. Effective heating value of wood residue from Scots pine and birch pulpwood as a function of moisture content (on the fresh mass basis) in Finland.

Source of residue	Tree species	Basic density	Moisture content					
			0	40	60	0	40	60
			MJ kg^{-1} dry mass			MJ m^{-3} solid		
Wood	Pine	405	19.3	17.7	15.6	7 817	7 169	7 318
	Birch	480	18.6	17.0	14.9	8 928	8 160	7 152
Bark	Pine	265	19.5	17.9	15.8	5 168	4 744	4 187
	Birch	480	22.6	21.0	18.9	10 848	10 080	9 072
Branches	Pine	415	20.2	18.6	16.5	8 383	7 719	6 848
	Birch	500	19.7	18.1	16.0	9 850	9 050	8 000

The volume of wood fuels is determined in practice in loose measure. Consequently, variation in the solid content or bulk density also requires consideration. The composition of biomass, method of comminution and loading, distance of transport, etc., effect the solid content. The solid volume content usually ranges for comminuted wood and bark residues from 0.2 to 0.5.

3 Nonwood fibers as raw material for pulp and paper

In 1993, the global production of mechanical and chemical pulps for papermaking was 171 million Mg. Although the most abundant source of pulp fibers is wood, fibers for production of good quality paper are available from a large number of vascular plants as Table 22 shows.

Table 22. Approximate dimensions of wood and nonwood pulp fibers.

Source of pulp	Fiber length, mm	Fiber diameter, µm	Slenderness ratio
Softwood	3.0	30	100
Hardwood	1.0	16	62
Wheat straw	1.5	13	120
Rice straw	1.5	9	170
Esparto grass	1.1	10	110
Reed	1.5	13	120
Bagasse	1.7	20	80
Bamboo	2.7	14	
Cotton	25.0	20	1 250

The fibers must be conformable – capable of forming a stable and uniform sheet, and they must develop fiber-to-fiber bonds. Mechanical treatment or beating with pulping improve these two properties of fibers. Unfortunately, few plant fibers can be grown, harvested, transported, and processed economically. To be useful for papermaking, a furnish must be available in large quantities at a reasonable cost.

The use of nonwood fiber for papermaking has been steadily increasing. During the last 10 years, the production capacity of wood pulps expanded 22% but that of nonwood pulps expanded 133%[60]. In 1994, the global production of nonwood pulps was 16 million Mg or 9% of all pulp. Excluding the mechanical pulps, the proportion of nonwood pulps was 12%.

CHAPTER 4

Nonwood pulps primarily come from regions of wood shortage in developing countries. The raw material base includes agricultural residues such as straw and bagasse, natural growing plants such as bamboo and reed, and plants grown for fiber production such as kenaf and hemp. Table 23 shows the production by regions in 1994.

Table 23. The global production of nonwood pulp in 1994[60].

Region	Production of non-wood pulp in 1994, million Mg
Africa	0.2
North America	0.2
Central America	0.2
South America	0.4
Asia	14.5
Oceania	0.0
Europe	0.2
Total	15.6

In 1994, China alone produced 12 million Mg or 75% of the world production of nonwood pulps. Only a small fraction of all pulp in China comes from wood. The common furnish there is straw[61]. Production units are small and outdated, and they usually only serve local needs. According to the provisional forest products outlook[60], China's annual nonwood pulp production will double by 2010.

Straw residue from agriculture represents a large potential of fiber. It is presently the most important source of nonwood pulp. The annual world production of straw is over 1.2 billion Mg, but straw has many other uses such as fodder and fuel. In 1997, the production capacity of straw pulp in China was over 9 million Mg. The total in all other countries was 1.2 million Mg. India, Pakistan, Turkey, Spain, and Egypt were other notable producers of straw pulp.

Wheat and rey straw are better pulp raw materials than oats, barley, and particularly rice straw. Compared with wood, straw has certain disadvantages as a raw material for pulp. Since the harvesting season is short, straw becomes available only intermittently and requires storage for long periods. Straw is liable to decay unless stored in a dry condition. Bulkiness increases the cost of collection and transport, and a high content of silica and ash is harmful in the recovery systems. In rice straw, over 15% of the dry mass may be ash. This makes the recovery of black liquor very difficult, but methods of removing the silica are under development.

The length of straw fiber is approximately 1.5 mm, and the slenderness or length-to-diameter ratio is 100:1. Straw pulp also has many thin-walled small fibers that increase the drainage resistance of the pulp. The runnability on the paper machine is poor at high speeds. Although the content of lignin is not as high as in wood, the yield of pulp is lower as Table 24 shows.

Table 24. The yield of chemical pulp from straw and wood[62].

Type of pulp	Straw pulp	Wood pulp
	Yield, %	
Semichemical	55–58	65–70
Chemical, unbleached	40–42	48–52
Chemical, bleached	35–38	45–50

Bagasse is the residue left when extracting or pressing sugar from cane (*Saccharum officinarum*). It is another important nonwood source of pulp. Since bagasse accumulates as process residue in large quantities in sugar plants, the cost of delivery to pulp mills is usually cheap. This is especially true in the integrated production of sugar and pulp. Sugar mills operate only during 75–225 days per year increasing the cost of storage. Fiber losses during storage are high.

The global production of bagasse is almost 100 million Mg dry mass per year. It is therefore a large reserve of fiber, but it also has uses for fodder, fuel, and especially in Brazil for alcohols to substitute for oil-based motor fuels. In 1997, the world production capacity of bagasse pulp was 3.1 million Mg. India and China were the major producers. In addition, Venezuela, Colombia, Mexico, and Argentina also have substantial production[60].

Before pulping bagasse, one should remove approximately one-third of the mass comprising the thin-walled epidermic and parenchymatous tissues of the pith due to their unsuitable cell composition. Washing and screening of pulp are more difficult than with hardwood pulp. The fibers are approximately 1.7 mm long and have a slenderness ratio similar to that of softwoods. The amounts of silicate and extractives are high, but the level of lignin is approximately the same as in hardwoods. Bagasse can make tissue, corrugating medium, wrapping, newsprint, and printing and writing papers.

Bamboos cover large areas in the tropical, subtropical, and temperate regions in Asia, Africa, and South America. There are 45 genera and 750 species of bamboo. Some of these grow over 30 m high. Bamboos grow fast and give a good yield in plantations. Cultivation is problematic because entire stands of bamboo may sporadically and unforeseeably flower and then die after the ripening of the seeds in large areas. The annual potential is approximately 30 million Mg of biomass.

The global production capacity of bamboo pulp was 1.3 million Mg in 1997. The major producer countries are India and China with Brazil and Thailand following on a much smaller scale.

Bamboos are perennial grasses with a woody stem or culm. The culm contains 50% parenchyma cells, 40% fibers, and 10% conducting cells such as vessels and sieve tubes. In most species of bamboo, the average length of fiber is 1.7–3 mm and the slenderness ratio 1:150–1:250. The lamellated secondary wall of bamboo fiber differs remarkably from the cell wall of wood fiber in the organization and orientation of

CHAPTER 4

microfibrils. Thin cell walls and a high content of hemicelluloses result in fast beating time. Dense and highly lignified nodes cause problems in pulping. The content of ash at 5–6% causes problems in chemical recovery in alkaline pulping. Bamboo fibers find use for the production of high-quality bleached papers.

Reed (*Phragmites*) forms dense vegetations in delta areas of large rivers in Eastern Europe, Africa, and Asia. The annual dry mass production may be 5–10 Mg per hectare. The world production of reed may be 30 million Mg of dry matter. Barges and amphibious boats harvest reed in shallow waters. Reed pulps have use in small quantities in China, Romania, Russia, Turkey, and Iran for printing and other papers.

Esparto grass (*Stipa tenacissima*) grows naturally in dry sandy soils in northern Africa and Spain. Great Britain used it for papermaking many years ago. The 1.5 mm slender fiber produces printing and dublicating papers with exceptional opacity and smoothness. Because of the high costs of harvesting and limited availability, esparto grass is presently of little importance as a source of pulp. The annual biomass potential is less than 0.5 million Mg.

Cotton fibers are an excellent raw material for dissolving pulp, since they are a very pure form of natural cellulose. Pulping requires only little chemical degradation. Raw material comes from rags, textile cuttings, and linters or the residue left behind from the seed after the removal of the staple cotton. Cotton fibers are 20 to 30 mm long, but the length requires reduction to 5–6 mm for papermaking. Cotton fibers make bulky, opaque, and long-lasting papers. They are particularly suitable for special grades of paper such as bonds, drawing paper, and electrical insulating paper. Blending of cotton fiber yarn with synthetic yarn in the textile industry makes it difficult to obtain cotton fiber in a pure form. Consequently, the use of cotton rags for papermaking is diminishing. Cotton lint has other competitive uses that contribute to its high price.

Linen fibers come from linen rags, textile cuttings, or directly from flax stalks. They are exceptionally long, have thick walls, and are difficult to fibrillate. Paper made of linen fibers is strong and bulky. Linen fibers alone or mixed with cotton fibers find use for making currency papers and cigarette papers. Availability is very limited and the cost of fiber high.

There are also many other plant fibers that find use in papermaking. Among them are manila hemp (*Cannabis sativa*), sunn hemp (*Crotolaria juncea*), ramie (*Boehmeria nivea*), sisal (*Agave sisalana*), jute (*Corchorus* sp.), and kenaf (*Hibiscus cannabinus*). Their biomass potential is not great making them suitable only for local use. Many European countries have studied the possibilities to reduce the surplus production of food by using agricultural land for growing nonwood fiber plants such as reed canary grass (*Phalaris arundinacea*). The problem is the cost rather than the quality of the fiber.

Total annual production in the world is over 2 billion Mg of nonwood biomass suitable for pulp and papermaking. Possibilities to expand the production are therefore great. By properly selecting the appropriate mixture of nonwood fibers and the proper pulping processes, one can produce any grade of paper and paperboard. If circumstances dictate, it is possible to make all grades without the addition of wood pulp. In most cases, nonwood fibers will have at least a small portion of wood pulp. A greater use of nonwood pulps awaits only the economic necessity[63].

While it is anticipated that non-wood fiber use will increase, the overall percentage of non-woods being used in the pulp supply will remain low. Realistic projections range from 12% to 15% for 2010[64].

CHAPTER 4

References

1. Young, H. E., Strand, L., and Altenberger, R., Maine Agr. Exp. Sta. Tech. Bull. 12:1(1964).
2. Hakkila, P., Utilization of residual forest biomass, Springer Series in Wood Science, Springer, Berlin, 1989, 568 p.
3. Young, H. E., Ribe, J. H., and Wainwright, K., Orono Life Sci. Agr. Exp. Sta. Miscell. Rep. 230:1(1980).
4. Standish, J. T., Manning, G. H., and Demaerschalk, J., Can. For. Serv. Pac. For. Res. Cent. Inf. Rep. BC-X-264:1(1985).
5. Hakkila, P., Kalaja, H., and Mäkelä, M., Folia Forestalia 240:1(1975).
6. Fegel, A. C., Bull. N.Y. State Coll. For. Tech. Publ. 55:1(1941).
7. Mork, E., Papier-Fabr. 26:741(1928).
8. Panshin, A. J. and de Zeeuw, C., Textbook of Wood Technology, McGraw-Hill, New York, 1980, pp. 62, 105, 107, 163, 243.
9. Ilvessalo-Pfäffli, M. -S., Fiber Atlas. Identification of papermaking fibers, Springer Series in Wood Science, Springer, Berlin, 1995, pp. 15, 18, 22, 28.
10. Bailey, I. W., Proc. Natl. Acad. Sci. 5:283(1919).
11. Fengel, D. and Grosser, D., Holz Roh-Werkst. 33:32(1975).
12. Browning, B. L., Methods of wood chemistry II, Interscience, New York, 1967, pp. 387, 388, 561, 562.
13. Timell, T. E., Adv. Carbohyd. Chem. 20:409(1965).
14. Timell, T. E., Compression wood in gymnosperms, vol. 1–3, Springer, Berlin, 1986, 2150 p.
15. Timell, T. E., Adv. Carbohyd. Chem. 19:247(1964).
16. Timell, T. E., Wood Sci. Technol. 1:45(1967).
17. Sarkanen, K. V. and Ludwig, C.H., Lignins: occurrence, formation, structure and reactions, Wiley, New York, 1971, 916 p.
18. Pettersen, R. C., in The chemistry of solid wood (R. M. Rowell, Ed.), Am. Chem. Soc., Washington, 1984, 614 p.
19. Larson, P. R., For. Prod. J. 16(4):37(1966).

20. Browning, B. L., Methods of wood chemistry I. Interscience, New York, 1967, pp. 75– 77.

21. Koch, P., U S Dep. Agr. For. Serv., Agr. Handbook 605, Vol. 1, U.S. Dept. Agr. For. Serv., Washington, 1985, pp. 400, 401, 435–450.

22. Hannus, K. and Pensar, G., Åbo Akad. Inst. Träkemi Cellul. B44:1(1970).

23. Hillis, W. E., Heartwood and tree exudates, Springer, Berlin, 1987, 268 p.

24. Hillis. W. E., in Wood extractives and their significance to the pulp and paper industries (W. E. Hills, Ed.), Academic Press, New York, 1962, 513 p.

25. Gardner, J. A. F. and Hillis, W. E., in Wood extractives and their significance to the pulp and paper industries (W. E. Hillis, Ed.), Academic Press, New York, 1962, pp. 307–403.

26. Sjöström, E., Wood chemistry. Fundamentals and applications, Academic Press, New York, 1993, pp. 114–161.

27. Hakkila, P. and Kalaja, H., Folia Forestalia 552:1(1983).

28. Ranby, B. G., in Fundamentals of papermaking fibers (F. Bolam, Ed.), British Paper and Board Makers Association, 1958, pp. 55–92.

29. Kollmann, F. F. P. and Côté, W. A., Jr., Principles of wood science, Springer, Berlin, 1968, pp. 18–64, 160–172.

30. Haygreen, J. G. and Bowyer, J. L., Forest products and wood science. An introduction, The Iowa State University Press, Ames, 1996, p. 53.

31. Côte, W. A., Jr., Wood ultrastructure. An atlas of electron micrographs, University of Washington Press, 1967.

32. Isenberg, I. H., in The chemistry of wood (B. L. Browning, Ed.), Interscience Publishers, New York, 1963, pp. 7–55.

33. Wagenführ, R., Holzatlas, Fachbuchverlag, Leipzig, 1996, pp. 453–459.

34. Ezpeleta, L. B. and Simon, J. L. S., Atlas de fibras para pasta de celulosa II parte, VOL. 1, Ministerio de Agricultura, Madrid, 1970, 89 p.

35. Thörnqvist, T., Swedish Council for Building Research. Document D13:1993, Stockholm, 1993, 109 p.

36. Timell, T. E., Wood Science and Technology 16(2):83(1982).

37. Timell, T. E., Svensk Pappertidn. 72:173(1969).

38. Koch, P., Agr. Handbook 420, vol. 1, U.S. Dept. Agr. For. Serv., Washington, 1972, pp. 87–94.

39. Koch, P., Lodgepole pine in North America, Forest Products Society, Madison, vol. 2, 1996, pp. 513–532.

40. Hakkila, P., Commun. Inst. For. Fenn. 66(8):1(1968).

CHAPTER 4

41. Hakkila, P. and Panhelainen, A., Commun. Inst. For. Fenn. 73(1):1(1970).

42. Hakkila, P., Commun. Inst. For. Fenn. 75(1):1(1971).

43. Anon., Wood Handbook: Wood as an engineering material. Agriculture handbook 72, U.S. Department of Agriculture, Washington, 1987, 451 p.

44. Stamm, A. J. and Sanders, H.T., Tappi 49:397(1966).

45. Stamm, A. J., Tappi 52:1498(1969).

46. Spurr, S. H. and Hsiung, W. Y., Journal of Forestry 52(3):191(1954).

47. Paul, B. H., Journal of Forestry 37:478(1939).

48. Conzales, J. S., Wood density of Canadian tree species, Information Report NOR-X- 315, Forestry Canada, 1990, 130 p.

49. Rendle, B. J., Quarterly Journal of Forestry 3:116(1959).

50. Hakkila, P., Commun. Inst. For. Fenn. 61(5):1(1966).

51. Nylinder, P., Om ved- och trädegenskapers inverkan på råvolymvikt och flytbarhet. I. Tall. Summary: Influence of tree features and wood properties on basic density and buoyancy. I. Scots pine (Pinus sylvestris), Kungl. Skogshögskolan, institutionen för virkeslära, Uppsatser R 36, 1961, 63 p.

52. Hakkila, P., Commun. Inst. For. Fenn. 96(3):1(1979).

53. Hakkila, P., Kalaja, H., Saranpää, P., The Finnish Forest Research Institute, Research Papers, 582:1(1995).

54. Megraw, R. A., Wood quality factors in loblolly pine, TAPPI PRESS, Atlanta, 1985, 88 p.

55. Paavilainen, L., Influence of fiber morphology and processing in the softwood sulphate pulp fiber and paper properties, Ph.D thesis, University of Technology, Helsinki, 1993.

56. Kibblewhite, R. P. and Bawden, A. D., New Zealand Journal of Forest Science, 10(3):533(1990).

57. Kollmann, F., Technologie des Holzes und der Holzwerstoffe, Springer, Berlin, 1951, 1050 p.

58. Nurmi, J., Acta For. Fenn. 236:1(1993).

59. Howard, E. T., Wood Science 5:194(1973).

60. Anon., Pulp and paper capacities, FAO, Rome, 1996, 201 p.

61. Anon., China National Statistical Yearbook, China Statistics Publishing House, Beijing, 1995.

62. Jeyasingam, J. T., TAPPI 1988 Pulping Conference Proceedings, TAPPI PRESS, Atlanta, p. 571.

63. Atchison, J. E., TAPPI 1988 Pulping Conference Proceedings, TAPPI PRESS, Atlanta, p. 25.

64. Pande, H., Unasylva 49(2):50(1998).

CHAPTER 5

Forest inventory and planning

1	**Forest inventory**	**187**
1.1	Introduction	187
1.2	Tree measurement	188
1.3	Measurement of forest stand	189
	1.3.1 Stand and compartment	189
	1.3.2 Site characteristics	192
	1.3.3 Forest stand plot	192
	1.3.4 Compartment inventory	194
1.4	Large area inventory	195
	1.4.1 Classification of methods	195
	1.4.2 Sampling concepts	195
	1.4.3 Simple random sampling	197
	1.4.4 Systematic sampling	198
	1.4.5 Stratified sampling	199
	1.4.6 Cluster sampling	200
	1.4.7 Two-phase sampling	201
2	**Forest management planning**	**202**
2.1	Introduction	202
2.2	Cutting budget methods	203
2.3	Planning based on stand economy	203
2.4	Numerical optimization	204
	2.4.1 Techniques available	204
	2.4.2 Linear programming	205
	2.4.3 Goal programming	207
	2.4.4 Heuristics	208
	2.4.5 The Finnish planning case	209
	2.4.6 Forest regulation	211
	2.4.7 Controlling sustainability	212
	References	**215**

CHAPTER 5

Timo Pukkala

Forest inventory and planning

1 Forest inventory

1.1 Introduction

Forest inventory provides information for forest management planning and forestry decision-making. The manner of inventorying a forest varies with the specific needs of the particular planning situation. Timber management planning requires that the standing volume, commercial volume, and volume increment of the stand or forest are known. These characteristics are very difficult, if not impossible, to measure directly on standing trees. Instead the combined use of easily measurable variables and models indirectly determines the volume, biomass, increment, and other important characteristics.

    A complete forest inventory system consists of measurement and calculation techniques. Figure 1 shows that calculation combines the information collected in field measurements with information contained in models. The models express the relationships between the measured properties and the variables considered in planning and decision making. Examples of models used in forest inventory are *volume tables*, *biomass functions*, and *height curves*. Tables and graphs are the most common types of models in manual computation. Computerized calculations use equations because they are more accurate and flexible than tables.

    A model that is valid for a restricted forest area or a specific forest inventory is a *local model*. If the model applies to an entire country, it is a *global model*.

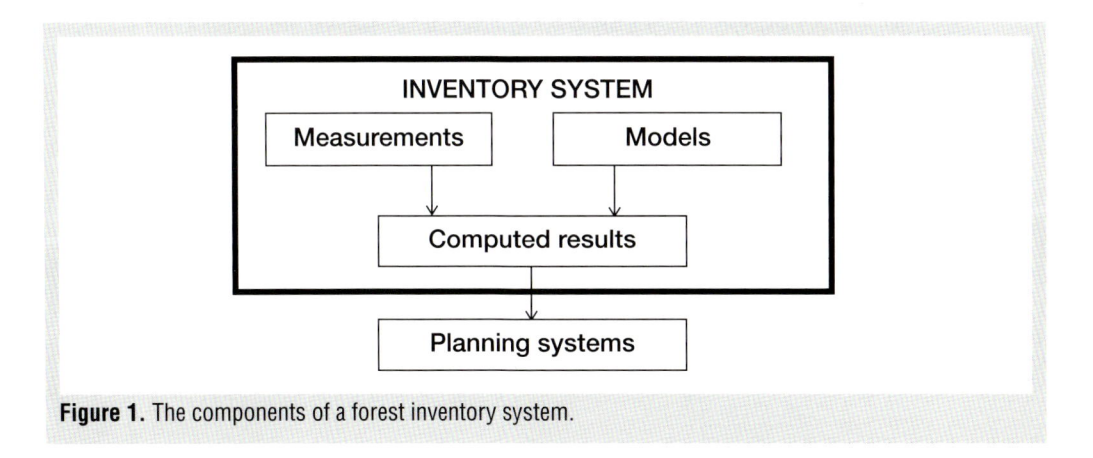

Figure 1. The components of a forest inventory system.

CHAPTER 5

1.2 Tree measurement

The divisions of a tree are roots, stump, stem, branches, and leaves. The most common object of measurement is the stem with occasional measurement of branches and leaves.

Forest mensuration rarely concerns roots and stumps. The different dimensions of a tree are *tree characteristics*.

Tree age is the number of years that the tree has grown since germination (seedling stand) or since sprouting (coppice stand). The tree age can sometimes be measured from the core of an increment bore by counting the number of annual growth rings. If the tree species produces one whorl annually and the whorls are distinguishable, the whorls will denote age. Sometimes the year of stand establishment is known.

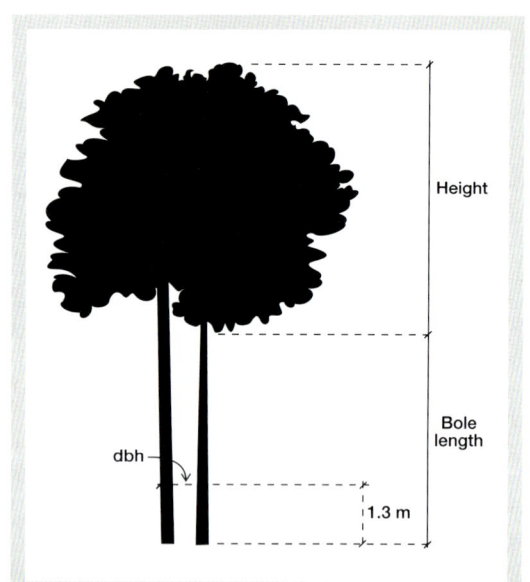

Figure 2. Definitions for diameter at breast height (dbh), tree height, and bole length.

The most common description of the thickness of the stem is the *diameter at breast height* (*dbh*). This is 1.3 m from the ground level for a seedling stand or from the starting point of the stem for a coppice stand. *Dbh* is measured with diameter calipers or diameter tape.

Tree height is the distance from the ground level to the tree top. The *bole length* of Fig. 2 gives the length of the clear (stem without branches) or almost clear part of the stem below the tree crown. *Commercial height* is the length of the marketable part of the stem. Tree height measurement uses a hypsometer or a measuring pole. *Bark thickness* is the distance from the cambium to the surface of the stem at the height of 1.3 m. Measurement uses a bark gauge.

Diameter increment is the increase in diameter for a specific number of years such as one or five. *Height increment* is the corresponding increase in height. If the annual growth rings are visible, the diameter increment can be measured from a radial core bored at the breast height. With conifers, the height increment can sometimes be measured as the distance between the whorls. If there are no clear annual rings or whorls, it is necessary to measure the diameter and height twice at intervals of one year or five years.

The *tree basal area* is the area of the transection of the stem at the breast height. *Stem volume* is the volume of the stem with bark from the stump level to the top of the tree. The stem volume of a standing tree cannot usually be measured directly but estimated using a volume table or volume equation with *dbh* and height as predictors.

Forest inventory and planning

The *form factor* is the ratio of the stem volume to the volume of a cylinder with the same *dbh* and height as the tree. It is approximately 0.45 for several species. Form factor is a "quick-and-dirty" estimator of the stem volume (stem volume = form factor × cylinder volume).

In fiber, pulpwood, fuel wood or fodder production, the masses of the stem, branches, and leaves are important characteristics. The mass (or biomass) can be expressed as fresh mass or dry mass (airdry or ovendry). The dry mass is a more indicative characteristic than the fresh mass since it expresses the energy and dry matter content of a tree better than the fresh mass.

Multiplying the stem volume by the density of wood also calculates the mass. To obtain dry mass, the volume is multiplied by the dry density or basic density. To obtain fresh mass, the volume is multiplied by the fresh density. The fresh mass of small felled trees can be weighed directly in the field. In this case, fresh mass converts to dry mass when multiplied by the dry matter content – the ratio between the dry and fresh masses.

1.3 Measurement of forest stand

1.3.1 Stand and compartment

A *forest stand* is a homogenous sub-area of the forest. In this book, *forest compartment* is the smallest management unit of the forest. A compartment may be smaller than a stand but not larger. A large, uniform stand may have two or more compartments managed differently. The compartment may also be a forest inventory unit for which the results are computed separately.

Forest compartments should be reasonably homogenous. A change in any of the following variables can differentiate a compartment from its neighbors:

- site quality and therefore growth rate
- species composition
- stand age
- management objective such as protection forest or production forest
- stand density
- stand structure – even-aged or uneven aged
- slope and aspect – level plateau, southern slope, northern slope, etc.
- accessibility
- administrative boundaries.

Figure 3 is an illustration of tree stands that are even-aged or uneven-aged. Tree plantations are examples of even-aged stands. Uneven-aged stands have diameter distributions in which there are much more small trees than large ones. The management schedule of an even-aged stand consists of regeneration, thinnings, and final felling once the stand age reaches the *rotation length*. Uneven-aged forests are typically managed as a *selection forest* by retaining the uneven-aged structure. The relevant management parameters are the *cutting cycle* (time interval between successive cuttings in the same forest) and the manner of harvesting different species and diameter classes.

CHAPTER 5

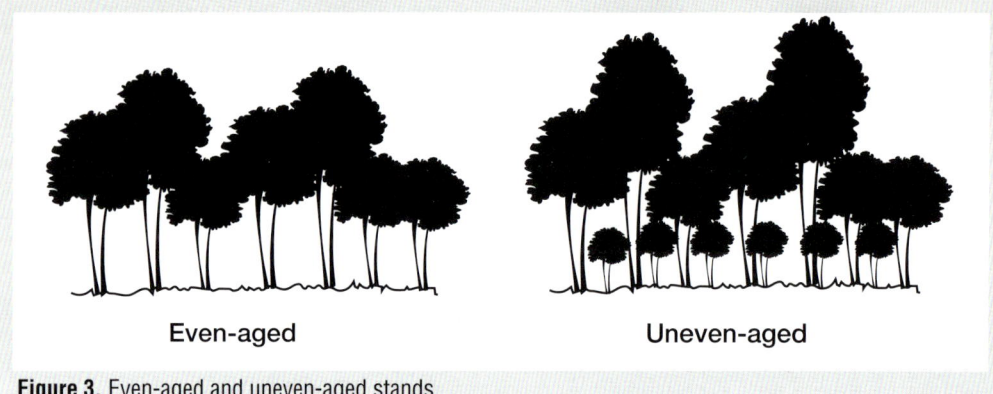

Figure 3. Even-aged and uneven-aged stands.

The measurements of forest inventory may vary with the stand structure. The relevant stand characteristics are different for even-aged and uneven-aged forests. For example, stand age and dominant height are very important for the description of an even-aged stand but not particularly relevant in uneven-aged forests.

Stand characteristics

Stand characteristics describe a tree population such as all trees of one plot, one stand, or one compartment. *Stocking* is the total number of stems per hectare. *Stand basal area* is the sum of the basal areas of individual trees. *Stand volume* is the sum of the tree volumes, and *stem mass, branch mass*, and *leaf mass* are the sums of the corresponding tree characteristics.

Calculations of the mean diameter, height, and age of trees usually use the tree basal area as a weight variable. If ordinary unweighted means were used, the weights of small trees would be the same as big trees, although the contribution of small trees to the total volume, mass, or growth is negligible.

Dominant height (H_{dom}) is the average height of the 100 largest trees per hectare. Dominant height at given age is a frequently used measure of site productivity.

Current annual increment (*CAI*) is the total volume growth of all trees in one year (previous or next year). *Mean annual increment* (*MAI*) of an even-aged stand is the average annual volume increment of the stand since seed sowing or previous coppicing. In a young stand, *CAI* is usually much greater than *MAI*. In an even-aged stand, *CAI* and *MAI* relate to each other such that when *CAI* falls below *MAI*, the *MAI* has reached its maximum (abbreviated as *MMAI*). The age at which this occurs indicates the rotation length that maximizes the volume production of the stand as Fig. 4 shows.

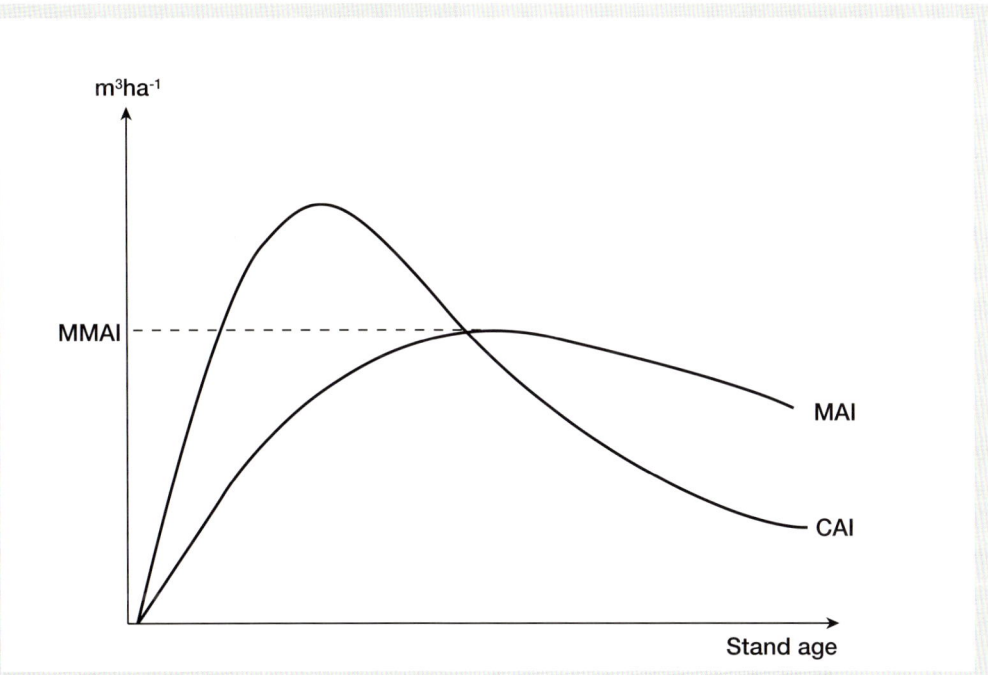

Figure 4. The temporal course of the current annual increment (CAI) and mean annual increment (MAI) of an even-aged stand. The maximum mean annual increment (MMAI) describes site productivity.

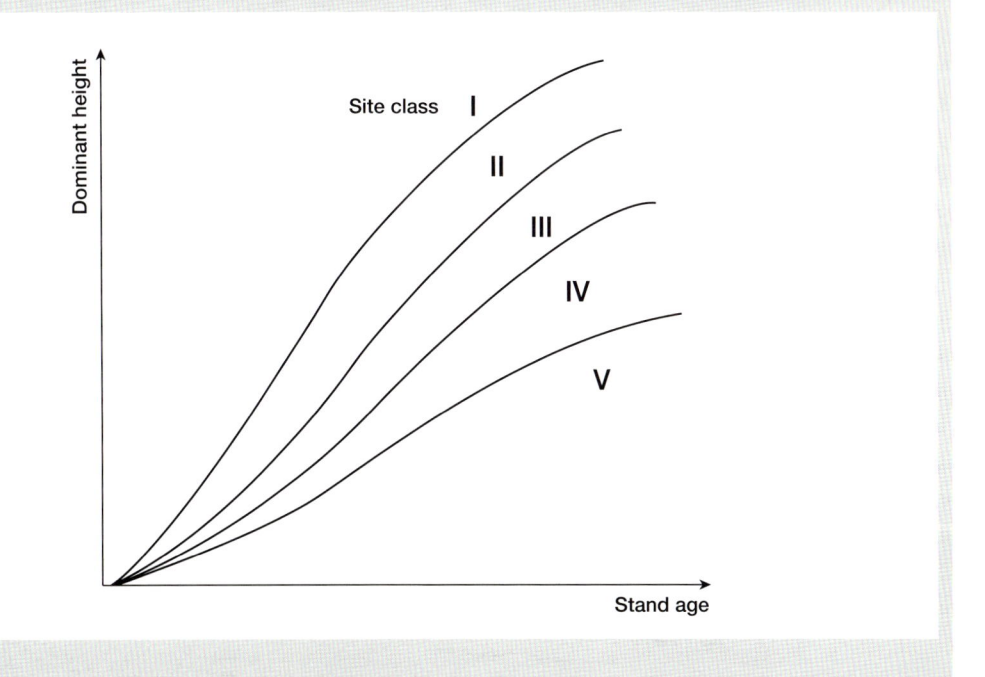

Figure 5. Site classification using stand age and dominant height.

CHAPTER 5

1.3.2 Site characteristics

Other variables that characterize forest compartments are *site characteristics* such as site productivity, soil type, slope, aspect, and accessibility. The most important of these is *site productivity*. Common measures for this are site class, site index, yield class, and forest site type.

The most common way to determine the *site classes* uses dominant height and stand age. A better site has a greater dominant height at a given age. Figure 5 shows a family of curves for the site classification system. These curves indicate the boundaries of site classes. This method to determine site fertility applies only to even-aged stands that are unthinned or thinned from below.

The site classes can have names according to the dominant height at a given stand age. These are *site indices*. For example, if dominant height at 30 years is 20 m and 30 years is the reference age, the site index of this forest is 20 m.

It is also possible to convert the site classes and site indices into *yield classes*. Yield class is the mean annual volume increment with the optimal rotation length. For example, if the maximum *MAI* is 12 m^3 ha^{-1}, the yield class is 12 m^3 ha^{-1}. The yield class shows the combined productivity of the site and the management schedule applied to the stand in terms of the MMAI.

Forest site types use ground vegetation. The vegetation changes with site fertility, and certain species are therefore useful indicators of site productivity.

Other site characteristics include slope, aspect, accessibility, stoniness, soil type, annual rainfall, mean temperature, altitude, etc. *Slope* is the average inclination of a sloping terrain in percentage or degrees. The *aspect* of the terrain indicates the main compass direction of the downward slope. The terrain may face south, southwest, east, etc.

1.3.3 Forest stand plot

It is usually too expensive and time consuming to measure all the trees of a compartment or stand to obtain the stand characteristics. Instead, a sample measurement represents the whole compartment. Usually the sample consists of one or several *sample plots*. The plots are *temporary, semipermanent,* or *permanent*. Temporary plots are the most common in forest inventory. They are not usually marked in the field meaning they cannot be found and visited later.

Stands of large trees and irregular stand structures require a large plot. In a homogenous young stand, 100 m^2 may suffice. If the trees are large, the stocking is low, and the spatial distribution of trees is irregular, the plot size should be 3 000 m^2 or more. The desired accuracy and precision of the results also affect the plot size. Higher requirements necessitate a larger plot. The location of inventory plots is mostly systematic or otherwise objective.

Figure 6 shows the four basic types of forest stand plots. These are *fixed-radius circular plot* or ordinary circular plot, *variable-radius circular plot*, *rectangular plot*, and *relascope plot*. The circular plot has a center and radius, the rectangular plot has corners, and the relascope plot has a center and a *basal area factor*.

Ordinary circular and rectangular plots are suitable for even-aged stands. The variable-radius circular plot in which the plot radius varies according to the dbh is suitable for uneven-aged forests where the number of small trees is many times higher than that of large trees. Relascope plots are the most suitable for stands with good visibility, regular stem form, and high variation in tree size.

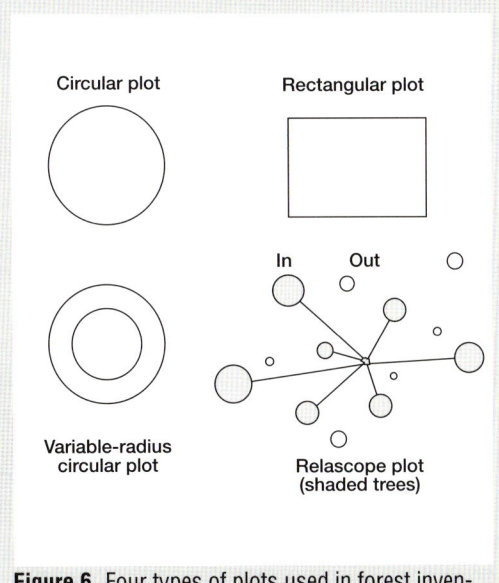

Figure 6. Four types of plots used in forest inventory.

If stem volume, biomass, and value of the growing stock are the most important stand characteristics, it is advantageous with limited resources to concentrate the measurements on big trees. Using relascope plots or variable-radius circular plots can accomplish this. With the same number of measured trees, the estimates for most stand characteristics are more precise with relascope and variable-radius circular plots than with ordinary circular or rectangular plots.

Trees that belong to the relascope plot are selected with a *relascope*. The simplest relascope consists of a stick and a plate with a hole attached to one end. All trees that fill the hole when aiming the stem at the height of 1.3 m belong to the plot. An important concept connected to the relascope plot is the *basal area factor, q* [m^2 ha^{-1}], of the relascope. The basal area factor gives the basal area [m^2 ha^{-1}] represented by one tree of the plot. Eq. 1 then gives the basal area of the stand, G [m^2 ha^{-1}]:

$$G = nq \qquad (1)$$

where n is the number of trees on the plot.

Relascope plots have use in forest inventory in two different ways. In the *visual compartment inventory*, the relascope estimates the stand basal area with no measurement of the tree diameters. The stand basal area with the mean tree height estimates the stand volume. Measuring the height of one or more trees of average size by using a hypsometer provides an estimate of the mean height.

CHAPTER 5

Relascope plots can also be used similarly to ordinary circular or rectangular plots. This technique requires measurement and recording of the diameter of every tree that fills the hole of the relascope. A disadvantage of the relascope plot is that checking of boundary trees is more complicated than with circular plots.

Measurement of a forest stand plot varies according to plot type and results needed. The measurement and calculation of results are slightly different when using ordinary circular or rectangular plots, variable-radius circular plots, or relascope plots.

Two types of trees commonly measured within the plot are *tallied trees* and *sample trees*. All trees are tallied or enumerated. This means determining the tree species, stem diameter, and probably stem quality and bole length. Some trees may be selected as sample trees for more detailed measurements of height, commercial length, past growth, bark thickness, etc. Preparing local models that relate height, stem volume, commercial volume, or other characteristics to stem diameter uses the sample trees.

1.3.4 Compartment inventory

Inventory of the growing stock of a stand or compartment may occur in several ways. One way is *total enumeration* with measurement of all trees. If an accurate estimate is necessary for timber sales, total enumeration is a choice for consideration. In old stands consisting of large trees that are few in number and valuable, total enumeration may be justified. In most instances, total enumeration is far too costly.

Another method is *visual compartment inventory*. Here the surveyor directly estimates the most important stand characteristics using a relascope, hypsometer, tables, and other tools. The method is quick, but it requires a high level of professional skill and experience. Visual inventory usually gives a rather low accuracy. If the purpose of inventory is to obtain a rough volume estimate only, a visual inventory may be satisfactory if skilled surveyors are available.

The most common type of compartment inventory is *sampling* with exact measurements taken over part of the compartment only. The measured trees comprise the sample. The sample usually consists of circular plots of fixed or variable radius, or relascope plots. Besides plots, *inventory strips* are a possibility. It is typical to use larger plots in an old or sparse forest than in a young or dense forest. The applied sampling ratio depends on the required accuracy and precision of the results, heterogeneity of the compartment, and the time and funds available.

The compartment inventory is a special case of *stratified sampling* in which each compartment is a separate *stratum*. Stratification is a means of improving the precision of inventory when the variation in stand characteristics between the strata is large compared with the variation in the stratum. Small variation in stand characteristics within a compartment increases the precision of inventory. This is a reason for delineating homogenous compartments.

Mapping of compartment boundaries is an integral part of compartment inventory. Mapping is usually done by visual interpretation of aerial photographs.

1.4 Large area inventory

1.4.1 Classification of methods

The number of different inventory methods is large. Classification therefore helps obtain a total picture of the methods. They fall into the following three categories:

- *total enumeration* measuring all trees of interest
- *sampling* with sample plots
- *compartment inventory* with delineation of compartments and then a separate inventory of each compartment using total enumeration, visual estimation, or sampling.

When using sample plots in the inventory, the location of plots may be random, systematic, clustered, or some combination of these patterns. In addition, the sample may be nonstratified or stratified. The plot size and type also vary. The most common plot types include ordinary circular plots, rectangular plots, relascope plots, variable-radius circular plots, and inventory strips.

A previous section described compartment inventory. Total enumeration means tallying or enumerating every tree of interest separately by recording its species and diameter and possibly some other characteristics such as bole length, quality class etc. Trees of interest may include stems marked for harvesting or all trees larger than a certain limit. Total enumeration resembles the measurement of an individual forest stand plot with tally of all trees and measurement of sample trees to develop a height curve. Total enumeration is the most accurate inventory method, but it is usually too expensive for large area inventories. Total enumeration is useful when measuring trees for timber sales or when all trees of interest are very valuable. Rare and exceptionally valuable tree species may be inventoried by total enumeration while the other species are inventoried by sampling.

1.4.2 Sampling concepts

The units of the forest selected for sampling are *sampling units*. Examples of sampling units are individual trees and sample plots. The sample may consist of every tenth tree (tree is the sampling unit) or circular plots placed at 200-m intervals (plot is the sampling unit). In most forest inventories, plot is the sampling unit, and the sample consists of sample plots.

The sample plots can be distributed over the target area in different ways. One possibility is to measure one large plot in each compartment or stratum. If the forest is not homogenous, the plot may not represent the entire area well. For precision, a general rule is that a large number of small plots is better than a small number of large plots. Measuring a large number of small plots requires considerable walking from plot to plot. An optimum plot size exists therefore that depends on the stand structure, the required precision, and the amount of time and funds available. The precision and accuracy of the results are usually best if the plots are as far from each other as possible. A systematic grid of sample plots is then the best pattern.

CHAPTER 5

The *accuracy* of the inventory expresses the similarity of the inventory result to the correct value of the characteristic. Accuracy depends on the *precision* and *systematic errors* (*bias*). Precision refers to the degree of certainty about the correctness of the results with measurement by the *standard error of mean* and *confidence limits* of the characteristic in question. Precision tells how certain the inventoried mean stand volume differs less than 10 m^3 ha^{-1} from the true mean volume, as an example.

The precision of the inventory depends on the variation in stand characteristics within the inventoried area with smaller variation giving higher precision. Improved precision is a reason why it is often beneficial to divide the inventoried area into homogenous compartments or strata.

The precision of the inventory is different for different stand characteristics. Usually, the stand volume is the characteristic for which the standard error and confidence limits are calculated when describing the precision of the inventory.

The first step of computing precision is to compute the *standard deviation* of the variable in question. Equation 2 provides the standard deviation:

$$s = \sqrt{\sum_{i=1}^{n}(x_i - \bar{x})^2/(n-1)} \qquad (2)$$

where s is the standard deviation, x_i is the value of the characteristic in sampling unit i, $\bar{x}$ is the mean of the characteristic calculated from all sampling units, and n is the number of sampling units (sample size).

The standard deviation indicates how much the results of sampling units vary around the average results. A larger standard deviation indicates greater variation between sampling units. The square of the standard deviation is the *variance* denoted by s^2.

The standard error of mean (s) describes how similar the results would be in repeated, independent inventories. The equation for s is as follows:

$$s_{\bar{x}} = s/\sqrt{n} \qquad (3)$$

Confidence limits give the range within which the correct value of a characteristic falls with a given probability. Equation 4 computes the confidence limits:

$$\bar{x} - t_\alpha s_{\bar{x}} \leq \mu \leq \bar{x} + t_\alpha s_{\bar{x}} \qquad (4)$$

where μ is the correct mean value of the stand characteristic, $\bar{x}$ is the corresponding estimate calculated from the inventory results (mean of the sampling units), t is the value of Student's distribution, and α is the probability that the correct value is outside the confidence limits.

The above equation indicates that the correct mean value (μ) of volume or some other characteristic is usually between $\bar{x} - t_\alpha s_{\bar{x}}$ and $\bar{x} + t_\alpha s_{\bar{x}}$ with a probability of α that the error is greater than $t_\alpha s_{\bar{x}}$. Parameter t depends on the number of sampling

Forest inventory and planning

units and the selected probability that the correct value is outside the confidence limits. Parameter t comes from statistical tables.

The *efficiency* of sampling technique measures the sampling error with a given sample size. The sampling technique that gives the smallest sampling error with a given sample size is the most efficient.

The ratio of the measured area and total area, or the ratio of the number of measured trees and total number of trees, is the *sampling intensity* or *sampling ratio* of the inventory. The sampling technique and sample size determine the *sampling design* of the inventory.

1.4.3 Simple random sampling

The essential items to know about a sampling technique are the calculation of the mean volume and the standard error or mean. Means are necessary for computing the totals, and standard error is necessary for computing the precision of the inventory.

The simplest sampling technique is *random sampling without stratification* or *simple random sampling*. Figure 7 shows that this method involves placing the sample plots randomly. Equation 5 computes the mean:

$$\bar{x} = \sum_{i=1}^{n} x_i / n \quad (5)$$

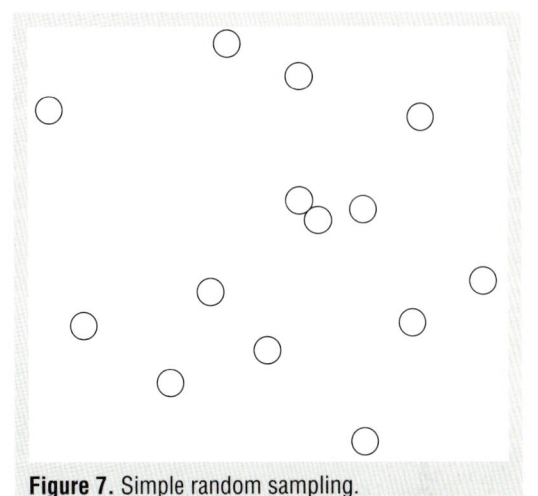

Figure 7. Simple random sampling.

If the sample is taken with replacement – the same sampling unit can be measured twice – or the sampling ratio is small, the standard error of mean comes from the equation given above as follows:

$$s_{\bar{x}} = s / \sqrt{n} \quad (6)$$

The precision of the results depends on the variation between the sampling units and sample size.

When planning a forest inventory, knowing how many plots are necessary for a certain precision is useful. Calculating the required sample size requires knowing the standard deviation of the characteristic in question. This is usually stand volume obtained from previous inventories or estimated in a preliminary survey by measuring some sampling units. Deciding the tolerable error denoted by E and the probability of the error denoted as α being greater than E is necessary. Calculating the required sample size n then uses the following:

CHAPTER 5

$$n = \frac{t_\alpha^2 s^2}{E^2} \tag{7}$$

where s^2 is variance, and t is the value of Student's distribution.

The number of sampling units belonging to different forest categories can estimate the proportions of these categories. The proportion of category i results from the following:

$$p_i = n_i/n \tag{8}$$

where n_i is the number of sampling units in category i, and n is the total number of sampling units. The variance of the estimated proportion s_p^2 is as follows for a small sampling ratio:

$$s_p^2 = p(1-p)/n \tag{9}$$

The benefits of simple random sampling are simplicity and the ease of estimating the precision. The disadvantage is that a random sample is difficult to draw in the forest. In addition, simple random sampling is inefficient. Other techniques often provide better precision with the same sample size.

1.4.4 Systematic sampling

Systematic sampling selects the sample units systematically rather than randomly such as plots measured at 100-m intervals as Fig. 8 shows.

Estimates of the means and the proportions of land categories use the same technique as simple random sampling. The formulas for standard error, confidence limits, and sample size do not apply to systematic sampling. No general formula exists for computing the standard error of systematic sampling.

Systematic sampling usually gives better precision than simple random sampling. The above formulas for simple random sampling can be used to approximate the precision of systematic sampling, but they usually underestimate the precision.

Figure 8. Systematic sampling.

If the variation among sampling units is completely random, systematic sampling is as good as simple random sampling. If there is trend-like variation in the forest as is often the case, systematic sampling is better than random sampling. If the forest properties vary periodically and systematically, systematic sampling may give biased results.

The prime advantage of systematic sampling is that the sample is easy to select in the field. Its precision is usually better than with simple random sampling. The disadvantage is the difficulty to estimate the standard error and the possible bias caused by periodical variation in the forest.

1.4.5 Stratified sampling

Stratified sampling divides the forest into homogenous categories called *strata* with a sample taken from each stratum systematically or randomly as Fig. 9 shows. The computations proceed in two steps. One first computes the parameters such as mean, standard deviation, and variance for the strata and then uses these results to compute the corresponding parameters for the total area.

Eq. 10 provides the mean of a stratified sample computed as a weighted average of the stratum means:

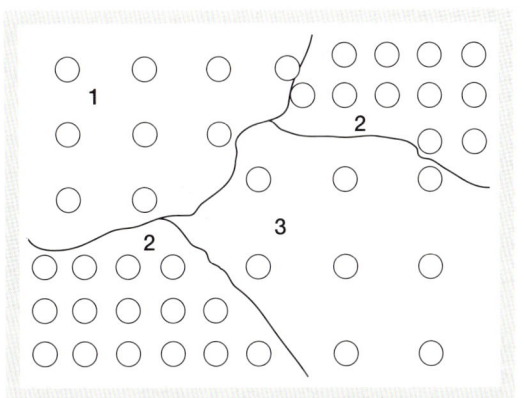

Figure 9. Systematic stratified sampling.

$$\bar{x}_{STR} = \sum_{j=1}^{m} p_j \bar{x} \qquad (10)$$

where m is the number of strata, and p_j is the proportion of stratum j of the total area. Eq. 11 provides the variance of the mean computed as the weighted mean of the corresponding variances of the strata:

$$s^2_{\bar{x}_{SRT}} = \sum_{j=1}^{m} p_j^2 s_{\bar{x}}^2 = \sum_{j=1}^{m} p_j^2 s_j^2 / n_j \qquad (11)$$

where n_j is the number of sampling units in stratum j.

Stratification improves the efficiency of sampling if the within-stratum variation is small compared with the between-stratum variation. The target of stratification is therefore to create the most homogenous strata possible.

One decision required in stratified sampling is the distribution of sampling units between the strata. The simplest method is *proportional allocation* in which the sample size in each stratum is directly proportional to the area of the stratum. This is not the most efficient method if the variation in stand characteristics is not equal in different strata. In such a case, it is better to take a bigger sample from the nonhomogeneous strata and a smaller one from the homogenous strata by using *optimal allocation*. In optimal allocation, the sample size in stratum j results from the following:

CHAPTER 5

$$n_j = n \frac{p_j s_j}{\sum_{j=1}^{m} p_j s_j} \qquad (12)$$

where n is the total number of sample plots.

The benefit of well performed stratified sampling is improved precision compared with unstratified sampling. Calculation of the mean and the precision are easy. The disadvantage is that stratification incurs additional work and costs.

Stratified sampling is very common in large-area forest inventory. Stratification typically uses aerial photographs or satellite images. With aerial photographs, *visual interpretation* draws the boundaries of different strata manually. When using satellite images, a computer can make the stratification by *numerical interpretation*. Examples of strata are different forest types, age classes, stand densities, and their combinations.

1.4.6 Cluster sampling

Cluster sampling means that the sampling units do not have a uniform distribution. Figure 10 shows that it uses bundles or *clusters* of plots located close to each other. Cluster sampling is an example of *two-stage sampling*. The first step involves selection of the subareas for measurements of the plots followed by the second step – measuring sample plots in each selected subarea. In both stages, selection of the sample can be systematic or random.

Clustering decreases the precision of the inventory compared with the same number of systematically or randomly selected sampling units. Clustering greatly decreases the time needed to reach the sampling units. Since this time can be used to measure additional units, clustering may improve the precision of the inventory for a given budget.

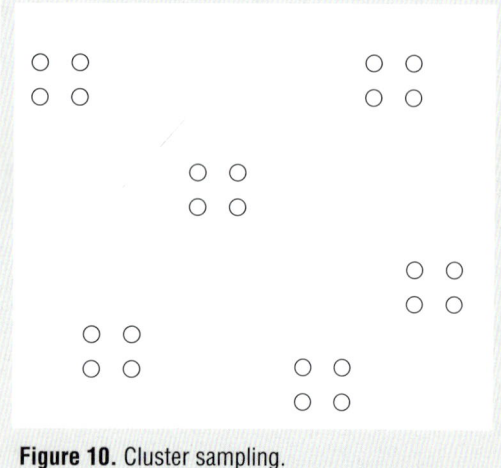

Figure 10. Cluster sampling.

In cluster sampling, the results are computed first for the clusters and then used to derive parameters for the whole sample. Computation of the mean uses the following:

$$\bar{x}_{CLU} = \sum_{k=1}^{c} \bar{x}_k / c \qquad (13)$$

where $\bar{x}_k$ is the mean of cluster k, and c is the total number of clusters. The variance of mean comes from the variation of the cluster means around the overall mean:

$$s_{\bar{x}_{CLU}}^2 = \sum_{k=1}^{c} \frac{(\bar{x}_k - \bar{x}_{CLU})^2}{c(c-1)} \qquad (14)$$

If the total number of clusters is small, this estimate requires multiplication by a correction factor 1-c/C where C is the total number of clusters and c is the sampled number of clusters. With a small sampling ratio, c/C is very small, and the correction factor approaches one.

When planning cluster sampling, the target is to find a scheme with much variation within the clusters and little variation between clusters. The sample plots of one cluster should not be too similar to each other, and the distance between the plots within one cluster should not be too short.

The efficiency of cluster sampling depends on the scale in which variation in the forest occurs. If most of the variation is within short distances, this variation can be gathered by a cluster, and cluster sampling is efficient. If little small-scale variation exists with only gradual and large-scale changes, cluster sampling is inefficient.

It is very common in large area inventory to stratify the sample and group the plots into clusters located along *inventory tracks*. The number of plots in each cluster usually allows measurement in one day. Sometimes the clustering may be two-fold with a group of clusters reachable from one camp.

1.4.7 Two-phase sampling

Two-phase sampling involves measurement of sampling units in two phases. The first phase involves measurement of a large sample. In the second phase, a smaller sample is selected from the same sampling units. The two samples of two-phase sampling are measured using different techniques.

Two-phase sampling has common use in forest inventory so that the first phase sample consists of many plots interpreted on aerial photographs or satellite images. The second phase involves measuring part of the sample plots in the field. The characteristics interpreted for the first-phase plots may be used to stratify the plots into homogenous groups or strata. In this case, the method is two-phase sampling with stratification. The difference from normal stratified sampling is the possibility to decide the strata after the photographic interpretation. After stratification, at least one first-phase plot in each stratum is measured in the field. The field plots usually have a larger number of stand characteristics than the first-phase plots.

The proportions of different strata are typically estimated from the first-phase sample. Because these proportions contain some error, the precision of two-phase sampling with stratification is poorer than in normal stratified sampling with the same number of field plots. The results are biased if the estimated proportions of strata are incorrect.

The mean stand characteristics of the field plots of a stratum are given to all first-phase plots of that stratum. All the final characteristics estimated for the various strata are therefore computed from the field plots.

In *two-phase sampling with regression*, the characteristics measured in the first and second phases are the same. In a typical case, stand volume is the principal characteristic determined in both phases. The second-phase sample plots for which the stand characteristics have been determined on the photograph and in the field are used to determine the relationships between the first- and second-phase results. These relationships can then predict the "true results" or field results of those first-phase plots not visited in the field. In both modifications of two-phase sampling, all first-phase plots will eventually receive the stand characteristics on the basis of field plots.

Two-phase sampling with regression can be done separately for different strata. Inventorying only part of the forest by this technique is also possible. Two-phase sampling is usually preferable if the measurement of a sampling unit in the first phase is cheaper than in the second phase and a correlation exists between the first- and second-phase measurements.

2 Forest management planning

2.1 Introduction

The aim in forest management planning is to find a plan to meet the production and other targets for the forest. Planning supports forestry decision-making by producing information on the consequences of alternate decisions and helping a manager to rank these alternatives. The task of forest management planning is to show the optimum way to use forest resources. The decision maker selects the criteria. They are not necessarily economic and material.

Planning is strategic, tactical, or operational. *Strategic planning* helps in setting the main direction of forestry so that the forest and its use contribute in the best possible way to the welfare of the owner. Examples of forest strategies are maximization of even timber flow, maximization of profitability, and nature conservation. Strategic planning helps to specify the management objectives of tactical planning.

Tactical planning answers the question of what to do in the forest to satisfy the production targets and other management objectives derived in strategic decision-making. Tactical planning typically produces a list of operations in the forest to meet the management objectives. The term forest management planning usually refers to tactical planning.

The third type of forest planning, *operative planning*, plans the implementation of the various treatments and other activities prescribed in tactical planning and subsequent decision-making.

Forest inventory and planning

2.2 Cutting budget methods

The first planning methods developed for timber production belong to the category of *cutting budget methods*. The idea is to calculate the *allowable cut* or *sustainable annual cut*. Several methods compare the present forest with the *fully regulated forest*.

In even-aged forestry, the fully regulated forest has a uniform age class distribution. The oldest age class regenerates every year. Management of each stand uses the optimum treatment schedule. The fully regulated forest therefore maximizes the production and income provided timber price remains constant. In addition, the total drain from the forest, the harvested volumes of different timber assortments, income, costs, and labor requirements are also the same every year. This makes the management easy once the forest has reached a fully regulated state.

Computation of the mean stand age, standing volume, and volume increment of a fully regulated forest is easy if the stand development under the optimum management schedule is known. These characteristics computed to the fully regulated forest form the basis of many cutting budget methods.

An example of cutting budget methods is the *Austrian formula* that computes the allowable cut (H) from the volume of the present forest (V), volume (V_r) and volume increment (I_r) of the fully regulated forest, and the length of the transition period (T) over which the forest will reach a fully regulated state:

$$H = I_r + (V - V_r)/T \qquad (15)$$

The problem with cutting budget methods is that they do not indicate the optimum way to change the forest into the regulated sate. In addition, the methods do not specify which compartments are logged and in which order the compartments are treated. A logical question is whether a regulated forest maximizing even timber flow coincides with the actual objectives of the decision-maker. Due to these and other problems, other planning approaches are replacing the cutting budget methods.

2.3 Planning based on stand economy

Planning methods based on stand economy evolved with the development of the concepts of tree stand and forest compartment. The idea of these approaches is to plan the management of individual compartments independently of the other compartments. The planning uses management instructions and silvicultural guidelines. These guidelines show the cutting regime – thinnings and final felling – that maximizes the timber production or net income of the stand. The guides may show the stand basal area to exceed before prescribing thinning treatment and the minimum basal of the residual stand. Guidelines may also exist on the minimum stand age or tree size for final felling.

With this planning approach, the allowable cut of the forest is the sum of the cutting volumes proposed for different compartments. A common practice is to check this figure against a cutting budget method. If the proposal deviates too much from the result of the cutting budget method, the treatment prescriptions of compartments require adjustment.

Timber management planning using stand economy produces clear treatment proposals for individual compartments and provides accurate estimates of the harvestable volume. Although the plan implementation is easy and straightforward, the methods do have serious limitations. The objectives of the forest owner may not agree with the targets used when preparing the management guides. Instructions for high timber production may not show the proper management for a forest owner interested in profitability or biodiversity. Another problem is that evaluation of some management criteria need simultaneous consideration of all compartments. This is the case when one objective is even timber flow. With the even-flow target, the optimum management of a compartment becomes dependent on the other compartments.

2.4 Numerical optimization

2.4.1 Techniques available

Forest management planning is always optimization, even when numerical optimization techniques are not used. The purpose is to maximize the welfare or utility of the forest owner within the limits set by forestry legislation. To accomplish this task, planning requires two kinds of information as Fig. 11 shows:

(1) preferences of the decision-maker that specify the criteria for the comparison of decision alternatives

(2) information on decision alternatives.

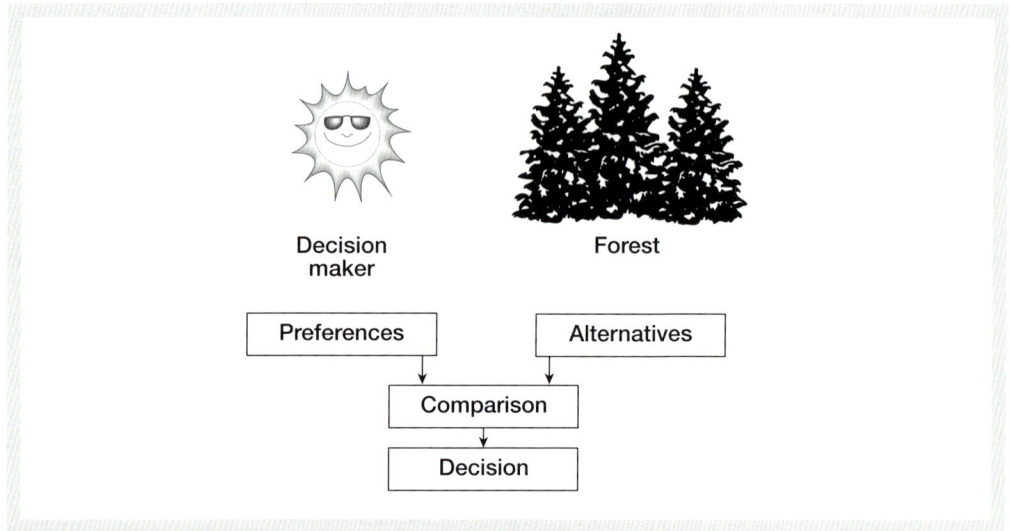

Figure 11. The elements of planning and decision making.

When evaluated against this basic task of forest planning, disadvantages of the cutting budget methods are the lack of clear and justified criteria for the selection of a forest plan and uncertainty about the optimality and efficiency of the plan. A disadvantage of the methods using stand economy is that they omit the influence that other compartments may have on the optimum treatment of a specific stand.

In many planning situations, the use of quantitative optimization can overcome these problems. This planning proceeds in the following steps:

- define the goals and constraints
- define decision alternatives and produce information about alternatives
- prepare a planning model
- solve the model
- test the solution and do sensitivity analysis.

The technique used to solve the planning problem determines the type of the planning model, i.e., how to formulate the planning problem. Several methods are available. The most common are *linear programming* (LP), *goal programming* (GP), and *heuristics*.

2.4.2 Linear programming

With linear programming, the planning problem is formulated by maximizing

$$\sum_{j=1}^{n} c_j x_j = z \tag{16}$$

subject to

$$\sum_{j=1}^{n} a_{ij} x_j \leq b_i \quad i = 1, \ldots, m \tag{17}$$

$$x_j \geq 0 \quad j = 1, \ldots, n \tag{18}$$

where z is an objective function, b_i are constraints, x_j are decision variables (activities), n is the number of decision variables, m is the number of constraints, and c_j and a_{ij} are constants. A computer usually solves the problem using a program based on the simplex algorithm.

In forest planning, the activities are typically treatment schedules of compartments or other units. The aim is to find areas for different schedules to maximize the objective variable while meeting the constraints. Constant c_j indicates how much one area unit (hectare) managed according to schedule j produces the objective variable. Constant a_{ij} gives the quantity of constraint i produced or consumed by one area unit under schedule j. Besides the "less than or equal" constraints, the problem formulation may contain "equal" or "greater than or equal" constraints. The LP model assumes that

CHAPTER 5

the totals of the objective and constraining variables result from adding the respective variables of the activities included in the solution. Another assumption is that the objective and constraining variables are linear functions of the decision variables.

The following example helps to understand how to use LP in production planning. A private landowner has 20 ha suitable for tree seedling production. Growing pine seedlings and growing spruce seedlings are the only relevant activities. Production of spruce seedlings requires 60 man days of labor per hectare annually, and pine production requires 30 man days. The total number of man days available is 900 per year. The landowner wants maximum net income. The net annual income is FIM 20 000 ha^{-1} in pine seedling production and FIM 30 000 ha^{-1} in spruce seedling production.

The activities and decision variables are the areas in hectares allocated to the production of pine seedlings (x_1) and spruce seedlings (x_2). The task is to find the optimum values for x_1 and x_2. The objective function expresses the dependence of net income on x_1 and x_2 by maximizing

$$20\,000 x_1 + 30\,000 x_2 = z \tag{19}$$

The first constraint states that the total area in seedling production must not exceed 20 hectares as follows:

$$x_1 + x_2 \leq 20 \tag{20}$$

The second constraint restricts the labor consumption to a maximum of 900 man days per year as follows:

$$30 x_1 + 60 x_2 \leq 900 \tag{21}$$

The LP problem is then to maximize

$$20\,000 x_1 + 30\,000 x_2 = z \tag{22}$$

subject to

$$x_1 + x_2 \leq 20 \tag{23}$$

$$30 x_1 + 60 x_2 \leq 900 \tag{24}$$

$$x_1, x_2 \geq 0 \tag{25}$$

Solving this problem with the simplex method gives $x_1 = 10$ ha and $x_2 = 10$ ha. This means using the entire area for seedling production with equal areas allocated for pine and spruce.

2.4.3 Goal programming

The problem with linear programming is that it allows only one objective variable. The forest owner often has several simultaneous goals. When using linear programming in this situation, one goal variable is the objective variable with the others controlled through constraints. This formulation does not correspond exactly to the actual problem. The constraints referring to the other goals are not as strict as the problem formulation suggests. The objectives may have target levels and desirable values, but reaching these values is not absolutely necessary.

To make the planning model correspond to the actual situation, an alternative problem formulation called goal programming model is availabale. In this model, the objective function consists of the deviations (d) from the target values of objective variables as follows:

$$\min \sum_{i=1}^{m} d_i^+ + \sum_{i=1}^{m} d_i^- \qquad (26)$$

where d_i^+ indicates how much the target value of objective i is exceeded (surplus), and d_i^- is the quantity by which target i falls short (slack). The target levels (b_i) are given through the *goal constraints* with one constraint per objective as follows:

$$\sum_{j=1}^{n} a_{ij}x_j + d_i^- - d_i^+ = b_i \quad i = 1, \ldots, n \qquad (27)$$

These constraints with the objective function mean that levels b_i are pursued. Deviations are regrettable and minimized, but they are tolerated since the constraints are flexible.

If falling short of the target is harmful but exceeding it is not, the surplus variable (d_i^+) is omitted from the objective function and goal constraint. If only exceeding is to be avoided, the slack variable (d_i^-) is omitted. It is also possible to multiply the deviation variables by constants (v_i and w_i) that reflect the relative importance of the respective objective variables as follows:

$$\min \sum_{i=1}^{m} v_i d_i^+ + \sum_{i=1}^{m} w_i d_i^- \qquad (28)$$

The goal constraints remain unchanged.

Another improvement to the model is to scale the deviation variables to the same range of variation – usually 0–1 – so that the units of measurement do not affect the actual weights of different goals. Dividing the deviations by the largest possible value of the goal variable (single-objective maximum) accomplishes this.

CHAPTER 5

$$\min \sum_{i=1}^{m} v_i(d_i^+/Q_i^{\max}) + \sum_{i=1}^{m} w_i(d_i^-/Q_i^{\max}) \tag{29}$$

The goal constraints remain unchanged. The goal programming model can be solved with the same software and algorithms as the ordinary LP model.

2.4.4 Heuristics

LP and GP are mathematical programming techniques. Another category of planning methods uses heuristics to reach the solution. Heuristics are methods that rely on iteration, human interaction, and gradual improvement of the plan. Heuristics have the disadvantage that they do not always find the global optimum – the best forest management plan. Some reasons for using heuristics despite this shortcoming include:

- The planning problem is so complicated that mathematical programming will not solve it. This may be the case when the assumptions about additivity and proportionality are not met.

- Mathematical programming is possible but too tedious – the model becomes too large or the solution too slow.

- The solution algorithms of heuristics are easier to understand and develop than those of mathematical programming.

Forest planning almost always uses heuristics as the main method or combines it with other techniques. *Model manipulation* is one kind of heuristics commonly used with mathematical programming. In model manipulation, modifications are made in the planning model (LP model or GP model), and the model is solved after every change. The solution and problem formulation most satisfying to the decision-maker is finally selected. Model manipulation resembles sensitivity analysis and should always be a component of forest planning that relies on mathematical programming.

Another kind of heuristics is *local improvement*. Forest planning using stand economy and compartment-wise treatment prescriptions often uses this kind of heuristics. A plan consisting of stand-wise prescriptions is evaluated at the level of the whole forest. If it is not satisfactory, prescriptions for individual compartments undergo alteration resulting in local improvements until reaching an acceptable plan.

An example of numerical heuristics based on local improvement is the HERO method[1]. With this method, the *utility function* of the decision-maker is estimated using paired comparisons of the importance of objective variables and paired comparisons of the priorities of different values of an objective as follows:

$$U = \sum_{i=1}^{m} a_i u_i(q_i) \tag{30}$$

where U is utility index, and a_i, u_i and q_i are the relative importance, partial utility function, and quantity of objective i, respectively. This utility function is maximized by first randomly selecting one treatment alternative for each compartment. The values of the partial utilities produced through different objectives are computed as is the total utility. Then one compartment at a time is considered to see whether another treatment would increase total utility. If increase is detected, the previous treatment is replaced by the one that increased utility. When all treatment alternatives of all compartments are studied in this way, the process is repeated starting with the first compartment. Iteration continues until no treatments occur that would increase utility.

2.4.5 The Finnish planning case

The exact way to formulate a forest planning problem varies widely depending on the decision-making situation. This means that no single model for forest planning exists. The following examples provide three typical planning situations with the formulation of the corresponding planning model. All the models are LP models and assume even-aged stand management. Planning models compatible with numerical optimization also exist for uneven-aged forestry[2], but the models and planning methods developed so far for uneven-aged forestry are more simplistic and less realistic than those for even-aged forestry.

The first example is a typical but simplified Finnish forest planning case. The forest of a private landowner has been divided into two compartments of 30 ha and 40 ha in area. Both compartments have three treatment alternatives during the coming 10-year period: (1) no treatment, (2) thinning, and (3) clear felling with regeneration by planting. The forest owner wants maximum 10-year harvest. He also requires that the volume of the remaining growing stock by the end of the 10-year period must be at least 7 000 m³.

The decision variables of the planning model are the areas of compartments 1 and 2 that are left without any treatment, treated by thinning, or clear felled. Let us use x_{ij} as symbols for the decision variables: x_{ij} is the area of compartment j treated with method i. Table 1 shows the harvest removals and remaining volumes pertaining to different activities computed by computer simulation.

Table 1. Symbols of decision variables and coefficients of the LP model in the first planning example.

Parameter	Compartment 1 Treatment			Compartment 2 Treatment		
	1	2	3	1	2	3
Area treated	x_{11}	x_{21}	x_{31}	x_{12}	x_{22}	x_{32}
Harvest, m³ ha⁻¹	0	55	200	0	65	230
Remaining volume, m³ ha⁻¹	240	195	10	260	200	5

CHAPTER 5

Formulating the LP model corresponding to this planning problem gives:

$$0x_{11} + 55x_{21} + 200x_{31} + 0x_{12} + 65x_{22} + 230x_{32} = z \tag{31}$$

subject to

$$240x_{11} + 195x_{21} + 10x_{31} + 260x_{12} + 200x_{22} + 5x_{32} \geq 7000 \tag{32}$$

$$x_{11} + x_{21} + x_{31} = 30 \tag{33}$$

$$x_{12} + x_{22} + x_{22} = 40 \tag{34}$$

The first equation is the objective function and expresses how the harvested volume depends on the areas of treatments. The first constraint states that the remaining volume must be at least 7 000 m^3. The second and third constraints state that the total area of the treatment alternatives must be exactly 30 ha for compartment 1 and 40 ha for compartment 2.

Commercial software can solve the problem with the simplex algorithm. The optimal solution is as follows: x_{21}=3 ha, x_{22}=0.487 ha, x_{32}=3.513 ha, and the other x_{ij} are 0 ha. This means that compartment 1 and 0.487 ha of compartment 2 should be treated by thinning (alternative 2). The remaining 3.513 ha of compartment 2 should be clear felled and planted.

2.4.6 Forest regulation

In the second case, the aim is to maximize harvest from a degraded forest and simultaneously convert the forest into a productive and regulated state. There are two categories of poorly managed eucalypt coppice forest containing 12 000 ha in category 1 and 18 000 ha in category 2. This forest is to be changed into fully regulated eucalypt forest in 15 years – the rotation length of an even-aged eucalypt stand. The coppice is clear felled, and new seedlings are planted on the cleared areas.

For planning purposes, the 15-year transition period has three 5-year sub-periods. The decision variables are the areas of categories 1 and 2 that are clear felled and regenerated during the three periods. The term denoting these areas is x_{ij}. This is the area of forest category j clear felled during period i. The harvested coppice volume is used as fuel wood. Table 2 shows the prediction of harvested biomass by growth models.

Table 2. Predicted removal (Mg of dry mass per ha) when clear-felling is done during the first, second, or third 5-year period.

Forest category	5-year period		
	1	2	3
Category 1	16	23	33
Category 2	24	32	45

Using the data from the table, the LP becomes as follows:

Maximize

$$16x_{11} + 23x_{21} + 33x_{31} + 24x_{12} + 32x_{22} + 45x_{32} = z \tag{35}$$

subject to

$$x_{11} + x_{21} + x_{31} = 12\ 000 \tag{36}$$

$$x_{12} + x_{22} + x_{32} = 18\ 000 \tag{37}$$

$$x_{11} + x_{12} = 10\ 000 \tag{38}$$

$$x_{21} + x_{22} = 10\ 000 \tag{39}$$

$$(x_{31} + x_{32} = 10\ 000) \tag{40}$$

CHAPTER 5

The objective function tells how the total harvest depends on the areas of activities. The first two constraints guarantee that the entire areas of coppice in categories 1 and 2 are regenerated during the three 5-year periods. The remaining three constraints specify the terminal state of the forest. The total area regenerated during each 5-year period must be exactly 10 000 ha to give a final forest in which the areas of the three different 5-year age classes are exactly 10 000 ha. The last constraint (in parenthesis) is unnecessary because the first four constraints already ensure that 10 000 ha will be regenerated during the third 5-year period.

The optimum solution to this problem is x_{11} = 10 000 ha, x_{21} = 2 000 ha, x_{22} = 8 000 ha, and x_{32} = 10 000 ha ha (other x_{ij} are 0 ha). This yields the regeneration areas shown in Table 3.

Table 3. Optimal clear-felling areas (ha) during three successive 5-year periods in forest categories 1 and 2.

Forest category	5-year period		
	1	2	3
1	10 000	2 000	0
2	0	8 000	10 000

2.4.7 Controlling sustainability

In the third example, the aim is to maximize sustainable timber harvest so that the timber flow is even – the drain remains constant over time. The forest consists of industrial pine plantations. Thinning treatments are not used. The only treatment is clear felling with regeneration by planting. In the total area of 5 000 hectares, 2 000 ha are in age class 21–25 years (class 1), and the remaining 3 000 ha are in age class 41–45 years (class 2). The planning horizon is 15 years divided into three 5-year management periods. The decision variables are the areas that are clear felled in age classes 1 and 2 during management periods 1, 2, and 3. They are again denoted by x_{ij} with subscript i referring to the 5-year period and j to the age class.

Since a rotation length of 30 years maximizes the mean annual wood production of the pine stand, this rotation will be applied to the present forest. Sustainability of harvests is controlled through the clear felling area (*area control*). With 30-year rotation, the 15-year clear felling area may not exceed 50% of the total area of the forest. In the present forest of 5 000 hectares, the maximum 15-year cutting area is 2 500 ha.

The harvestable volume comes from a yield table of the pine species in question shown as Table 4.

Forest inventory and planning

Table 4. Predicted removal (m³ ha⁻¹ of stem wood) when clear-felling is done during the first, second, or third 5-year period.

Age class	5-year period		
	1	2	3
1 (21–25 years)	120	230	250
2 (41–45 years)	310	320	350

Formulating the LP model yields:

Maximize

$$120x_{11} + 230x_{21} + 250x_{31} + 310x_{12} + 320x_{22} + 350x_{32} = z \quad (41)$$

subject to

$$x_{11} + x_{21} + x_{31} + x_{12} + x_{22} + x_{32} \leq 2500 \quad (42)$$

$$x_{11} + x_{21} + x_{31} \leq 2000 \quad (43)$$

$$x_{12} + x_{22} + x_{32} \leq 3000 \quad (44)$$

$$230x_{21} + 320x_{22} - 120x_{11} - 310x_{12} = 0 \quad (45)$$

$$250x_{31} + 350x_{32} - 230x_{11} - 320x_{22} = 0 \quad (46)$$

The objective function maximizes the 15-year harvest. The first constraint is the area control restricting the cutting area of the 15-year planning period to a maximum of 2 500 ha. The second and third constraints are necessary to guarantee that a maximum of 2 000 ha are felled in age class 1, and a maximum of 3 000 ha are felled in age class 2. The last two constraints are the even-flow controls. The first of these two equations shows that the harvests of the 2nd 5-year period ($230x_{21} + 320x_{22}$) must be equal to the harvests of the first 5-year period ($120x_{11} + 310x_{12}$). The difference of these harvests must be zero. The last constraint specifies that the harvests of the second and third 5-year periods must be equal.

The optimal solution to this problem is $x_{12} = 876$ ha, $x_{22} = 848$ ha, and $x_{32} = 776$ ha (other x_{ij} are 0 ha). This yields the cutting schedule in Table 5.

CHAPTER 5

Table 5. Optimal clear-felling areas (ha) during three successive 5-year periods in age classes 1 and 2.

Age class	5-year period		
	1	2	3
1 (21–55 years)	0	0	0
2 (41–45 years)	876	848	776

The total cutting area is exactly 2 500 ha. The harvested volumes of the successive 5-year periods are the same, approximately 27 000 m^3.

Another way to control sustainability is through *volume control*. In this case, the first constraint (area control) is replaced by a constraint stating that the total harvested volume must not exceed a certain limit (50% of the total volume of the forest, the total volume being computed for the middle of the 15-year period). Otherwise volume control does not alter the problem formulation.

CHAPTER 5

References

1. Pukkala, T. and Kangas, J., *Scandinavian Journal of Forest Research* 8(4):560(1993).

2. Buongiorno, J. and Gilles, J. K., *Forest management and economics*, Macmillan, New York, 1987, pp. 89–103.

CHAPTER 5

CHAPTER 6

Management of forest ecosystems

1	**Introduction**	**219**
2	**Forest succession and management**	**219**
3	**Management measures and the response of different tree species**	**222**
4	**Management of forests for optimal productivity**	**223**
5	**Regeneration processes and measures for natural regeneration**	**226**
5.1	Establishment of new growth and properties of seedling stand	226
5.2	Techniques and application of natural regeneration	230
6	**Regeneration processes and measures in plantation forestry**	**232**
6.1	Introduction	232
6.2	Selection of tree species	232
6.3	Provenance selection	233
7	**Selection of regeneration method for plantations and technical applications**	**234**
8	**Genetic improvement**	**236**
8.1	Origin	236
8.2	Production	239
8.3	Traits	240
9	**Processes and measures to grow established trees**	**243**
9.1	Effect of stand density on tree growth and mortality	243
9.2	Stand density and differentiation of tree size	246
9.3	Thinning and development of tree stands	247
9.4	Mechanisms behind thinning effects: thinning and availability of resources	248
	9.4.1 Solar radiation	248
	9.4.2 Soil water	250
	9.4.3 Nutrient availability	251
9.5	Thinning regimes and selection of trees for removal	252
9.6	Impact of thinning on stand growth and productivity	253
9.7	Thinning during the rotation	256
9.8	Length of rotation	258
9.9	Thinning program over a rotation	258
10	**Processes for improving timber quality**	**259**
10.1	Effect of spacing and thinning on knots in wood	259
10.2	Effect of pruning on knots in wood	263
10.3	Effect of spacing on basic density of wood	264

11	**Management of nutrient resources and site fertility**	**264**
11.1	Concepts	264
11.2	Measures to manage nutrient availability	265
	11.2.1 Site preparation	265
	11.2.2 Ditching	267
	11.2.3 Prescribed burning	268
	11.2.4 Thinnings	270
	11.2.5 Fertilization	271
11.3	Impacts of timber harvesting on nutrient resource of forest soils and their implications for forest management	277
11.4	Nutrient management and sustainable forest production	278
12	**Ecology and management of forest pests**	**282**
12.1	Introduction	282
12.2	Forest pest concept	282
12.3	Characteristics of pest outbreaks	284
12.4	Herbivorous species attacking trees	286
	12.4.1 Tree-distribution related herbivore abundance	286
	12.4.2 Host plant size and growth form	286
	12.4.3 Taxonomic affinities of host plant	286
	12.4.4 Insect recruitment by introduced plants	287
	12.4.5 Role of introduced herbivores in forest ecosystems	288
12.5	Resistance mechanism of trees against herbivores and pathogens	289
	12.5.1 Constitutive resistance	289
	12.5.2 Induced defense	291
12.6	Models explaining variation in chemical defense between plants	295
12.7	Management of forest pests	297
	12.7.1 Tree resistance	297
	12.7.2 Biological control	298
	12.7.3 Examples of biological control	299
	12.7.4 Chemical control	301
	12.7.5 Behavior-modifying chemicals	301
	References	**305**

CHAPTER 6

Seppo Kellomäki, Veikko Koski, Pekka Niemelä

Management of forest ecosystems

1 Introduction

Forest management manipulates the structure of forest ecosystems to make them function as desired. Any kind of forest-based production, *P(i,j)*, is a result of the interaction between the environment, *E(j)*, and genotype, *G(i)*:

$$P(i, j) = G(i) + E(j) + G(i) \cdot E(j) \tag{1}$$

Manipulation of the genetic properties of tree populations, changing the properties of habitats occupied by populations, or both can maintain or increase production. Ultimately, sustainable management aims to modify the interaction of both factors to optimize the need to conserve the functioning and structure of forest ecosystems and satisfy production needs. This occurs by controlling the long-term functional and structural development of forest ecosystems (succession) to induce them to produce the items determined in the management goals of Fig. 1. Management directly or indirectly influences the regeneration, growth, and mortality of tree populations.

2 Forest succession and management

In forest management, the measures enhancing regeneration return the successional development of the ecosystem to the initial or earlier phases of succession. Enhanced growth depends on maintaining optimum availability of resources for growth over the rotation. Suppression of herbivore and pest attacks reduce tree mortality. The timing and intensity of the various measures comply with ecological limits set by the biological and physical properties of forest ecosystems.

Disturbance dynamics of forest ecosystems determine the management. Two categories of disturbance are possible – autogenic disturbances and allogenic disturbances that drive succession (autogenic, allogenic, or both). In *autogenic succession*, the regeneration, growth, and death of single trees – minor disturbances related to the life cycle of trees – drive the structural dynamics of the forest ecosystem. The death of single trees and the consequent gaps in the canopy release resources and enhance regeneration and growth. The concept of autogenic succession applies to the form of selection forestry or uneven-aged management where single trees are the basic management objective. This management system occurs widely in the temperate zone in regions where autogenic succession predominates the dynamics of forest ecosystems.

CHAPTER 6

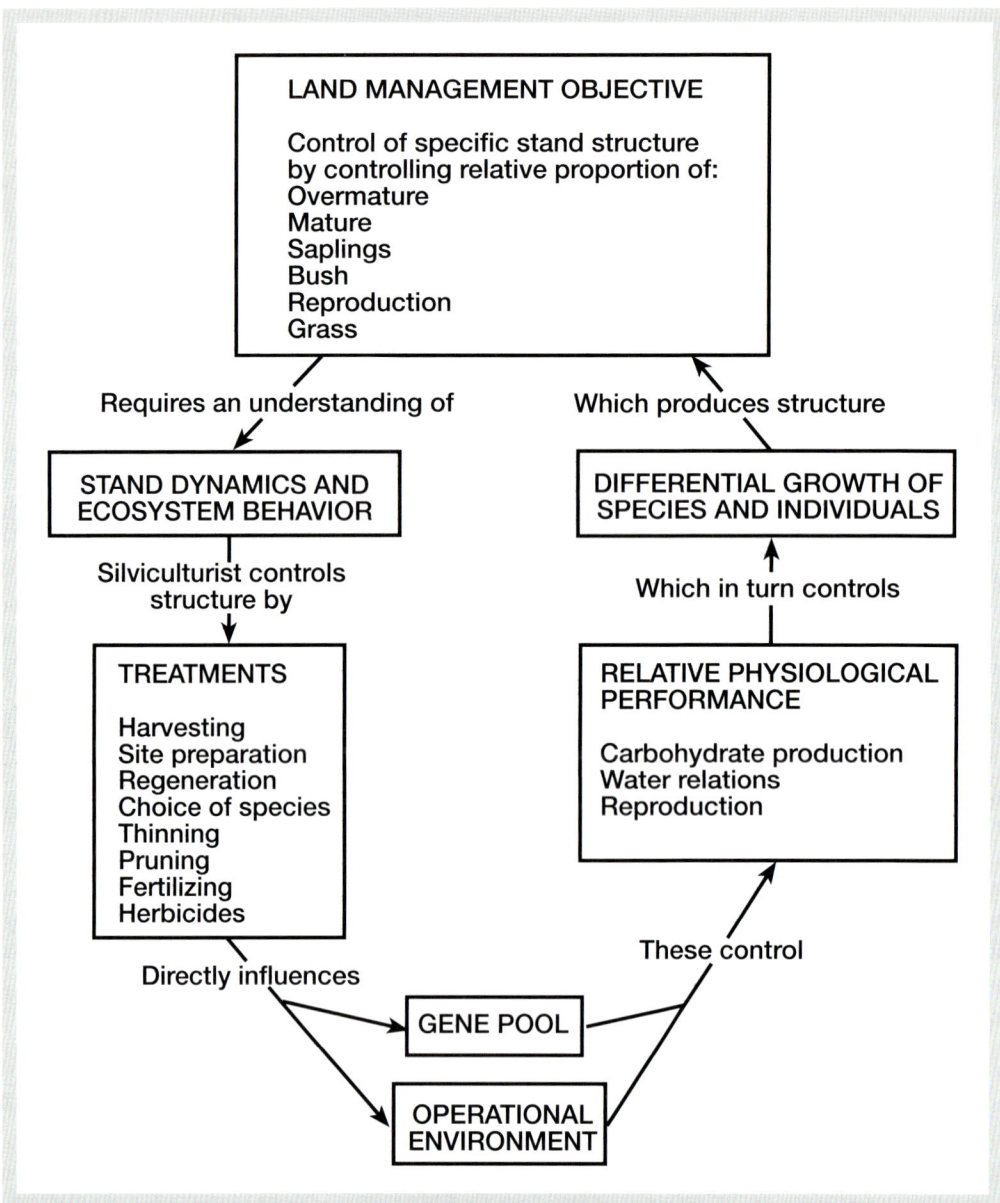

Figure 1. Interaction between management goals, dynamics of forest ecosystems, and the management process in controlling the successional dynamics of forests ecosystems subject to management[1].

Allogenic succession refers to the dynamics driven mainly by major disturbances induced by fires, gales, and excessive snowfall. Because of major disturbances, several trees die at the same time with a release of space for trees to regenerate and grow as a population with individuals of approximately the same age. In allogenic succession, a single stand is the basic unit driving the structural dynamics of the forest ecosystem. In management, the concept of allogenic succession is the form of standwise forestry or even-aged management where single stands form the basic object of management. This management system has wide use in the boreal zone in regions where allogenic succession predominates in the dynamics of forest ecosystems.

In practice, the distinction between autogenic and allogenic succession is not very clear. Both normally act at the same time. The difference between these two forms of succession lies in scale when ignoring the driving factors behind the successional processes. The applicability of different management regimes primarily depends on the ecological and physiological properties of trees – how they respond to the size of the gaps and the subsequent amount of resources liberated for regeneration and growth. In this context, classification of tree species uses the categories of pioneer, intermediate, and climax species indicated in Table 1. This classification applies widely to the management of forest ecosystems.

Table 1. Some principal properties of pioneer and climax species.

Property	Pioneer species	Climax species
Properties related to regeneration biology		
Age at onset of seed crop	Early	Late
Length of seeding period	Short	Long
Seed yield	Big	Small
Frequency of seeding	Variable	Repeated
Seed mass	Small	Big
Distribution of seeds	Wide	Limited
Storing of seeds	Yes	No
Vegetative regeneration	Significant	No significance
Properties related to growth biology		
Life span	Short	Long
Growth rate	Fast	Slow
Total yield	Small	Big
Tree size	Small	Big

Pioneer species are dominant early in succession driven by major disturbances. They are often characterized by shade-intolerance, quick regeneration and fast growth, and short life span under a substantial supply of resources. Many broadleaf trees in the temperate and boreal zones are pioneers. Contrary to pioneer species, *climax species* are dominant late in succession, and they can take over from pioneer species even with a scarcity of resources. Many climax species are characterized by shade-tolerance, longer life spans, and slower growth rates than pioneer species. Climax species include both broadleaves and conifers. *Intermediate species* is the term for species falling between pioneer and climax species. They can flourish on forest sites in all stages of succession.

CHAPTER 6

The ecological properties of tree species imply that selection forestry or uneven-aged forestry applies whenever the practice of forestry uses species adapted to the scarcity of resources (climax species). Even-aged forestry or standwise forestry needs species adapted to exploiting the substantial supply of resources (pioneer species). Proper forest management can use the response difference between climax and pioneer species. The environmental conditions within a stand change during the succession and allow climax species to invade a stand established by pioneer species. This may lead to long-term cycling of stands of pioneer and climax species in a site.

3 Management measures and the response of different tree species

In management, several measures are used to control the supply of resources and the properties of tree populations. The measures to control the supply of resources fall into two categories relating to the disturbances driving the succession of the forest ecosystem – methods liberating resources to a large extent (allogenic measures related to allogenic succession) and methods liberating resources to a smaller extent (autogenic measures related to autogenic succession) as Table 2 shows.

Allogenic measures primarily control the properties of the soil system and subsequently the nutrient cycle and soil moisture with the largest enhancement of the growth and regeneration of pioneer species through the disturbance of the soil surface. Site preparation refers to the mechanical measures used to reduce the effects of ground vegetation on the seedlings and to control the physical and chemical properties of soil. In this context, ditching only lowers the ground water level and reduces excess soil moisture. Prescribed burning influences the chemical properties of soil and eliminates the effects of ground vegetation on seedlings. Fertilization adds nutrients to the site with substantial increase in the availability of nutrient for tree growth.

Autogenic measures primarily represent spacing with the enhancement of growth and regeneration for pioneer and climax species through thinning of seedling stands and more mature stands. Regenerative cutting refers to the terminal cuts intended to reforest the site through natural regeneration with the help of natural seeding or through artificial regeneration with the help of seeding or planting. The tending of seedling stands aims for proper spacing as also occurs in thinning. Tending of seedling stands also eliminates tree species not preferred in the management. Pruning removes dead or living branches from the lower crown to increase the amount of wood without knots obtainable from the stem.

Table 2. Schematic presentation of effects of different managing measures on the properties of site and tree populations. Legend: + = slight effect, ++ = moderate effect, +++ = substantial effect.

Measure	Effect on the site properties and communities				
	Nutrients	Moisture	Temperature	Soil physics	Species
Allogenic measures					
Site preparation	+++	++	+++	+++	+++
Ditching	++	+++	++	++	+++
Prescribed burning	+++	+	++	++	+++
Fertilizing	+++				
Autogenic measures					
Regenerative cuttings	++	++	++	+	+++
Tending of seedling stands	+	+	+	+	+++
Thinning	++	++	+	+	++
Pruning	+	+	+		+

In management, the functioning and structure of the ecosystem in relation to the management goals determine the time and intensity of measures as Fig. 2 shows. Soil preparation to enhance the availability of resources and subsequent restocking can precede regeneration. Tending of seedling stands results in the allocation of stand productivity to fewer trees demonstrating faster growth rates than would be possible without increased spacing. Regular thinning enhances the nutrient cycle and subsequent growth. Increased spacing in mature stands enhances the regeneration rate. Further enhancement can occur through regeneration cuttings and soil preparation. Timber, other items, and services are concurrently supplied throughout the successional cycle.

Economic goals of forestry are among the main factors influencing the intensity of management. Forest management is a long-term investment with a slow return rate. Management is therefore appropriate only when the increasing investment produces an equal increase in timber production or production of other commodities as Fig. 3 shows. The degree of intensity can be freely selected in an area limited by the trajectories of the functions for the management intensity vs. the amount of commodities and the trajectory for the rate of enhancement of management and the rate of enhancement of the production of commodities. Outside the area thus defined, intensifying management is profitable only to a limited extent.

4 Management of forests for optimal productivity

Among the key issues addressed in managing are those of optimum distribution of tree species and stand age or tree size over the forest area to maximize long-term growth. In uneven-aged forestry, the optimum size distribution represents the mean tree mass per tree equal to half the mass potentially supported by climatic and edaphic factors. The

CHAPTER 6

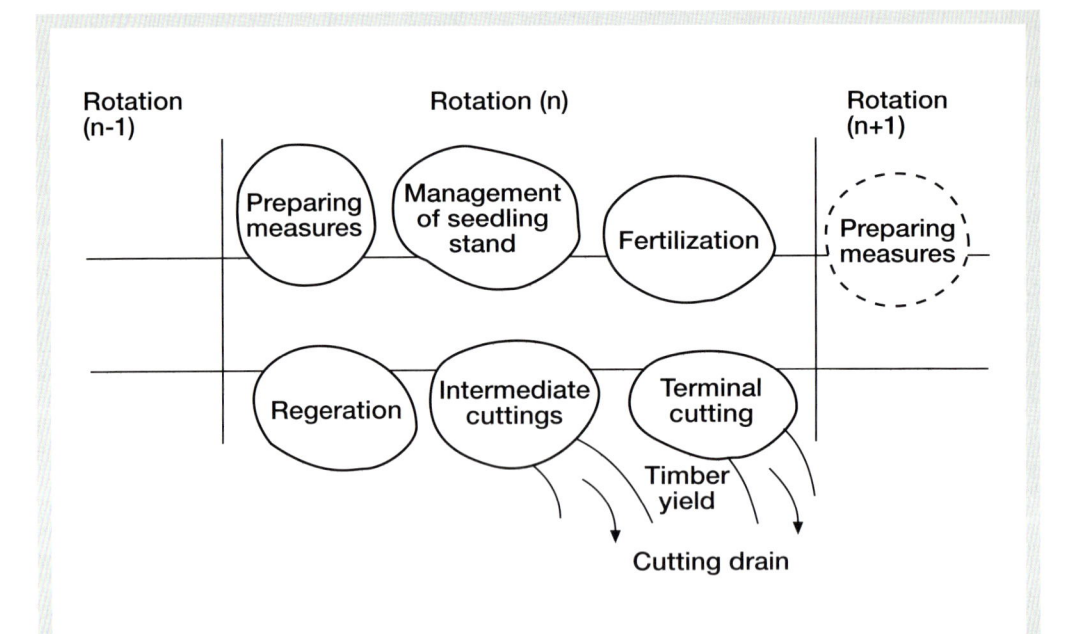

Figure 2. Schematic presentation of the interrelations between the long-term dynamics of forest ecosystems and the impacts of different management measures on forest-ecosystem structure and functions over the rotation.

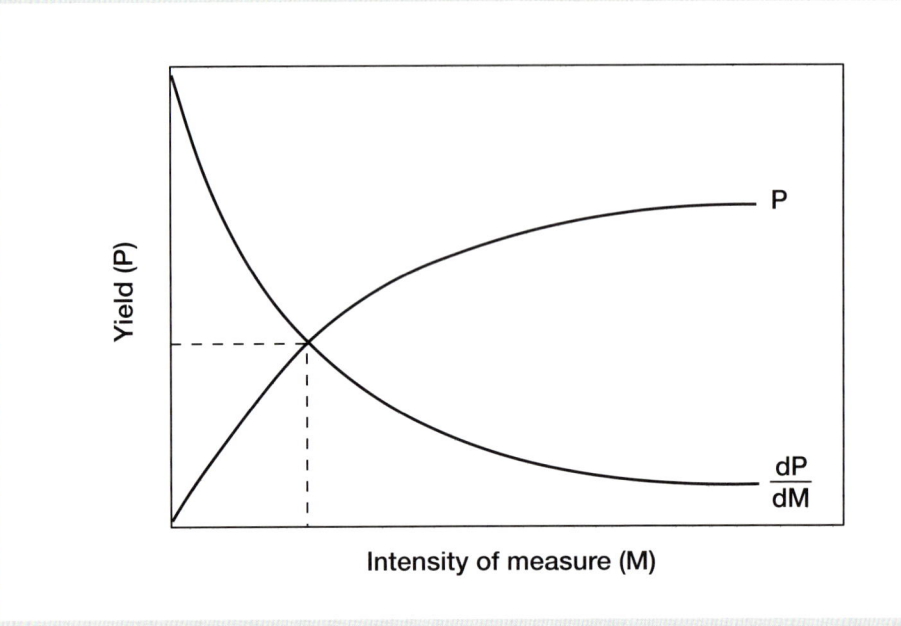

Figure 3. Schematic presentation of the relationship between the intensity of management and production of different commodities.

same is true in even-aged forestry but with the mean mass of trees per stand equaling half the mass potentially supported by climatic and edaphic factors.

Assume an example with six Scots pine stands representing the age classes 2, 10, 22, 38, 62 and more than 82 years used in Fig. 4. The growth during the 10 yr. was the highest for the stand with an initial age of 38 yr. – growth was five-fold compared with the stand with initial age of 85 yr. Assume further that a forest area of 100 ha is divided into six compartments allocated in different ways to initial age classes. Consequently, distribution skewed to the dominance of old stands represents the largest stocking, but growth is approximately 15% smaller than that for the normal distribution as Fig. 5 shows. The total tree growth in the forest area dominated by young stands was substantially smaller – one-third of that for the normal distribution – than in the other age distributions. For even distribution, the total growth was 70%–80% of that for the normal distribution.

Maximization of timber production in a forest area would require an age structure representing the normal distribution. Over a rotation, this age distribution will gradually lead to the dominance of old stands with increasing stocking and decreasing growth.

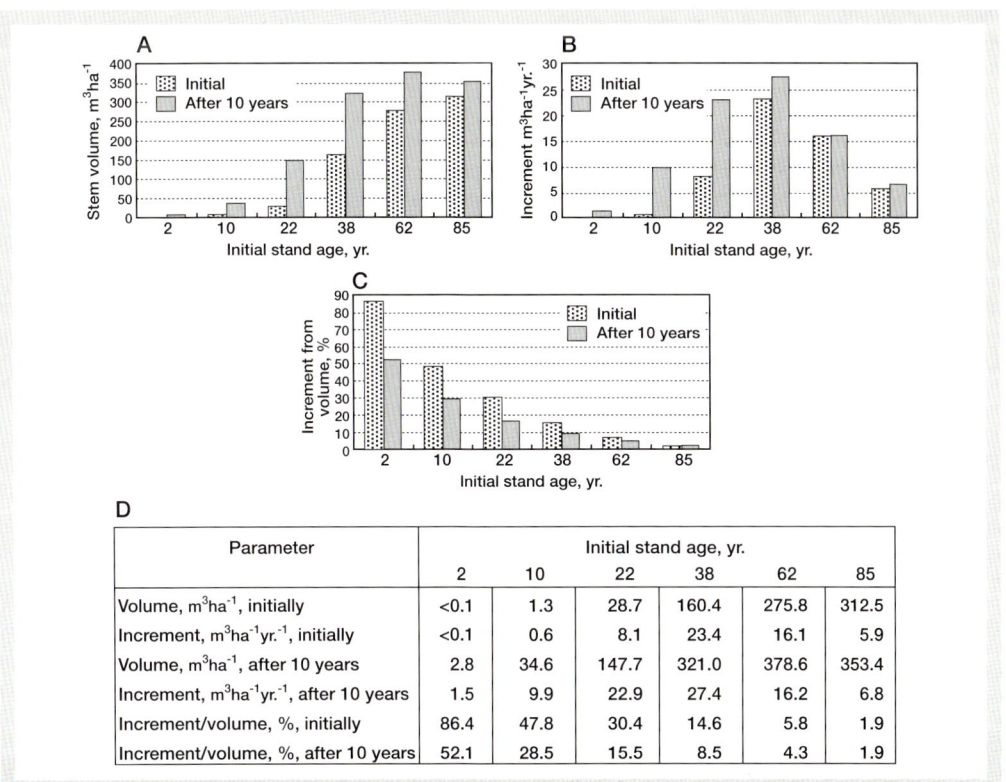

Parameter	Initial stand age, yr.					
	2	10	22	38	62	85
Volume, m^3ha^{-1}, initially	<0.1	1.3	28.7	160.4	275.8	312.5
Increment, $m^3ha^{-1}yr.^{-1}$, initially	<0.1	0.6	8.1	23.4	16.1	5.9
Volume, m^3ha^{-1}, after 10 years	2.8	34.6	147.7	321.0	378.6	353.4
Increment, $m^3ha^{-1}yr.^{-1}$, after 10 years	1.5	9.9	22.9	27.4	16.2	6.8
Increment/volume, %, initially	86.4	47.8	30.4	14.6	5.8	1.9
Increment/volume, %, after 10 years	52.1	28.5	15.5	8.5	4.3	1.9

Figure 4. Example of how stand age affects growth and stocking of a tree stand. A: Stocking at different ages, B: growth over 10 yr. assuming initial stocking as in Fig. A, C: relative growth over 10 yr. assuming initial stocking and growth as in Figs. A and B, and D: description and statistics for the calculation example. Regardless of initial age and volume, each stand grew for 10 yr. with no management measures applied.

CHAPTER 6

Later, young stands become dominant when a large area of stands previously of intermediate age matures. They will be cut at the end of the rotation with a shift of the age distribution to dominance by young stands. Now the forest area represents low stocking and low absolute growth but high relative growth. Maximization of timber production over several rotations in a sustainable manner therefore requires dividing the forest area into sections representing an age distribution with the dominance of young, intermediate, and old stands so that the total age distribution for the forest area represents the dominance of intermediate stands with high stocking and high growth.

5 Regeneration processes and measures for natural regeneration

5.1 Establishment of new growth and properties of seedling stand

In natural regeneration, the establishment of new growth (restocking) occurs through seeding or sprouting. Seeding refers to sexual regeneration through seeds, and sprouting refers to vegetative (asexual) regeneration through activation of dormant buds. Natural regeneration through seeding is the only real choice for conifers, since

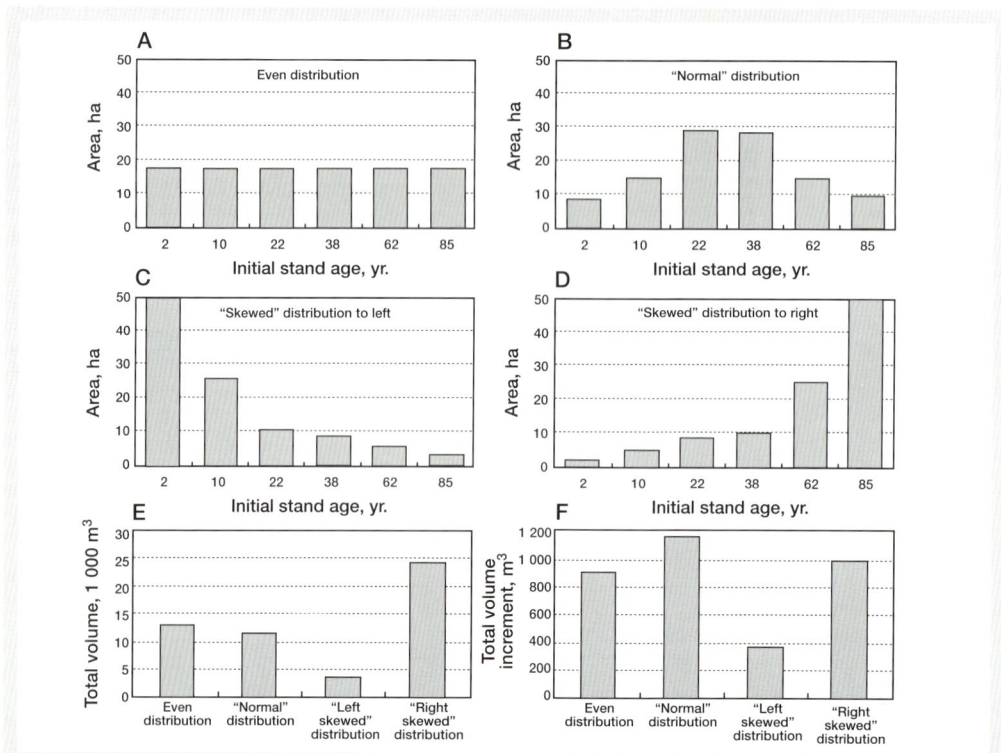

Figure 5. Computational example of how age distribution in tree stands can influence the productivity of a forest area. The distribution of tree stands with an overall area of 100 ha. A: Dominated by no single age class (even distribution), B: dominated by intermediate age classes (normal distribution), C: dominated by young age classes (distribution skewed to the left), and D: dominated by old age classes (distribution skewed to the right). Figures E and F indicate, respectively, stocking and growth over the forest area for the various distributions.

they cannot regenerate vegetatively. Many broadleaved species regenerate vegetatively, but sexual regeneration often produces more offspring with regular growth over the whole rotation.

The quantity and quality of the seed crop, the success of germination, and the establishment of seedlings on the site primarily influence the success of sexual regeneration as Fig. 6 shows. Typically, the magnitude of the seed crop increases gradually after regeneration cutting, and the total seed crop increases to a limited extent with an increasing number of parent trees per unit area. The majority of seeds are shed to fall in the immediate vicinity of the parent trees with a small amount dispersing over greater distances. This leads to considerable variability in the spatial distribution of seedlings. The probability of a seedling surviving the early years is very small compared with the survival of more mature seedlings. The growth of natural seedlings is very slow in the early years. This means a longer period for major hazards to influence the survival of seedlings.

Over a regeneration period (time from the regeneration cut to the time when a seedling stand fulfills given criteria), the seed crop repeats annually, but only few germi-

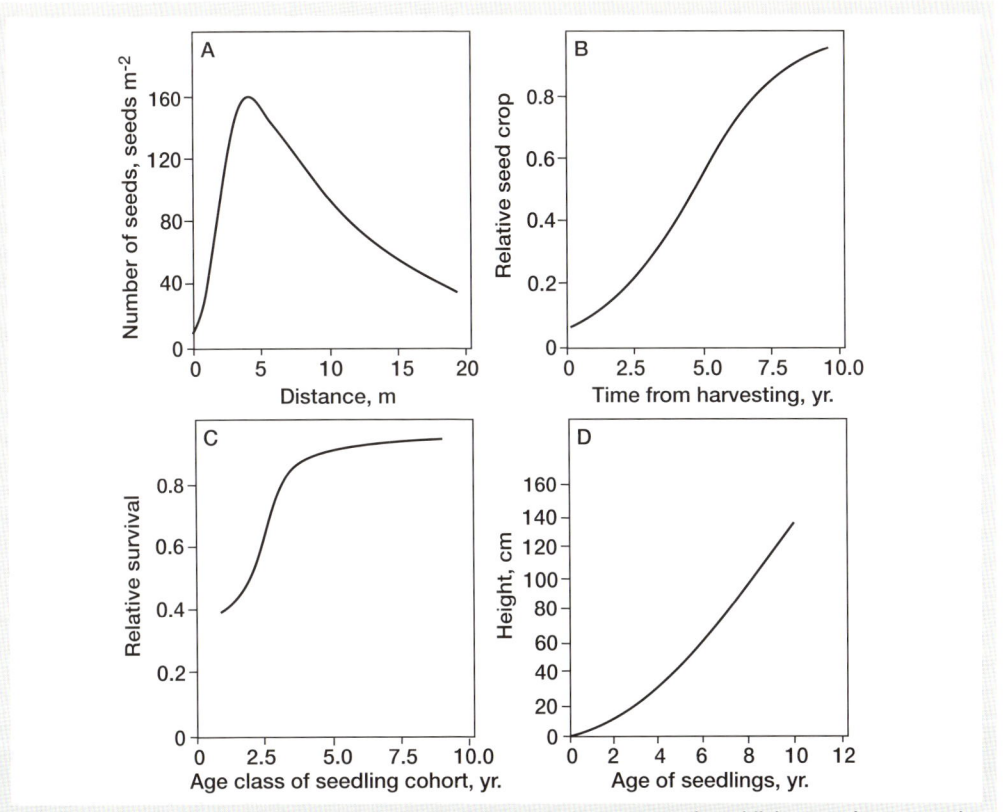

Figure 6. A schematic presentation of the factors impacting the success of establishment of new growth (restocking) following sexual and vegetative regeneration. A: Number of seeds per unit area as a function of distance from parent tree, B: selected seed crop as a function of time from regeneration cutting, C: survival of seedlings as a function of age, and D: seedling height as a function of age^2.

CHAPTER 6

nated seeds from each crop develop to form an established seedling as Fig. 7 shows. The density of a seedling stand at a given moment therefore represents the accumulation of seedlings over the years since the regenerative cut. Because the forest soil provides a heterogenous seed bed with variable conditions for gemination of seeds, a large variability in the spatial structure of a seedling stand occurs. Natural regeneration produces a seedling stand representing a skewed structure of sizes dominated by a large number of smaller seedlings unevenly distributed over the regeneration area.

The regeneration process has a large probability of only limited success in establishment of seedlings. Success of regeneration is therefore the probability of obtaining a seedling stand fulfilling given criteria when influenced by several random processes representing the main phases of the regeneration process[4]. Equation 2 shows the mathematical relationship:

$$Seedlings(t) = Seeds(t) \cdot SURF(t) \cdot FULLS(t) \cdot \ldots \qquad (2)$$
$$\ldots \cdot MATS(t) \cdot GERS(t) \cdot DISTS(t)$$

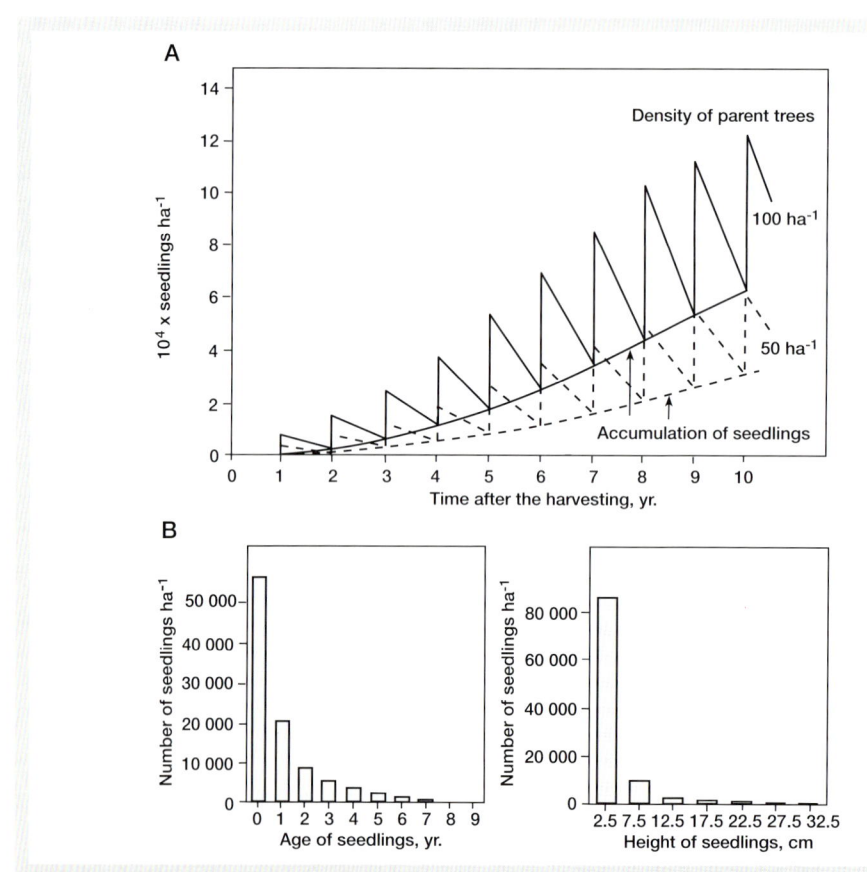

Figure 7. A schematic presentation of A: accumulation of seedlings in a seedling stand over a given period in a stand regenerated naturally and B: age and size distribution of seedlings at the end of the regeneration period[3].

where *Seedlings(t)* is the number of seedlings established in the year *t*, *Seeds* is the number of seeds shed on the site, *SURF* is the share of stockable soil, *FULLS* is the share of full seeds, *MATS* is the share of mature seeds, *GERS* is the share of maturated seeds, and *DISTS* is the share of seeds attacked by herbivores and pests.

Over a given period, establishment of a seedling stand of Scots pine of a given density is more likely on a nutrient-poor site than on a fertile site where herbs and grasses reduce the establishment rate as Fig. 8 shows. In these conditions, soil preparation substantially enhances the establishment rate, i.e., the establishment rate is faster when soil preparation bares more mineral soil. Under disadvantageous temperature conditions, the probability of achieving a dense seedling stand is much lower than under favorable conditions. Even with favorable conditions, the number of parent trees per area unit should be sufficiently high to enhance the establishment rate. Increasing density of parent trees reduces the availability of resources for seedlings with a consequent slowing of the rate of restocking.

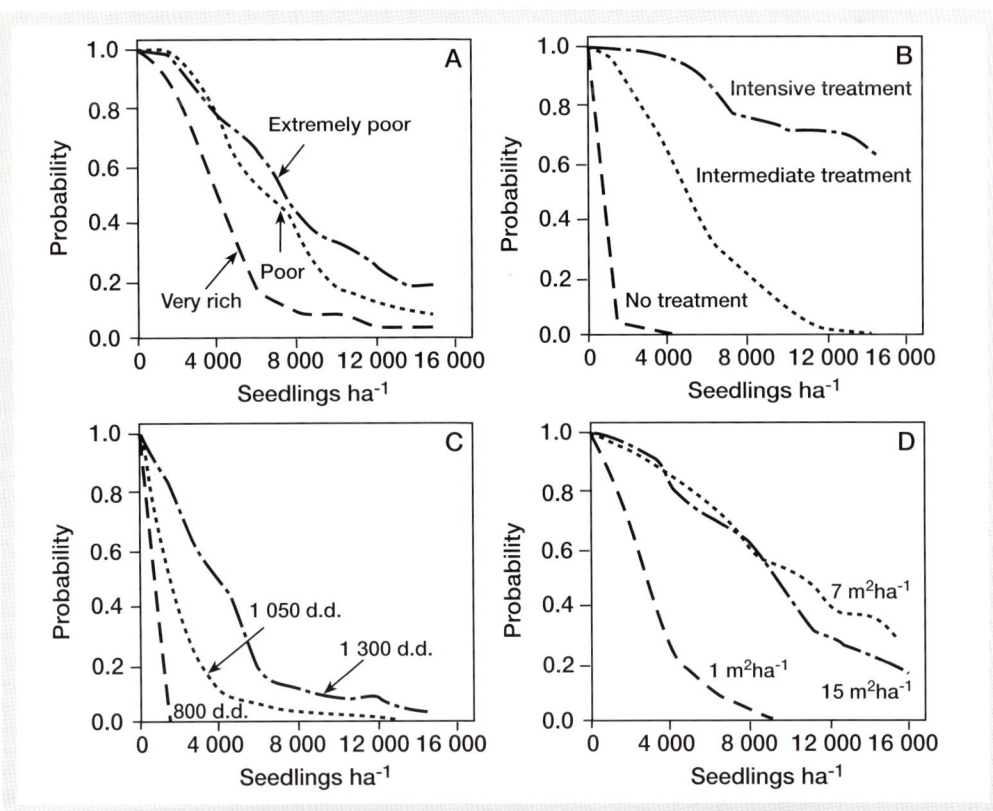

Figure 8. Effect of selected factors on success of establishment of a Scots pine stand in terms of probability of a given number of seedlings becoming established in the course of the regeneration period as a function of A: site quality, B: intensity of soil treatment, C: temperature sum, and D: density (spacing) of parent trees.

5.2 Techniques and application of natural regeneration

Using the density (spacing) of the parent trees, the technique of natural regeneration comprises the seed-tree 10–50 parent trees/ha method and shelterwood method 50–150 parent trees/ha indicated in Fig. 9. Optionally, the final cut can occur in open strips with seed dispersal provided by fully-stocked stands between the strips or seeds dispersed from stands around small clear-cut areas. When applying the seed-tree or

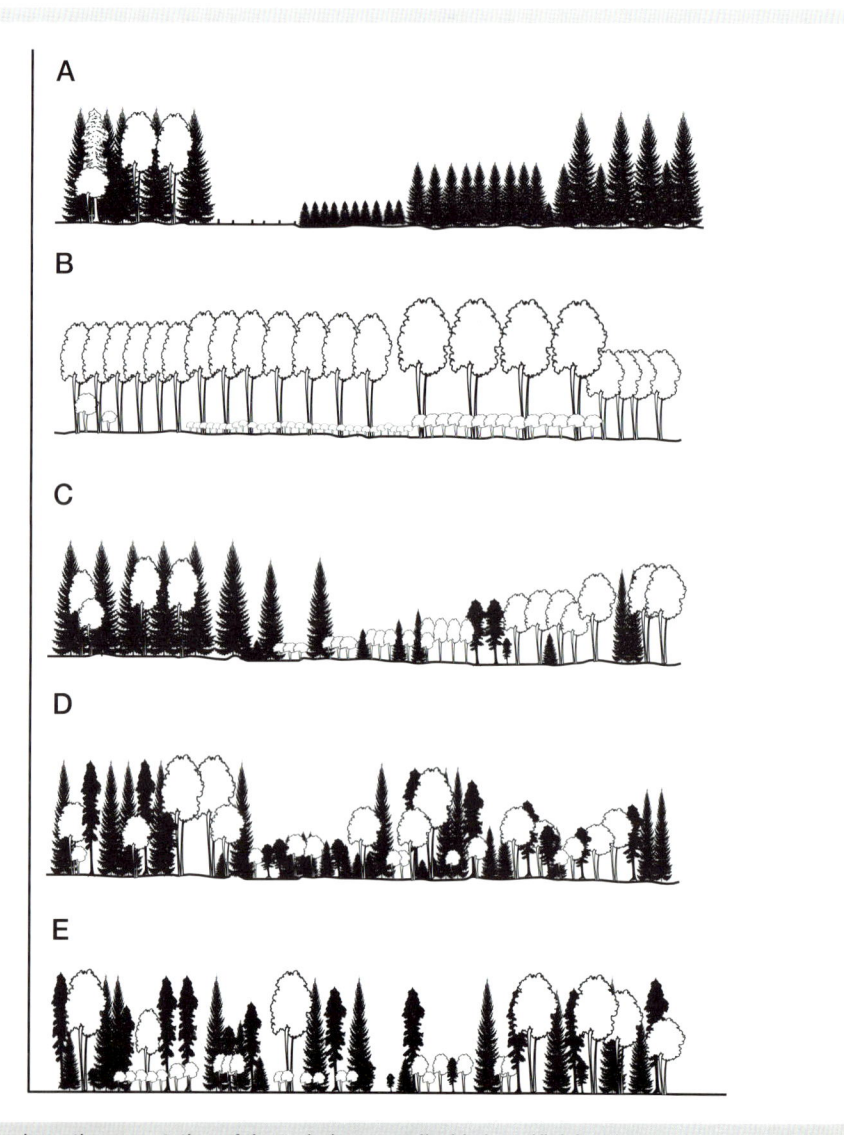

Figure 9. A schematic presentation of the techniques applicable in artificial and natural regeneration. A: Clear-cut with planting, B: natural regeneration with shelterwood or seed-tree method, C: natural regeneration with shelterwood or seed-tree method in mixed stand, D: selective cutting with small openings, and E: selective cutting.

shelterwood method, the phases to enhance growth and regeneration occur one after the other using the feature that seed crops increase with increasing maturation of trees. The separation of enhancement of growth and regeneration also allows more space in which to optimize the environmental conditions for regeneration and growth. In this respect, selection forestry represents a combination with regeneration and growth enhancement of trees occurring simultaneously. Removal of single mature trees creates space for establishment of seedlings. Selection forestry represents continuous regeneration through the shelterwood method with high stocking of parent trees and an uneven-aged structure of the tree population.

The seed-tree and shelterwood methods and their variations are most applicable when dealing with shade-intolerant tree species (pioneer species) capable of using resources substantially available in open stands. Table 3 shows regeneration methods. On fertile sites, grasses and herbs reduce the success of regeneration and require a combination of soil preparation with the seed-tree and shelterwood methods. On nutrient-poor sites, soil preparation also enhances the success of regeneration but to a smaller extent than on fertile sites. In selection forestry, regeneration is most successful with shade-tolerant tree species that adapt to a scarcity of resources. Disturbance of the soil surface enhances the regeneration rate of climax species also, but intensive soil preparation could be detrimental to the trees densely stocking the site.

Table 3. Regeneration methods.

Method	Sites with most success	Preparation methods
Clear-felling and seeding by adjacent stand	Dry upland forest sites, poor pine mires, poor spruce mires, spruce, birch	Clearing of felling area, soil preparation
Seeding felling	Dry upland forest sites, pine mires, moist upland forest sites, pine, birch	Clearing of felling area, soil preparation
Shelterwood felling	Moist upland forest sites spruce (pine) spruce mires	Clearing of felling area, soil preparation
Selection felling	Best sites, species with marked shade-tolerance (common beech, fir)	

CHAPTER 6

6 Regeneration processes and measures in plantation forestry

6.1 Introduction

Artificial regeneration means planting or seeding as used in plantation forestry. In planting, the establishment of the tree stand uses plants raised in nurseries, and seeding means sowing seeds in the area for regeneration. A forest plantation can be an area with a recently removed tree cover (*reforestation*) or an area without a recently removed tree cover (*afforestation*). Reforestation and afforestation use the same methods.

In plantation forestry, regeneration success depends on the properties of the seed and planting stock used and therefore on the seed source and nursery practices that Fig. 10 shows. The genotypic and phenotypic properties of seeds and seedlings should match the ecological conditions of the site for regeneration. In this context, the appropriate-to-the-purpose selection of tree species and provenance forms a solid basis for success. The planting stock should also be morphologically such that it can acclimatize to a variable supply of resources and to competition from grasses and herbs. Herbs are often predominant in the initial stage of plantation development on fertile sites.

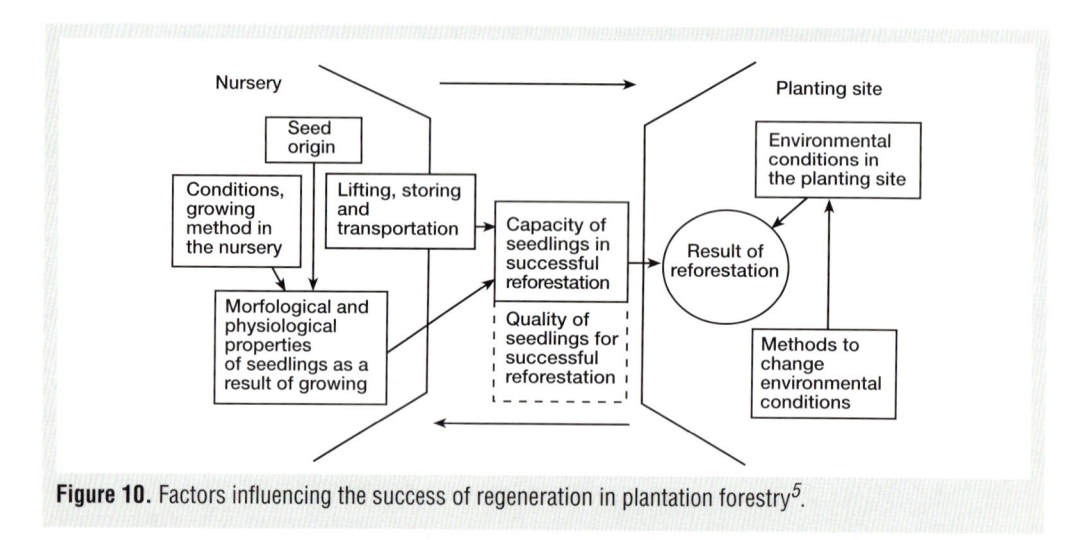

Figure 10. Factors influencing the success of regeneration in plantation forestry[5].

6.2 Selection of tree species

The growth and yield characteristics expected of tree species on a particular site primarily influence species selection. Differences in commercial value among tree species also influence selection of tree species to favor those enjoying the greatest demand in the marketplace. The applicability of different tree species when considering protection of the environment – control of soil erosion or improvement of soil fertility – is also an important criterion to note in species selection. Any tree species combining high growth and yield with high commercial and environmental value is of great value in forestry aimed at sustaining the multi-objective productivity of forest ecosystems.

In practice, only a few rules control selection of tree species. Whenever using local tree species, the main criteria are high survival and growth in relation to the prevailing conditions. In climatically homogenous areas, site fertility in terms of site type and soil texture has common application for selection of tree species as shown in Table 4. The species with the highest capacity to acclimatize and grow under specific conditions has preference. In the short term with an ample supply of resources such as nutrients and water, any species in the pioneer category has high probability to survive and grow. Over the long term, some climax species could also perform well. With a scarce supply of resources, it is primarily the climax species that have the highest ability to survive and grow regardless of the time horizon.

Table 4. An example of selection of tree species considering site fertility and soil texture in Finland.

Site fertility and soil texture	Tree species, first choice	Tree species, second choice
High site fertility		
Fine texture	*Betula* sp.	*Picea abies*
Coarse texture	*Picea abies*	*Betula* sp.
Intermediate site fertility		
Fine texture	*Picea abies*, *Betula* sp.	*Betula* sp., *Picea abies*
Coarse texture	*Pinus sylvestris*	*Picea abies*, *Betula* sp.
Low site fertility		
Fine texture	*Pinus sylvestris*	*Pinus sylvestris*
Coarse texture	*Pinus sylvestris*	*Pinus sylvestris*

6.3 Provenance selection

Local or northern provenances of tree species usually adapt well to the prevailing environmental conditions. They do this at the expense of growth that is usually higher for provenances representing a more southern location. In addition, trees of northern and local provenances have more ability to resist attacks by local pests than do trees of southern provenances. Improved growth with appropriate selection of provenance with reduced death risk requires balance whenever contemplating the establishing of plantations in forestry. Figure 11 demonstrates the importance of having an appropriate seed source. It presents the survival of Scots pine seedlings of southern and northern provenances on a particular site at 5, 7, 13, and 21 yr. since planting.

The transfer of the genotype by more than two degrees northward substantially increased mortality throughout the monitoring period. Mortality was high even when the southern provenance represented higher altitude than the planting site. This was contrary to the situation resulting from southward transfers that promoted survival. This beneficial effect was at its maximum when the northern provenance represented a higher altitude than the site of the plantation. Eliminating the effect of altitude shows that the increased survival of young Scots pines closely relates to the southward transfer of the genotype. In practice, the southward transfer should be limited, or the growth and yield will decrease too much.

CHAPTER 6

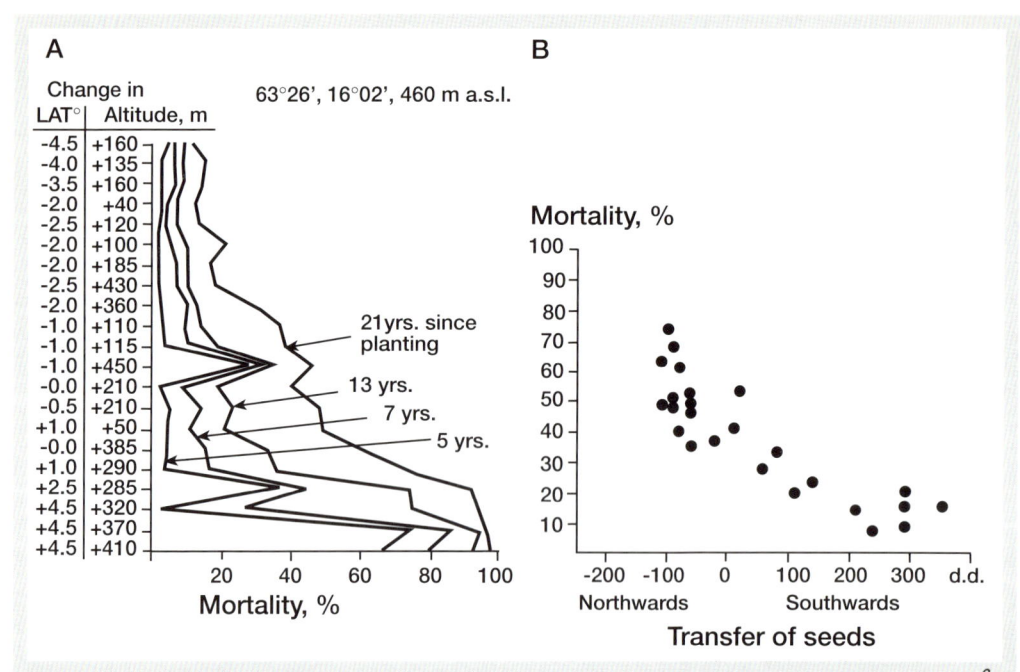

Figure 11. A: Survival of Scots pine seedlings of northern and southern origin and high and low altitude[6]. A decrease in latitude and altitude indicated transfer to the south or lower altitude and an increase indicated transfer in the opposite direction. B: Mortality of Scots pine seedlings in relation to south-north transfer of the genotype after elimination of the impact of altitude[7]. Transfer is expressed in terms of the temperature sum, i.e., negative values indicate cooler climate and positive values indicate warmer climate.

7 Selection of regeneration method for plantations and technical applications

In plantation forestry, the interactive processes between the trees and the environment can be controlled through proper management; i.e. appropriate selection of tree species and manipulation of the site properties enhance the survival and growth of trees in the plantation established on the site. In addition, the properties of the planting stock and seeds influence the outcome of regeneration and the subsequent growth. Furthermore, the management process relates closely to the technical and economic ability of the forest owner to manipulate the genotype-environment interaction.

In practice, the properties of the site, tree species and planting stock, and technical and economic resources provide a few rules for the selection of the regeneration method. Seeding is most applicable when dealing with nutrient-poor sites with coarse soil textures. Success of regeneration increases substantially whenever application of an appropriate soil treatment technique disturbs the litter and humus layers. In sites of intermediate and high fertility, planting is more successful than seeding. The outcome of planting improves when using increasingly larger planting stock with increasing site fertility and increasing disturbance of the litter and humus layers. An ample supply of resources eliminates seeding, and even the planting of pioneer species requires efficient soil management to limit the competitive impact of herbs and grasses. Lack of

Management of forest ecosystems

resources favors climax species, and only a minor degree of soil management is necessary to enhance regeneration through planting or seeding as Table 5 shows.

Table 5. Relationship between tree species selection, properties of site, cultivation methods, and site preparation for Finnish conditions. Legend: + = recommended, (+) = optional.

Tree species and growing site	Cultivation method		Prerequisites and explanations
	Sowing	Planting	
Scots pine			
Upland forest with grass-herb vegetation		(+)	Well prepared soil
Moist upland forest site	(+)	+	Well prepared soil
Medium dry upland forest site	+	+	Prepared soil
Dry upland forest site	+	(+)	Prepared soil, no planting
Spruce swamps		+	Prepared soil, large seedlings
Pine swamps	(+)	+	Prepared soil
Open peat lands	(+)	+	Prepared soil
Fields		+	Prepared soil, no poor soils
Norway spruce			
Upland forest with grass-herb vegetation		+	Prepared soil
Moist upland forest site	(+)	+	Prepared soil, no sowing
Spruce swamps		+	Prepared soil, the best
Fields		+	Prepared soil
Birch			
Upland forest with grass-herb vegetation		+	Prepared soil
Moist upland forest site		+	Prepared soil
Fields		+	Prepared soil, the best fields

CHAPTER 6

8 Genetic improvement

8.1 Origin

The goal in the genetic improvement of forest trees is to develop reforestation material resulting in higher yields in commercial forestry. This uses the enrichment of desirable genes controlling commercially important traits. High timber quality is typically the foremost trait sought, but high survival rate and fast growth rates are other characteristics of genetically-improved stock. The genetic component is always involved when establishing a new forest stand by artificial means. Even with natural seeding, the performance of the new tree stand depends on whether good or inferior parent trees were used.

Proper forest tree breeding uses the genetic variation among individuals of an adapted population. An adapted population may be native (autochthonous) or introduced. A large number of genes control all the important traits of survival, growth, and quality. Gene effects are largely additive. This means that the variation is continuous. A variable such as height can have any value between minimum and maximum. The environment also modifies the expression of quantitative traits. The desired result of the complex genetic background and the interaction with environment is that a tree population display the normal distribution of commercially important traits. With a certain probability, good trees produce better offspring than inferior trees – a high probability that a

Figure 12. Controlled crossing is the basic technique used in forest tree breeding. Currently, the Finnish Forest research Institute's Punkaharju Research Station is studying the possibility of using genetically tranformed pollen in controlled crossings. (Photo by Teijo Nikkanen).

moderate genetic gain can occur in each breeding cycle. Figure 12 shows an example of forest tree breeding. The genetic instruments available in forest tree breeding are selection and recombination. Conventionally, selection started with phenotypical selection in existing natural stands of good outward quality such as Fig. 13 shows. The very best individuals – one or two per hectare – were accepted as plus trees. In the early days of tree breeding more than 50 years ago, the main emphasis was on stem quality, straightness of the bole, and good natural pruning. Stem volume depicting the growth rate in even-aged stands became important later. Tree size and growth rate are readily measurable traits that are therefore suitable for the computation of various genetic parameters.

Estimation of the contribution of the genetic component especially that of additive genetic variation is possible only by progeny testing. Progeny testing means experimental cultivation of progenies (full-sibs or half-sibs) of many plus trees side-by-side on a common test site. Experimental design including replications and randomization allows computing the significance of the differences among the entries and the variance components. The primary purpose of progeny testing is to rank parent trees according to their breeding values. Genetic parameters such as heritability allow improvement of breeding methods and computation of economic consequences. Heritability, h^2, is the proportion of additive genetic variance, V_A, of the total phenotypic variation, V_P

$$h^2 = V_A / V_P \qquad (3)$$

Figure 13. Selection of plus trees in Finland started in 1947. Plus pine E138 (pictured) was selected in 1948. Attention focused on the quality of the tree, e.g., bole straightness, branch thickness, natural pruning, overall state of health of the tree, and its size. (Photo by Teijo Nikkanen).

CHAPTER 6

The results provided by progeny testing find use in backward and forward selection. Backward selection means screening phenotypically selected parent trees according to the performance of their offspring. Selection has direct impacts on management of a seed orchard. Forward selection usually involves selecting the best individuals of the best families. This kind of recurrent selection is considerably more effective than the original plus tree selection. The environment of the test sites is much more uniform than in practical forestry that reduces the environmental variance and subsequently improves the heritability value. In addition, one can evaluate family performance from many trees rather than the one used in plus-tree selection.

The rapid advances made in molecular genetics make it possible to anticipate the selection of trees carrying desirable genes by detecting certain pieces of DNA. For the present, marker-assisted selection is the first step. This means searching for favorable genes using proximate neutral marker genes. The difficulty in this approach is that most traits have several additive genes acting upon them, and the link between the target genes and the marker genes is not very tight. Due to the long life cycle of trees, DNA-based selection could accelerate this otherwise very slow process considerably.

Recombination means rearrangement of genes. With trees, recombination is a natural process that occurs in flowering and seed set. Trees are out-crossing plants – the seed-producing parent and the pollen-producing parent are different, unrelated individuals. In forest tree breeding, recombination normally means controlled crossing. Consider an imaginary example. If tree A has fast growth but poor quality and tree B has slow growth but excellent quality, crossing A x B could combine the good properties of both parents. The progeny could also include trees with slow growth and poor quality. Recombination by crossing requires following with selection and probably another crossing.

Since most tree species have a juvenile period of 10–20 yr. when they do not flower, the alternation of selection and recombination is a time-demanding stage. Efforts have therefore concentrated to induce flowering at an early age. Some species such as birch (*Betula spp.*) do respond to long-day, high-temperature, and elevated-CO_2 level (1 000 ppm) treatments and begin to flower a year after germination from seed. Other species such as *Thuja* spp. respond to hormonal treatments with gibberellic acid. Most commercially important tree species are still conservative, i.e., efficient measures to induce early flowering are not yet available for them.

Some efforts have occurred to recombine genes over species limits. Spontaneous hybrids of larches (*Larix*) and firs (*Abies*) are common in arboreta, and artificial species hybridization is possible when applying controlled crossing. Only a few have proven successful in forestry. Species hybrids of poplars and aspens (*Populus*) are common examples of successes in this endeavor. Recombination by artificial species hybridization is a matter of trial and error. The formation of a new variation is only the first step. The breeder must continue testing and selecting suitable variants. This kind of recombination breeding is like a lottery – the expected gains are high, but the probability of success is modest or low.

Management of forest ecosystems

Advances in molecular biology and gene technology offer revolutionary paths to recombination. Transferring only the gene that controls the desired trait might be possible without the ballast of all other genes. Study focusing on other organisms has shown that genes can be transferred even over kingdoms of organisms as the example, human insulin gene transferred to bacteria, shows. Transgenic trees would most likely meet very stiff ethical and emotional opposition. Genetic engineering is applied in the same regime as controlled crossing. The potential risk is negligible when simply transferring solitary, well known genes. Thousands of genes are transferred in controlled crossings.

8.2 Production

Breeding processes involve limited numbers of plants. Large-scale use of genetically improved planting stock requires propagation on the order of millions per year to achieve reasonable cost levels. Figure 14 shows the most common application – producing genetically improved seed in seed orchards. Selected parent trees cross with each other in a special population. The seed produced is in many respects therefore similar to natural seed formed in normal stands. The difference lies in the higher frequency of favorable genes. The seed-orchard method is adequate when implementing the response to selection of additive gene effects.

Figure 14. The results of forest tree breeding are exploited in artificial regeneration (planting or seeding). Genetically improved seed for artificial regeneration is produced in seed orchards. Most pine seed orchards in Finland are now in full production. They produce enough seed for nurseries in Southern and Central Finland. The picture shows Scots pine cones being collected in a seed orchard (No. 338. Pelkola) in Imatra. (Photo by Teijo Nikkanen).

CHAPTER 6

If the superiority of a product of breeding uses nonadditive gene effects such as a specific combination of certain genes, then mass propagation must occur by vegetative means. In sexual reproduction, the advantageous genetic composition would segregate. Grafting, root cuttings, tissue culture, etc., can achieve vegetative propagation of trees. In all cases, the basic individual gives a genetically uniform set of plants – a clone. Grafting has wide use in forest tree breeding, but it is too expensive for the production of commercial reforestation material. Rooted cuttings are common with poplars as an example and technically possible with spruce. Micropropagation by tissue culture is under development for certain species. This technique still requires further development and better cost efficiency to be competitive. Somatic embryos and artificial seed are potential large-scale applications. Note that cloning *per se* does not improve the genetic value of the plant material.

8.3 Traits

The performance and the commercial value of a tree consist of numerous components. Placing trees in a category of good or bad is not possible. Dealing with hundreds or even thousands of characteristics and biological processes is also impossible. To arrange and manage the breeding system, a certain realistic framework is necessary. The traits subject to genetic improvement therefore comprise three main categories: survival, growth rate, and quality.

Survival includes having the trait of adaptation to the ambient environment such as climate and soil and resistance to harmful abiotic and biotic factors. Adaptation links closely to suitable geographic provenance, but room for selection even within a local provenance exists. Harmful abiotic factors usually imply various pollutants. Unfortunately, genetic improvement has only limited measures to improve tolerance of toxic compounds. The problem is most severe in harsh environments where natural stress reduces the capacity to tolerate chemical loads.

Achieving resistance to pests and diseases is perhaps the most complex objective in forest tree breeding. The number of parasites for most species is high. This makes the goal of breeding trees totally resistant against all forms of injury unrealistic. The fundamental dilemma is that the counterpart or enemy is also a living organism. Pests and macro and micro organisms have much shorter life cycles than trees, and they can evolve during the life span of a tree generation. Instead of achieving complete resistance against any one damaging agent, forest tree breeding aims at an acceptable level of damage. The existence of genetic diversity and favoring of tolerant genotypes are concrete possibilities in this endeavor.

Growth rate occupies a central position among tree breeding objectives. As a rule, rapid juvenile growth is advantageous to stand establishment. Rapid start and vigorous height growth help planting stock win competition with grasses and the suppression caused by slash and escape from various herbivores. Success in stand establishment yields immediate savings through reduced need for weed and pest control. Accelerated growth rates shorten the time lapse from planting to harvesting. The actual figure for genetic gain in increment varies from case to case. A 10% relative

Management of forest ecosystems

Figure 15. Forest tree breeding aims to improve the properties of wood as a raw material by increasing the frequency of desirable genes and genotypes in commercial forests. Forest tree breeding can ensure continued production of high quality magazines such as House & Garden and Classic Cars printed on Finnish paper made of spruce pulp. (Photo by Teijo Nikkanen).

genetic gain is a reliable overall estimate. Higher figures have occurred and are possible. Some reported values may be questionable. Very high growth rates can negatively impact quality or can use a narrow genetic base. The previously mentioned heritability values of the growth rate are low at approximately 0.20. This figure indicates that environment including soil, tending measures, etc., has a very powerful impact on the growth rate.

Quality – especially stem form – is the original objective of forest tree breeding. The early pioneers in this area concentrated on the negative effect of selective cuttings and the future shortage of superior saw logs. Straight bole, small taper, and good natural pruning are still desirable characteristics. The difficulty with stem form is that no simple numeric descriptor of quality excists as it does for stem volume and height. Genetic gain means a higher frequency of superb trees rather than expressing any value for an average tree.

Wood quality, i.e., the physical and chemical properties of stemwood, is of special interest to the pulp and paper industry. Figure 15 indicates the need for high quality products. Wood density, ring width, and grain are important from the mechanical point of view. Wood properties vary widely even within species. The problem often lies in goal setting. What kinds of wood and timber will industries want in 50 years? Fiber length is one example. The long tracheids of spruce were once a favorite characteristic among paper engineers. Recently, the short and pale fibers of Eucalypts and aspen have become desirable in paper making. The evolution of technical processes is considerably faster than that of trees. Molecular biology provides fascinating possibilities for the genetic improvement of wood. Identification of the genes controlling lignin may allow reduction in the lignin content of wood. Natural impregnation of heartwood by decay-resistant compounds is another promising feature.

Genetically improved reforestation material is a little more expensive than wild or unselected one. However, the difference is negligible in relation to total costs of reforestation. Genetic improvement is a "soft", biological method. No additional inputs are needed during the whole rotation. On the other hand, genetic improvement is unable to compensate for mistakes made in silviculture.

Management of forest ecosystems

9 Processes and measures to grow established trees

9.1 Effect of stand density on tree growth and mortality

Over the rotation, the availability of resources and attacks of herbivores and other pests controls the dynamics of tree growth and mortality. Mortality and especially growth relate to the spacing of trees within the stand and the influence on tree size. Spacing or stand density indicates indirectly the availability of resources and aggregates the impacts of the various environmental factors on the physiological and ecological processes of tree growth. These density dependent processes control the dynamics of the tree population in the stand, and they provide a vehicle to control tree growth and properties through cuttings or removal of trees leading to decreased stand density.

Larger trees require a greater amount of resources for vigorous growth. This implies that the space available to individual trees should increase with increasing tree size. If this does not occur, Fig. 16 shows that trees will die. Pioneer tree species especially are sensitive to the crowding induced by the increased need for space necessary for growth to occur. The same pattern also holds for climax tree species. A young Pendula birch with stem volume of approximately 0.02 m^3 needs approximately 2 m^2 of growing space, and a mature birch with stem volume approximately 0.25 m^3 needs approximately 10 m^2. Similarly, a young Norway spruce needs approximately 1.5 m^2, but a mature one requires approximately 6 m^2. Spacing also influences the allocation of growth along the bole and therefore the stem form; growth, senescence, and death of branches; and the subsequent branching of the stem and knot content of the wood.

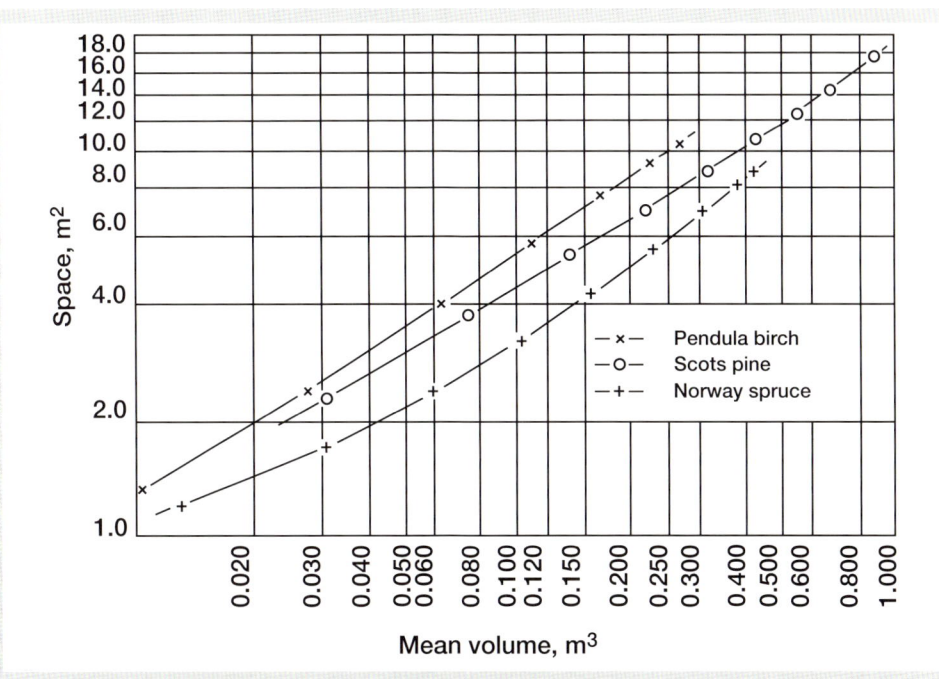

Figure 16. Space needed by Scots pine, Norway spruce, and Pendula birch for successful growth as a function of stem volume[10].

CHAPTER 6

The Yield-Density Model developed by Shinozaki and Kira can describe the relationship between spacing, tree growth, and mortality[11]. Assume *a set of tree stands of the same age but varying in terms of spacing, i.e., stand density.* Over a long time, total growth does not depend on the initial stand density:

$$Y \cdot \lambda^{a-1} = K = Constant \qquad (4)$$

where Y is the total stand growth, λ is the stand density, and a is a parameter with value $a = 0$ when $t = 0$ and $a = 1$ when $t \rightarrow \infty$.

The Yield-Density Model indicates that a high stand density or close spacing of small trees gives the same yield as a stand composed of only few widely-spaced trees. In other words, numerous small trees or a few large trees can fill a given space with the consequence that self- thinning will eliminate several trees from the stand leading to decreased stand density. The remaining trees will continue to grow with time as Fig. 17 shows.

The application of the Yield-Density Model to a *single stand* shows that the self-thinning of the stand over time determines the limits for the mass of trees of specific age that can exist on a site. With time, stand density will decrease, but the mass of individual trees will grow. The *development of a given stand* will form by the sequence where stands of different ages and densities follow one another. This process follows the Power Model of Yoda[13] that replaces the Yield-Density Model in stands of varying age:

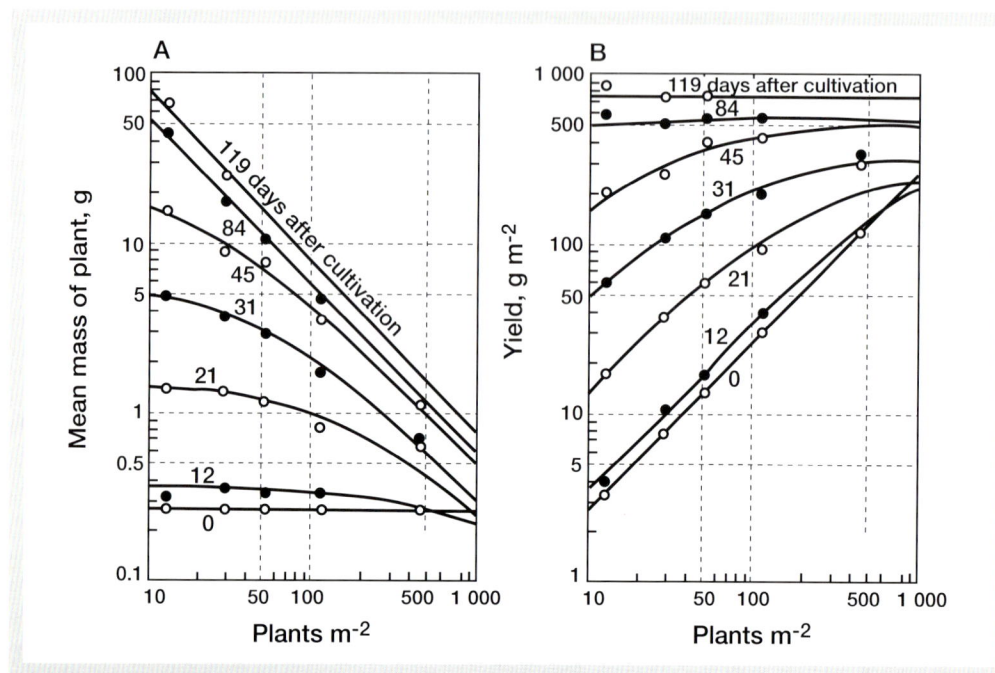

Figure 17. Outlines of the Yield-Density Model controlling the relationship between stand density and growth. A: Mean mass of plants as a function of stand density and time since planting and B: total plant yield as a function of stand density and time since planting[12].

Management of forest ecosystems

$$Y = C\lambda^{-\frac{1}{2}} \tag{5}$$

where Y is the mass of trees forming the stand, and C is a parameter, i.e., the mass of trees when $t \to \infty$ and only one tree forms the stand.

The value of -0.5 for the power is valid for a range of different tree species occurring in the boreal and temperate zones. It represents the relationship between space, s, and mass determining the growth of an individual tree in the stand:

$$s = \frac{1}{\lambda} \tag{6}$$

Let L be the dimension of an organ of a tree. Since the square of the dimension and the volume by the cube of the dimension determines the area, then

$$s \propto L^2 \propto (L^2)^{\frac{2}{3}} \propto (y^2)^{\frac{2}{3}} \tag{7}$$

Consequently,

$$s \propto \lambda^{-\frac{3}{2}} \Rightarrow Y = C\lambda^{-\frac{3}{2}} \tag{8}$$

In logarithmic coordinates, Fig. 18 shows that the Power Model forms a line with a gradient of -1.5 between stand density and the mass of a single tree (y, mean tree) assuming equal size for different trees. The line indicates how large the tree can grow under given spacing or stand density. Whenever trees grow larger than the size determined by the density with mass trajectory, trees die and the stand undergoes self-thinning.

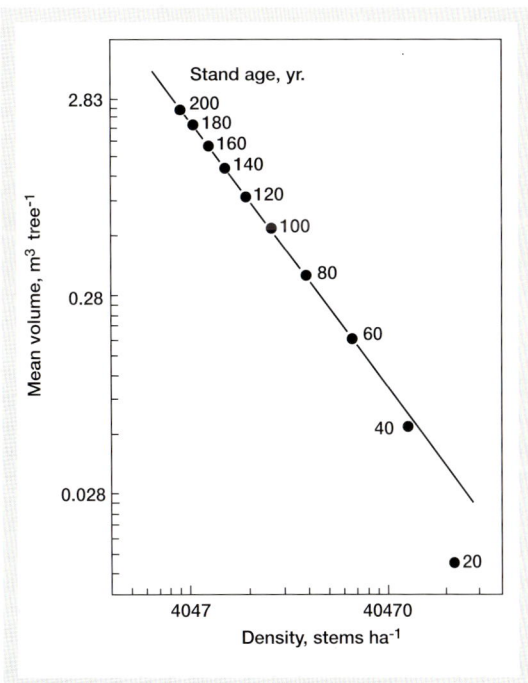

Figure 18. Dependence of the mean volume of trees on stand density when applied to stands of varying age [14].

9.2 Stand density and differentiation of tree size

The inverse relationship between stand density and tree size implies that decreasing stand density increases the mass of individual trees. This is true during the rotation, although growth rate decreases as trees mature. Simultaneously, the growth and subsequent accumulation of mass by individual trees is different due to genetic differences in growth rate and the heterogeneity of the environment. Consequently, size distribution of trees changes over time following the differentiation in tree growth as Fig. 19 shows. This differentiation is particularly clear in stands with close spacing such as dense seedling stands where most trees are small and tree size distribution is highly skewed. Small, suppressed trees represent that part of the population for elimination in the course of succession as large and dominating trees become even more dominant.

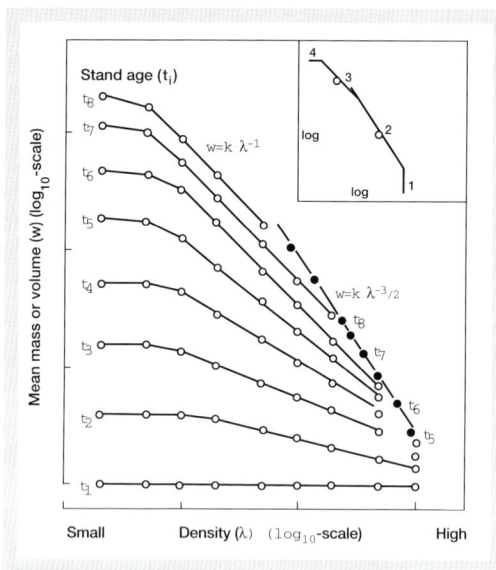

Figure 19. Mean mass or volume of trees as a function of stand density and time since stand establishment. The insert shows the course of stand development: 1=phase with no self-thinning, 2=differentiation of rapid self-thinning, 3=phase of slowing down of self-thinning, and 4=phase of very slow self-thinning[14].

Differentiation of tree size characterizes the growth and development of stands with variable initial densities as Fig. 20 shows. Over time, the frequency of large-sized trees is clearly higher in widely-spaced stands than in stands with close spacing. In the

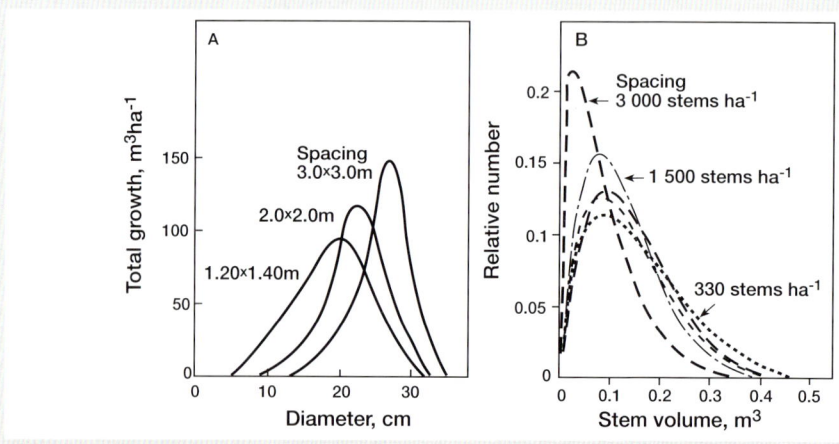

Figure 20. Examples showing initial stand spacing influencing the size differentiation of A: *Picea abies* and B: *Picea sitchensis*[15, 16].

latter case, the frequency of small-sized trees dominates over smaller mean sizes. Whenever the value of timber is differentiated by size such as pulpwood vs. saw timber, controlling spacing over the rotation is a prime measure available to forest managers to increase the value of timber. Proper thinning can allocate growth to selected trees to enhance their growth and shorten the time necessary to exceed specific dimensions.

9.3 Thinning and development of tree stands

During the natural succession, some initial trees in a given stand will die in the succession process through natural mortality, natural removal, and self-thinning. The current mass forming the stocking of the stand represents only part of the total mass produced since stand establishment indicated in Fig. 21. The mass otherwise lost through natural removal can be harvested through repeated thinnings. Simultaneously, the total mass of trees stocking the site remains lower than that determined by the self-thinning. The thinning removal is therefore larger than the natural removal alone. This results in the trees forming the stand having more growing space with concomitant growth enhancement. Stand growth becomes allocated to fewer trees, and these can grow larger than would be possible in a stand subject only to natural succession. In this context, thinning is comparable to a minor disturbance promoting autogenic succession in a stand.

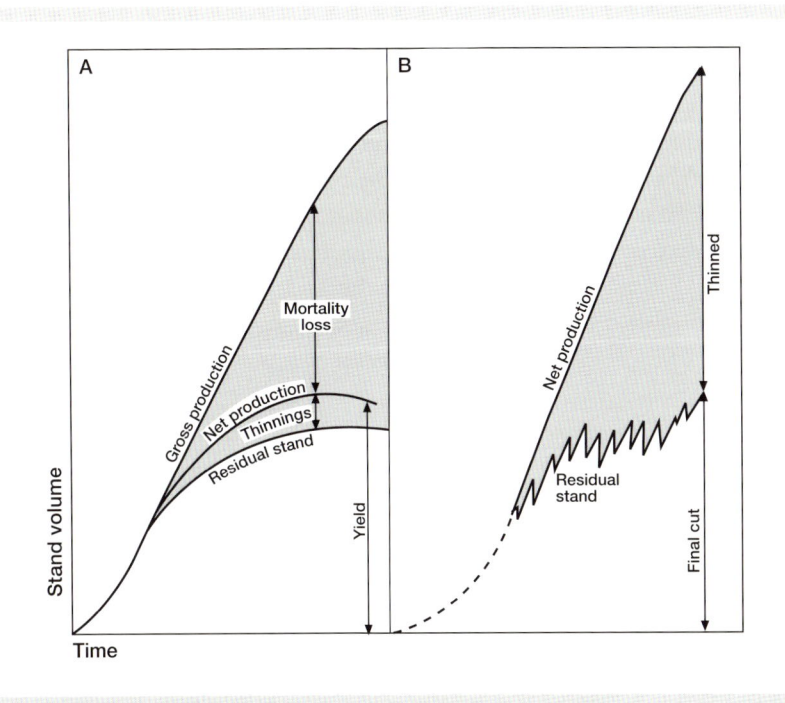

Figure 21. Outlines of thinning related to natural succession in a tree stand. A: Thinning related to tree growth and mortality and B: thinning related to the net production of trees[1].

CHAPTER 6

In a tree stand composed of a population of *a single tree species,* the thinning rule of differentiation of tree size holds over the whole rotation. First thinnings are normally precommercial, since no marketable timber can be harvested due to the small size of the trees. Precommercial thinnings are frequently necessary to widen the spacing and therefore enhance tree growth and shorten the time required for commercial thinnings. Precommercial thinnings also increase the mechanical strength of the remaining trees enabling them to resist damage from wind and snow loads.

In a tree stand composed of populations of *several tree species,* the situation is more complex. Then the self-thinning rule applies to the relationship involving density, mass, and size differentiation of the trees. Precommercial thinning today can also include the elimination or cleaning of tree species that have little or no commercial or environmental value. Even then, precommercial thinning includes the spacing of the base tree population and the spacing or cleaning of competing tree populations with the aim of improving the growth of trees belonging to the base population through enhanced availability of resources.

9.4 Mechanisms behind thinning effects: thinning and availability of resources

9.4.1 Solar radiation

Proper spacing can especially control the availability of radiation from the sun, soil water, and nutrients. In addition, the temperature conditions, wind force, and air humidity within the stand relate to spacing but not to the same extent as the first three. They are the major determining factors controlling tree growth within a stand and the productivity of the entire stand.

Considering the impact of spacing on the *solar radiation* received by a tree requires distinguishing within-tree and between-tree shading. Within-tree shading means that a leaf or needle in the crown receives less radiation than a leaf or needle on the crown surface. This is due to the interception of radiation by other leaves or needles or branches and stem. If $I(r)$ indicates the radiation incident on a point r inside the crown and I is the radiation outside the crown, then

$$I(r) = p(r) \cdot I \qquad (9)$$

where $p(r)$ is the probability of radiation to be transferred through the crown to point r (x,y,z). The probability $p(r)$ depends on the directional distribution of light and the density of mass in the crown space:

$$p(r) = e^{-p \cdot c \cdot t(r)} \qquad (10)$$

where p is the density of the mass in the crown space, $t(r)$ is the path length of the ray between the point r and the surface of the crown, and C is the light-extinction coefficient influenced by the properties of the plant material of the crown and the distribution of the incident radiation.

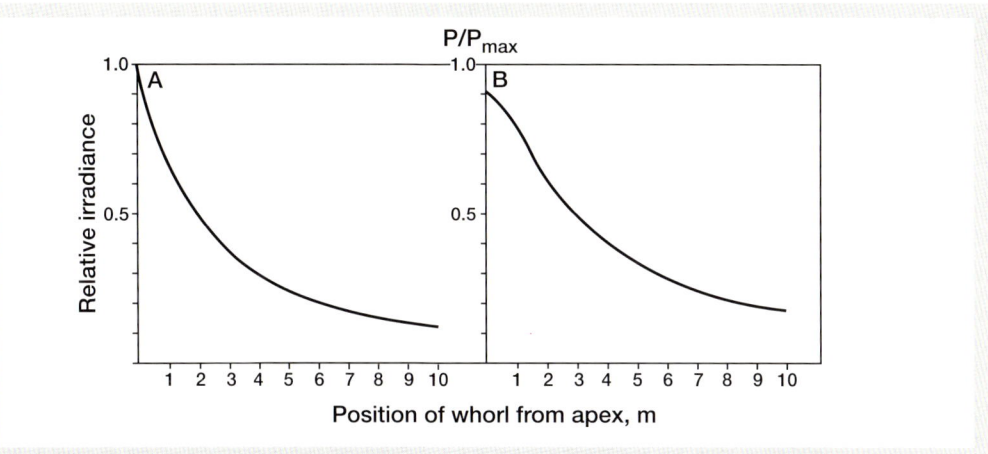

Figure 22. A: Proportion of radiation outside the crown incident on branch whorls (P) and **B:** proportion of photosynthesis of the maximum photosynthesis (P max) as a function of crown depth[17].

Consequently, the mean amount of radiation, incident on a branch whorl at height z is as follows:

$$\bar{I} = \frac{1}{K}\iint p(r)dxdy \qquad (11)$$

where K is the cross-sectional area of the crown at height z.

Over the crown, the proportion of radiation outside the crown incident on the lower crown is substantially less than that incident on the upper crown. The effect of within-crown shading on photosynthesis is less than that expected on the basis of shading, since the photosynthetic response to radiation is a saturating one as Fig. 22 shows.

Between-tree shading relates to stand density. If $r(x,y,z)$ is a point in the crown of a tree, $T(z)$ is the mean projection area of the tree crown at height z, A is the unit area, and N is the number of trees per A (stand density is λ), then the probability $(p(r))$ that no tree is shading point r is the following:

$$p(r) = \left(1 - \frac{T(z)}{A}\right)^N \qquad (12)$$

The total projection area of the crown is then

$$\iint 1 - e^{-p \cdot C \cdot T(r)} dxdy \qquad (13)$$

where V is the crown projection at height z.

The probability of point r not being shaded by itself or other trees is the following:

$$e^{-p \cdot C \cdot t(r)} \cdot e^{-p \cdot \lambda \cdot T(r)} \qquad (14)$$

CHAPTER 6

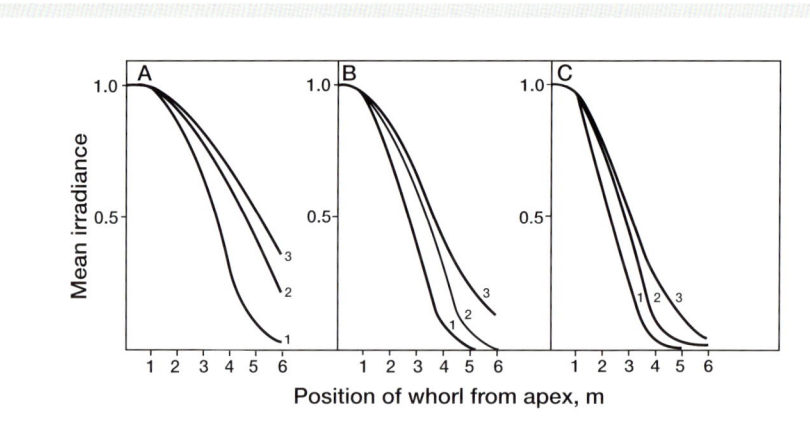

Figure 23. Effect of stand density on relative radiation received by a given tree as a function of the spatial distribution of trees[17]. Stand density (stems ha-1) A=1 000, B=2 000, C=3 000. The numbers in the figures indicate the spatial distribution of trees: 1= homogenous, 2=regular, and 3=random.

Consequently, the mean relative radiation incident on a branch whorl as influenced by within-crown and between-tree shading is then

$$\bar{I}(z) = \frac{1}{K}\iint e^{-p \cdot C \cdot t(r)} \cdot -e^{-p \cdot \lambda \cdot T(r)} dxdy \qquad (15)$$

The extinction of radiation in the upper crown is small regardless of stand density as Fig. 23 shows when compared to lower parts of the crown where the effect of stand density is quite distinct; i.e., the influence of neighboring trees on a given tree focuses primarily on the lower crown. This influence also depends on the spatial distribution of trees so that the influence is at its greatest in a homogenous distribution and is at its smallest in a random distribution.

9.4.2 Soil water

The effect of thinning on *soil water* relays through several mechanisms related to the reduction of foliage area following thinning. The interception of precipitation by tree crowns or stand canopy and subsequent evapotranspiration decrease substantially with an associated increase in throughfall as Table 6 shows. The amount of water in the foliage balances the incoming precipitation with the evaporated and infiltrated water. Thinning reduces interception with the result that water conditions in the soil balance at a higher level of groundwater, higher soil moisture level, and water availability throughout the soil profile.

Table 6. Model computations showing how thinning can influence the different components of the hydrological cycle in a Scots pine stand located 60°N. The computations focus on a stand with age 45 yr., mean stem diameter (dbh) of 12 cm, and mean height of 12 m. Pre-thinning stand density was 1 500, and post-thinning density is 1 000 stems per hectare. Soil moisture value indicates the volumetric moisture level of the top 10-cm layer of the soil. The values are the mean values for a period of 15 yr. for June, July, and August assuming a constant tree stand.

Component	Pre-thinning mm	Pre-thinning %	Post-thinning mm	Post-thinning %
Precipitation on site	101	100	101	100
Evapotranspiration	79	78	71	70
Surface flow-out from site	4	4	4	4
Flow deeper down from upper soil	18	18	26	26
Balance	0	0	0	0
Soil moisture, $m^3\ m^{-3}$	0.28		0.33	

9.4.3 Nutrient availability

Thinning influences *nutrient availability* primarily by improving the nutrient cycle, i.e., by adding to nutrient availability as harvesting residues decompose and enhancing the decomposition process of litter and humus in the soil following higher soil temperatures. The increase in nutrient supply for nitrogen is greater in young stands than in mature stands as Table 7 indicates. Young stands are more responsive to the addition of nitrogen than mature stands. Thinning therefore has a major impact on the growth and development of young stands.

Table 7. Example showing thinning influencing the amount of soluble nitrogen (ammonium and nitrate) in stands of Scots pine on sites of medium fertility. The post-thinning situation is two years after thinning. The computations use the Forcyte 10 model developed by Kimmins and Scoullar[18].

Thinning example	Stand age, yr.	Stand density, stems ha^{-1}	Stand volume, m^3 ha^{-1}	Nitrogen supply, kg ha^{-1} yr.$^{-1}$
Example #1				
Pre	40	1 568	232	38
Post	42	1 098	162	48
Example #2				
Pre	55	1 443	315	24
Post	57	1 082	236	35
Example #3				
Pre	70	712	365	20
Post	72	534	274	32

9.5 Thinning regimes and selection of trees for removal

The regimes applied in removing trees in thinning are selective or systematic. *Selective* thinning uses tree-by-tree selection to determine whether to remove a particular tree or allow it to grow further. In selective thinning, biological properties such as tree health or biological status or technical properties such as crookedness are the prime decision factors in the selection process. In *systematic* thinning, the removal of a particular tree primarily depends on its position in relation to other trees, the location of strip roads to be opened in the stand for entry of logging machines, or both. Only selective thinning relates to the prevailing stand structure with the specific aim of controlling the future growth and development of the tree stand. Mechanization of selective cutting is far more problematic than the mechanization of systematic thinning.

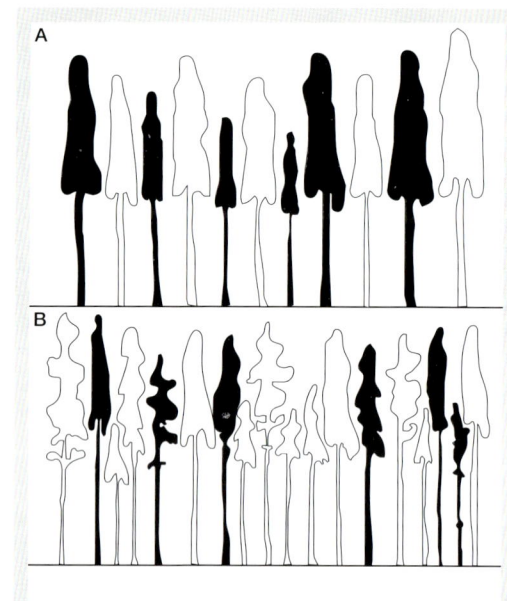

Figure 24. Schematic presentation of low thinning (upper) and high thinning (lower)[19].

Selection forestry or uneven-aged forestry uses selective thinning solely.

Selective thinning can occur from below (low thinning) or from above (high thinning) as Fig. 24 shows. In the former case, the trees for removal are suppressed individuals with the highest risk of elimination through natural mortality. In the latter case, the increased spacing for the remaining trees uses the removal of dominant or codominant trees. This allows suppressed trees to grow further. In both cases, the retention trees have the best growth and technical properties. In practice, selective thinning is a combination of low and high thinning aimed at maximizing the quantitative and qualitative growth of tree populations of single tree species or communities formed by several tree species occupying the site.

Figure 25 shows the similarities and differences between low thinning and high thinning. The figure shows the size distribution of the removed and retained trees for both thinning regimes. In low thinning, the trees to be removed consist solely of suppressed or intermediate trees – no dominants are removed. In high thinning, the trees to be removed are mainly dominants, but intermediate and suppressed trees are also removed. Typically, high thinning focuses on the entire tree population, although the probability for a tree to be removed is greater among dominants than among intermediate and suppressed trees. In low thinning, the probability for a tree to be removed is highest among suppressed trees.

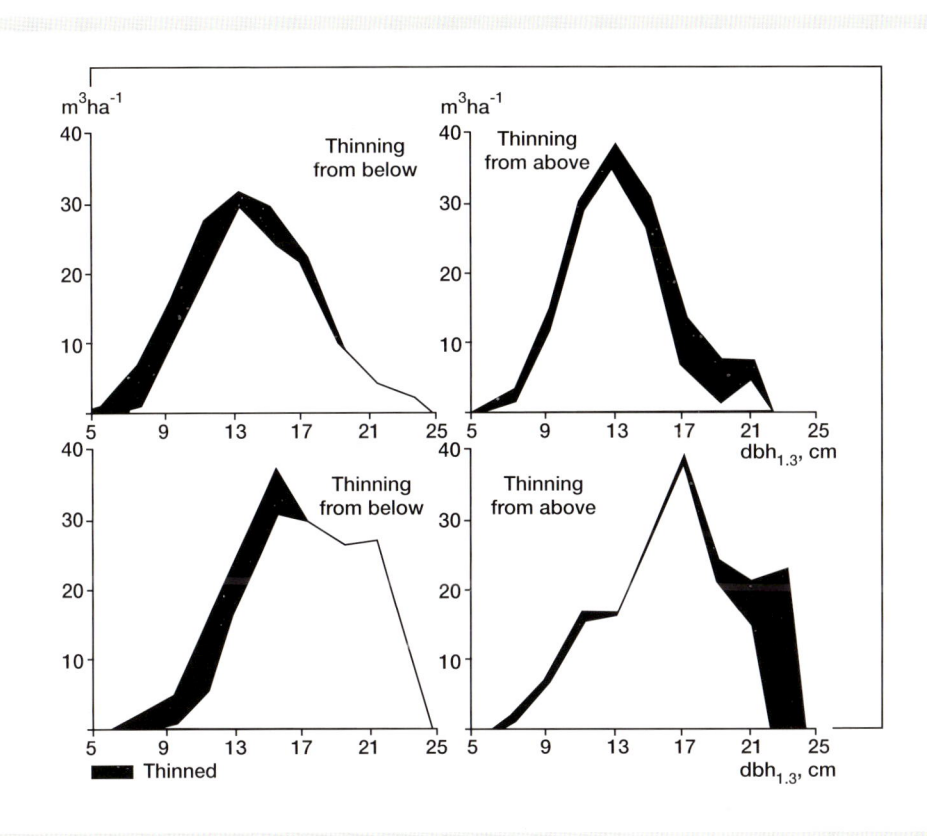

Figure 25. Example of how low thinning (left) and high thinning (right) focus on different segments of the tree size distribution on a site[20].

9.6 Impact of thinning on stand growth and productivity

Spacing or thinning have only minor impacts on the height growth of boreal and temperate trees. Typically, moderate thinning will slightly increase height growth in closely spaced stands. Height growth may even be impaired if spacing is executed to a great degree (wide spacing). This response pattern does not apply to radial growth or diameter growth. It is even sensitive to low degrees of thinning that result in the substantial post-thinning increases in growth of Fig. 26. This increase is maximum in the lower part of the bole making the bole more conical than it would be without thinning. Thinning also increases growth in other parts of the tree as a result of the allometry of growth. Thinning therefore promotes root growth and increases the mechanical strength and anchorage of trees to withstand the force of wind better. Intermediate and subdominant trees respond most vigorously to thinning, since the availability of resources for these trees increases more than with dominant and suppressed trees.

CHAPTER 6

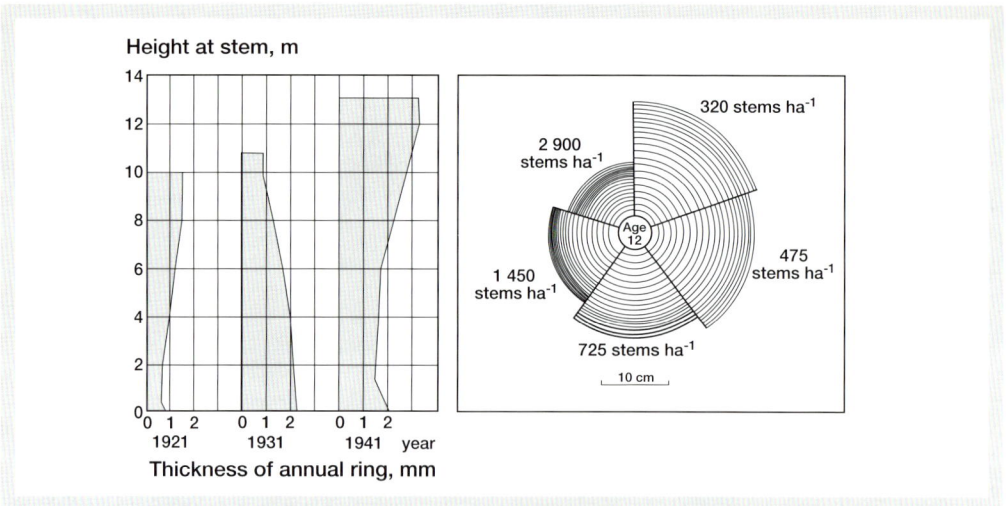

Figure 26. Distribution of radial growth along the bole (left) and the impact of spacing on radial growth (right)[21, 22].

At the stand level, spacing has only a minor impact on the productivity of the tree stand although the growth of individual trees can increase substantially. This is because post-thinning growth is allocated to fewer trees, and the growth increase of individual trees can only compensate for the reduction of growth due to the reduced number of trees indicated by the Yield-Density Model. The productivity of a tree stand is somewhat constant and independent of spacing as Fig. 27 shows. Constant growth is only valid if the mass of trees exceeds 60% of the maximum possible for the site. In practice, this occurs when the canopy is fully closed and branches of neighboring trees overlap. When the canopy is incompletely closed, growth can increase linearly with an increasing degree of canopy closure. This indicates that the high growth of individual trees in an incompletely closed stand cannot compensate for the reduction in total growth due to the reduced number of trees.

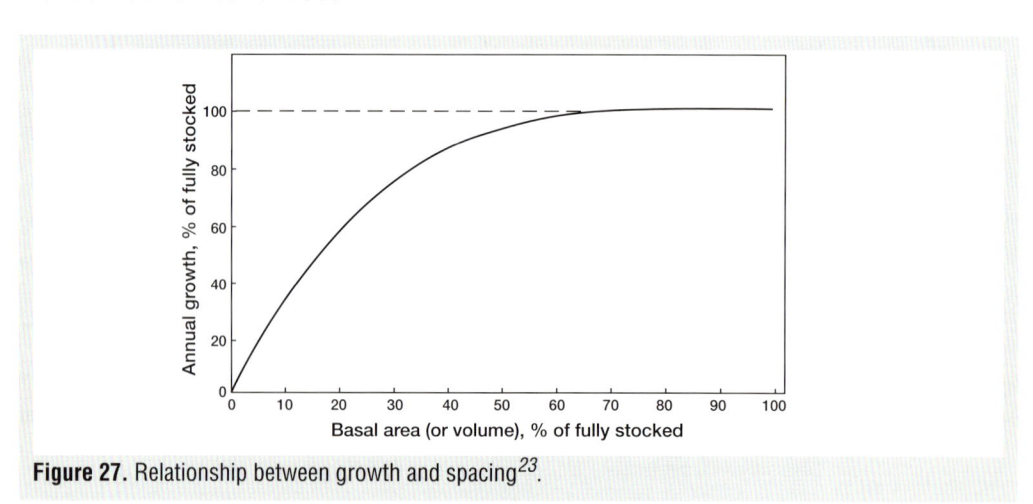

Figure 27. Relationship between growth and spacing[23].

Management of forest ecosystems

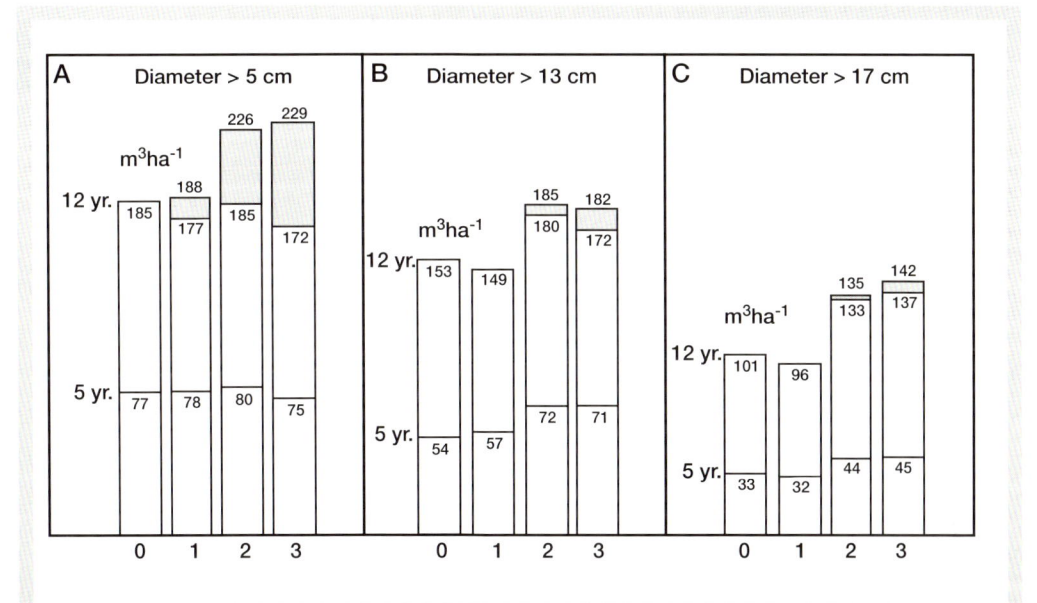

Figure 28. Impact of spacing on the timber yield in young Norway spruce stands in terms of commercial timber harvested during the first 5-yr. (lower part of columns) and 12-yr. periods (total height of columns). Minimum diameters for commercial timber A: >5 cm, B: >13 cm and C: >17 cm. Each column represents a single thinning intensity: 0=no thinning, 1=10%, 2=25%, and 3=40% removal of the basal area. Shaded area indicates precommercial thinning[24].

The impact of thinning on *timber yield* is opposite to that of stand productivity. The harvest of trees otherwise lost through natural mortality can increase the total timber yield over the rotation. In Finland, more than one-third of the total timber yield composed mainly of pulpwood comes primarily from repeated thinnings. Wider spacing can also enable a large number of trees to exceed the minimum dimensions expected for commercial timber assortments resulting in a greater timber yield over the rotation. Figure 28 shows this for a young Norway spruce stand.

Increasing removal rate resulted in increased timber yield (composed mainly of pulpwood) up to 40% compared with a nonthinned stand. With the widest spacing, stem dimensions exceed even those of saw logs resulting in a substantial increase of the value of timber. The thinning regime (low or high) appears to have no impact on the yield of pulpwood compared with saw timber that can be greater in high thinning than in low thinning. This is probably the case only if low thinning occurs early in the rotation, and high thinning occurs late in the rotation.

CHAPTER 6

9.7 Thinning during the rotation

Programmed thinning over the rotation includes timing and selection of the thinning rate. After the current growth rate of the tree stand, the main purpose of thinning varies according to Fig. 29. During the phase of accelerated growth, the main purpose in thinning is to select the trees retained for further growth to increase the level of stand stocking and the strength of the retained trees against the force of wind and snow and pest attacks. The culmination of growth functions as a turning point beyond which thinnings become increasingly aimed at harvesting timber. Simultaneously, the frequency and rate of thinning change so that frequent thinning with low removal rates is preferable before growth culmination. The thinning rate can then increase with subsequent lengthening of the thinning interval. No excessive mortality occurs as indicated by constant productivity in relation to stocking. The length of the thinning cycle and thinning rate show an inverse correlation.

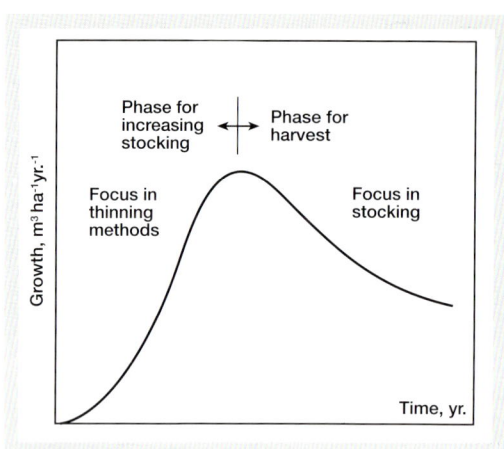

Figure 29. Application of thinning over the rotation[19].

Thinning rules are country-specific and relate to the functions that forests have and the practices applied in forestry. In Finland, the stand growth and development as indicated by the dominant height and basal area of the stand in Fig. 30 govern the timing and rate of thinning. The associated rules are species- and site-specific so that the relationship between dominant height and basal area in thinned stands follows the expected trajectory. Whenever the relationship falls above the given trajectory line, thinning is recommended at the given rate leading to a reduction in stocking to a level not lower than a given value. This cycle repeats until the trees forming the stand attain maturity for final cutting, i.e., clear cutting accompanied by artificial regeneration or seed-tree or shelterwood cutting relying on natural regeneration. The application of the relationship between dominant height and basal area to govern thinning is most successful in the latter phase of the rotation. During this phase, the basal area can be reduced by 20%–25% in each thinning operation.

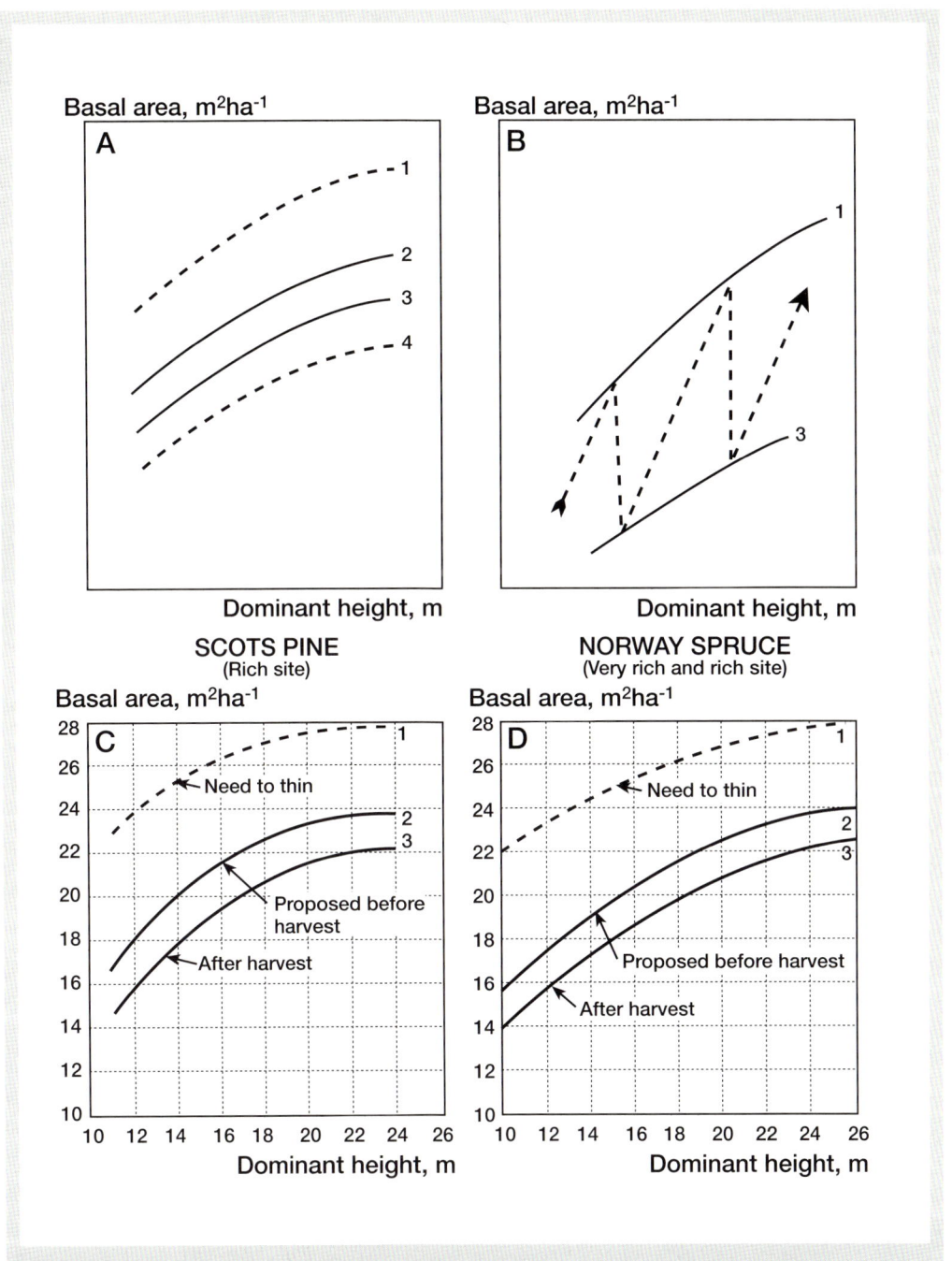

Figure 30. A and B: Application of dominant height and basal area in governing the timing of thinning. Thinning rules for C: Scots pine and D: Norway spruce as applied in Finland. Legend: 1=trajectory for thinning, 2=trajectory for basal area after thinning, 3=trajectory for basal area after logging, and 4=trajectory for lowest acceptable basal area after logging[19].

9.8 Length of rotation

Over the rotation, several thinnings are necessary before final cutting. In southern boreal conditions, this means two or three thinnings over a rotation of 100 yr. The need for thinning depends strongly on the growth rate of trees and the rotation length generally applied in forestry. The length of the rotation is often a compromise determined by ecological factors, economic and environmental values, and traditional and institutional factors governing forestry. One way to determine the length of the rotation is to use the intersection of current annual growth and mean annual growth to indicate the point in time of the biological maturity for final cutting as Fig. 31 shows.

Rotation is frequently longer than indicated by the intercept, and even a small lengthening of the rotation will increase the yield and quality of saw timber. In practice, the length of the rotation can be determined by resorting to a specific diameter value (dbh) beyond which trees are considered to have attained final-cutting maturity. In Finland, the threshold dbh for maturity of Scots pine is currently 28 cm, for Norway spruce it is 25 cm, and for Pendula birch it is 25 cm. This corresponds to rotation lengths of 80, 90, and 70 yr. on sites of medium fertility.

9.9 Thinning program over a rotation

The thinning program over a rotation depends on tree species, site fertility, harvesting method, specific aims of forestry, etc. The effects of these factors must balance so that sustainable productivity conforms to the economic profitability of thinning. This implies that biological and environmental factors impose limitations on the technical and economic solution applied in thinning. In biological terms, a low intensity of thinning associated with a short thinning cycle is the most preferable alternative. Such a thinning pattern would probably be uneconomical. Table 8 shows a set of thinning programs over a range of site fertility in Finnish conditions.

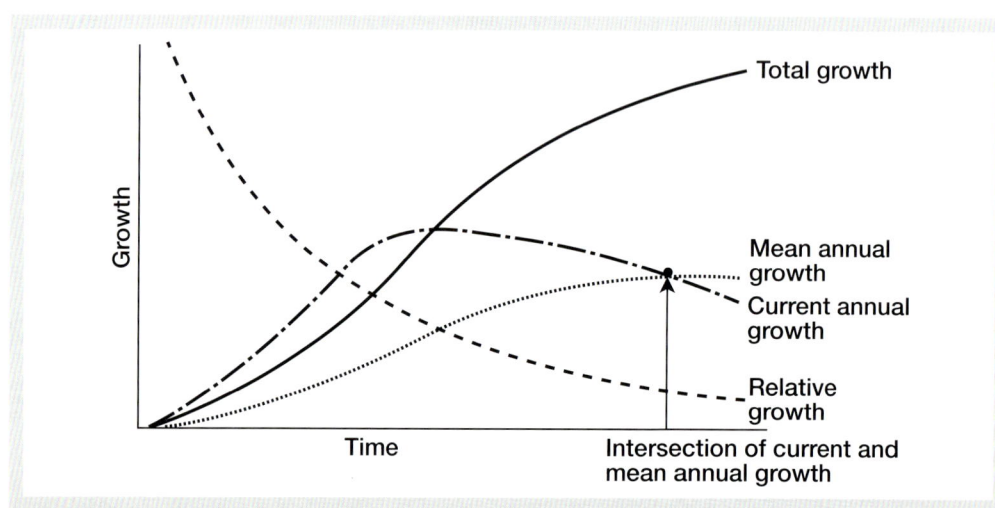

Figure 31. Course of current annual, mean annual, total, relative growth, and rotation length indicated by the intersection of current annual growth and mean annual growth.

Table 8. Number of thinnings, length of thinning cycle, and length of rotation for a range of site fertilities indicated by site classification based on dominant height.

Site classification based on dominant height, H_{dom} [a]	Length of rotation	Number of thinnings	Dominant height at time of thinning, m			Dominant height at time of final felling
			First thinning	Second thinning	Third thinning	
30+	70	3	13	18	22	26
27	80	3	13	17	21	25
24	90	3	13	17	20	23
21	100	2	12	17		21
18	110	2	12	16	19	
15	120	1	13		16–17	
12		0				
9		0				

[a] Dominant height, H_{dom}: corresponds to the average height of 100 largest trees per hectare

10 Processes for improving timber quality

10.1 Effect of spacing and thinning on knots in wood

Numerous indicators such as stem branching, quantity of knots in wood, basic density, etc., can depict timber quality. Timber quality relates therefore to tree growth and growth allocation among the different parts of the tree. Although genetic properties of the tree strongly influence timber quality, environmental conditions can be more important. The spacing of trees in a stand therefore plays an important role in quality. The dimensions and life spans of the branches and the dimensions of knots in the wood correlate to initial stand spacing and the spacing modified in thinning.

Diameter growth of Scots pine branches is very sensitive to increasing stand density at lower stand densities. At higher stand densities, this sensitivity decreases. In young Scots pine stands, a spacing of 1 500–2 500 trees per ha is sufficient to limit the diameter of the thickest branch to less than 15 mm. Stand densities over 10 000 trees per ha are necessary to limit the maximum branch diameter to less than 10 mm as Table 9 shows. Over the rotation, the impacts of initial spacing on branch growth gradually disappear. They are still evident throughout the main part of the rotation especially until canopy closure occurs.

CHAPTER 6

Table 9. Diameter of thickest branch in Scots pine stands of variable densities on sites of different fertility. These results are representative of young 20–25 yr. Scots pines growing in southern Finland.

Diameter of thickest branch, mm	Stand density, stems ha^{-1}		
	High fertility	Intermediate fertility	Low fertility
16	1 700	1 300	
15	2 300	1 700	
14	3 300	2 500	
13	4 500	3 500	1 100
12	6 500	5 000	1 700
11	9 500	7 500	3 500
10	14 000	11 000	3 800

Spacing and thinning primarily influence timber quality during the early years of the rotation when the inner parts of the lower bole (the future butt log) form and transmit the effect to the quality of saw timber. The structural growth and development of the crown can therefore have a substantial impact on timber quality through spacing. The impact of spacing on branch growth depends heavily on site fertility. Table 9 shows that the interaction of stand density and site fertility is obvious in the management of Scots pine for high-quality timber regarding knots in wood.

Figure 32 shows that initial seedling growth is characterized by the accumulation of living branches in the crown. During this phase, the life span of the branches is still short. For Scots pine, this is 10–15 yr. Dead branches occupy the bole for an additional 10–20 yr. until the bole wood is totally clear of live or dead branches, and only the surface wood in butt logs is clear of knots and yields clear wood. The inner parts of the bole contain wood with knots – live in the upper parts of the bole and live and dead knots in the middle and lower parts. Close spacing can strongly enhance the share of clear wood, but it also restricts stem diameter growth to a substantial degree compared with wider spacing. Management aimed at attaining high-quality timber – a large amount of clear wood – is a matter of striking a balance between increased stem diameter growth and reduced branch growth plus early branch senescence and death and early pruning of branches.

Thinning totally eliminates the effect of initial spacing on the growth of branches that have emerged after thinnings as branch growth immediately responds to wider spacing. This response is two-fold. The growth of existing branches improves, and branches increase in diameter. Additional branches may emerge in new branch whorls. The senescence and death of branches will decrease, and the branches will remain on the stem longer. Consequently, the proportion of clear wood will decrease, but stem diameter growth will simultaneously increase. In absolute terms, the amount of clear wood can increase considerably, although the proportion of clear wood could diminish as Fig. 33 shows. Enhanced growth can compensate for the effect that enhanced branch growth has on wood quality.

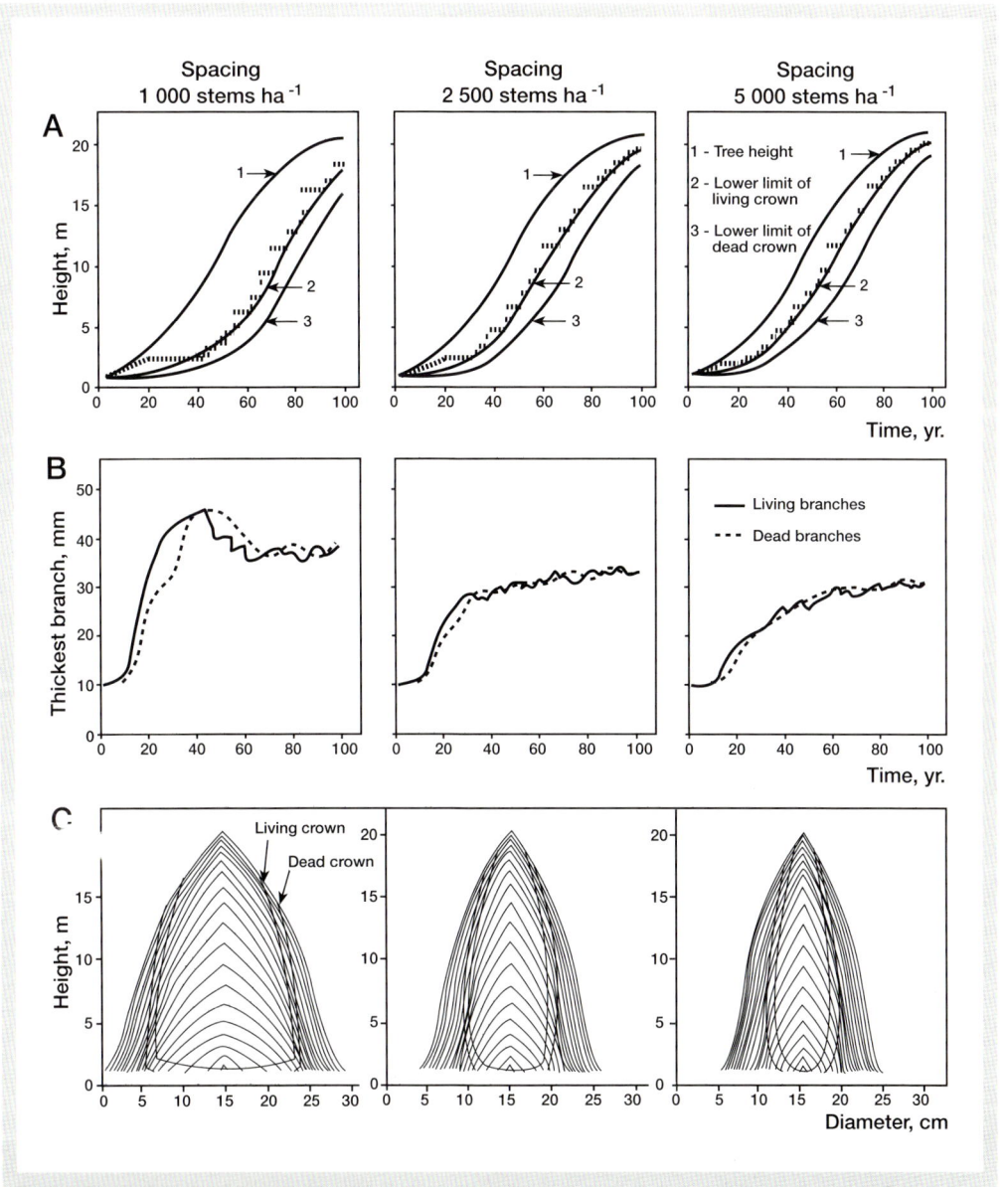

Figure 32. Impact of initial spacing with no management on crown growth and development of Scots pine over the rotation. A: Development of tree height (1) and position of lower limit of live (2) and dead crown (3), B: diameter of thickest live and dead branches on Scots pine over the rotation, and C: position of live and dead knots in stemwood of Scots pine over the rotation as a function of initial spacing.

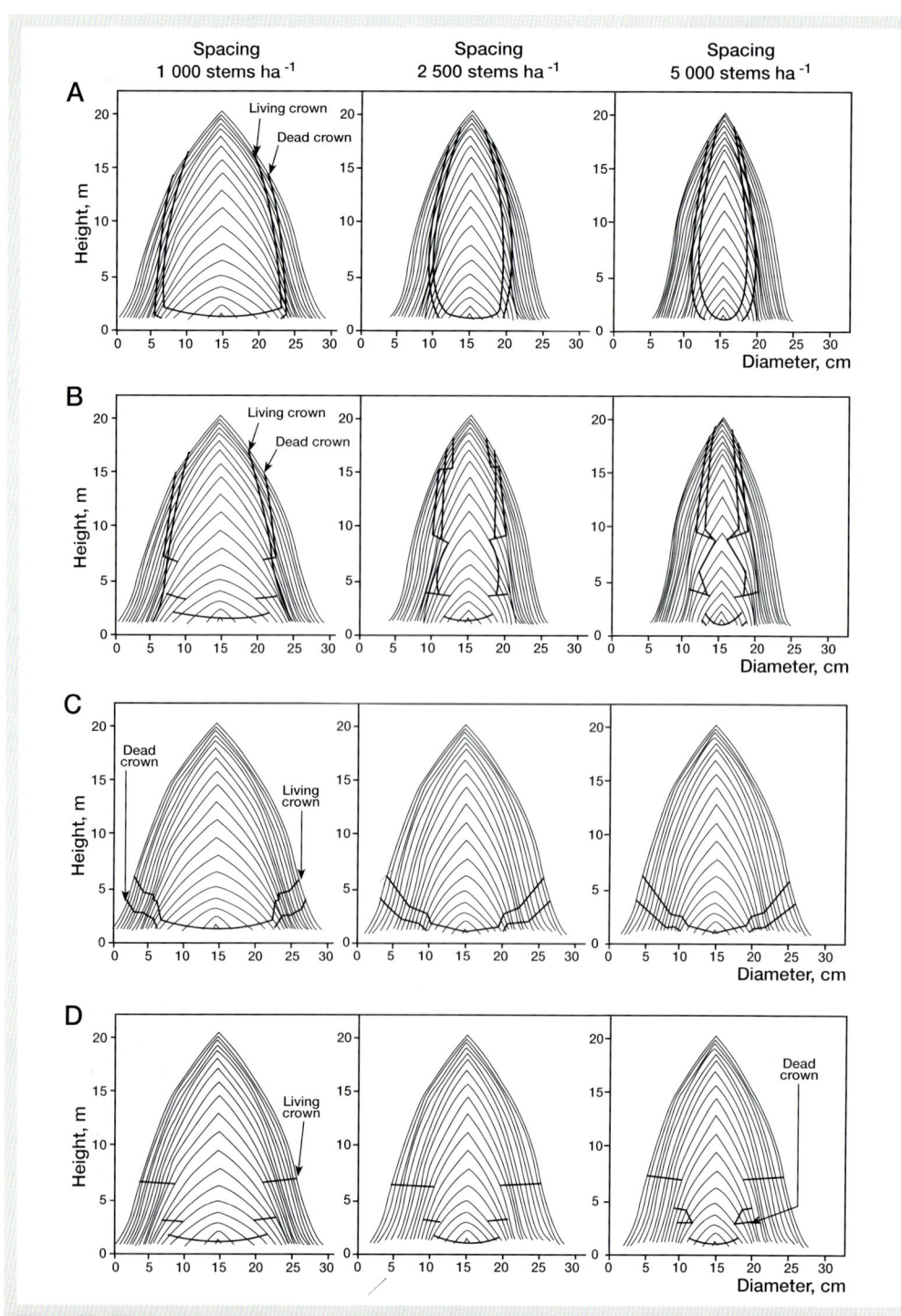

Figure 33. Effect of A: natural pruning, B: artificial pruning, C: thinning, and D: thinning and artificial pruning on knots in wood as a function of initial spacing in Scots pine stands.

10.2 Effect of pruning on knots in wood

Initial spacing is only one component of management for timber quality. Mechanical pruning is necessary to enhance the formation of clear wood. *Pruning* as a silvicultural measure means the removal of slow-growing lower branches and already dead branches still attached to the bole indicated in Fig. 34. This promotes the formation of clear wood compared with what happens naturally as branches die and are shed. The time available for stem growth free of branches and subsequent knots in wood will increase. This will result in more clear stemwood. Pruning can occur several times over the rotation to maximize the amount of clear wood. The removal of branches still alive will substantially reduce the number of knots in wood due to the elimination of dead branches attached to the stem.

The removal of branches showing reduced growth has no effect on the growth of the tree as a whole. The removal of still vigorously growing branches can have a substantial impact on stem growth by reducing height growth and diameter growth. Diameter growth especially can take several years to recover after excessive pruning. With Scots pine, the removal of 20% of the living crown has no impact or will increase diameter growth. Removal of 50% of the living crown will result in diminished growth lasting ten years. Optimum timing of pruning coincides with the tree age when growth is at its maximum. Forest managers should therefore consider combining pruning and thinning to enhance stem growth while reducing the wood-quality impairing effect of dying and dead branches.

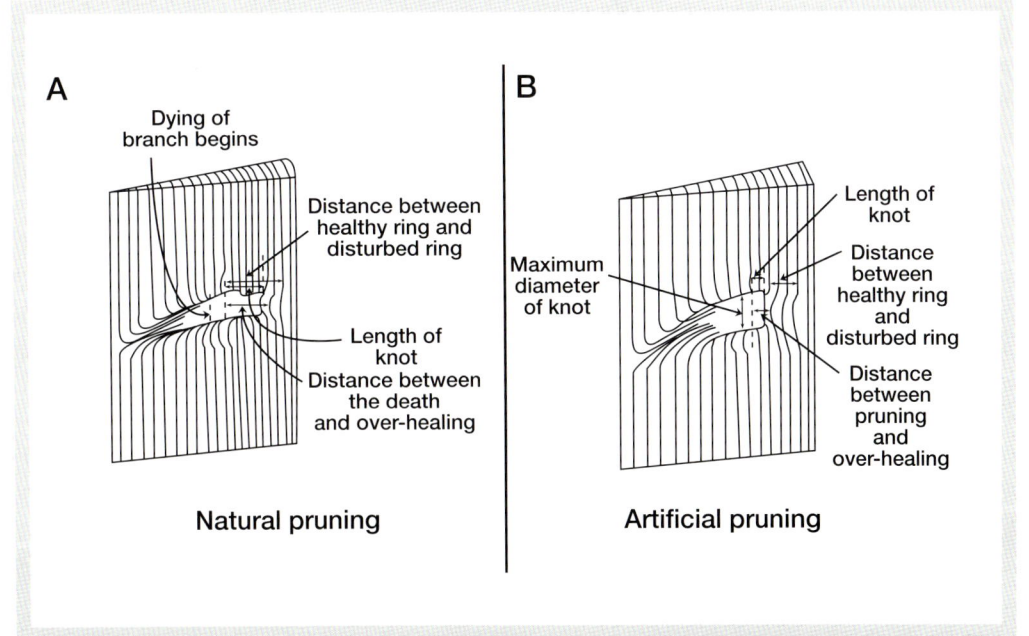

Figure 34. Effect of A: natural pruning, and B: mechanical pruning of dead branches on the formation of clear stemwood.

10.3 Effect of spacing on basic density of wood

The basic density of wood correlates with the width of annual growth rings and with stand spacing. With young Scots pines, basic density of the wood could be approximately 350 kg m^{-3} at a spacing of less than 2 000 stems per ha but approximately 400 kg m^{-3} when stand density is greater than 7 000 stems per ha. In very dense stands over 30 000 stems per ha, the basic density of Scots pine wood can be as high as 450 kg m^{-3}. This increase relates to the smaller proportion of early wood associated with increasing stand density. The inverse relationship between wood density and width of annual rings implies that the basic density of wood will be higher on nutrient-poor sites than on fertile sites. Basic density also depicts the chemical properties of wood and the quality of resultant pulp and paper. Stand spacing at the time of stand establishment and following thinning is a basic factor influencing the quality and mechanical strength of wood.

11 Management of nutrient resources and site fertility

11.1 Concepts

The wood-production capacity of forests closely relates to the nutrient cycle in the forest ecosystem. On a global scale, nutrients cycle through terrestrial and aquatic ecosystems under the control of atmospheric and geological processes. On a local scale, nutrients are bound by vegetation as plants take them from the soil. They eventually return to the soil in litter. This decomposes and releases nutrients into the soil for reuse by plant life. In this context, *nutrient management* means making nutrients available for tree growth as they cycle through the ecosystem with minimum leaching beyond the site. This is a matter of sustaining nutrient availability for plant growth and maintaining or improving the natural fertility of the site.

In boreal conditions, the soil profile is characterized by the accumulation of organic matter in the mineral soil in the form of litter with less decayed matter in the upper part and more decayed matter (humus) lower down, as shown in Fig 35. The organic layer clearly separates from the mineral soil. Many humus components carried by the downward flow of water deposit deeper in the soil. Nutrients leached from the surface layers of the soil similarly become intercepted deeper in the soil where the original parent material is only slightly weathered. The soil profile has podzolized due to prevailing climate with precipitation exceeding the evaporation with a consequent downward flow of water deeper into soil. The main task in nutrient management is to make the nutrients bound in the soil profile available for tree growth without undue loss of nutrients beyond the forest ecosystem.

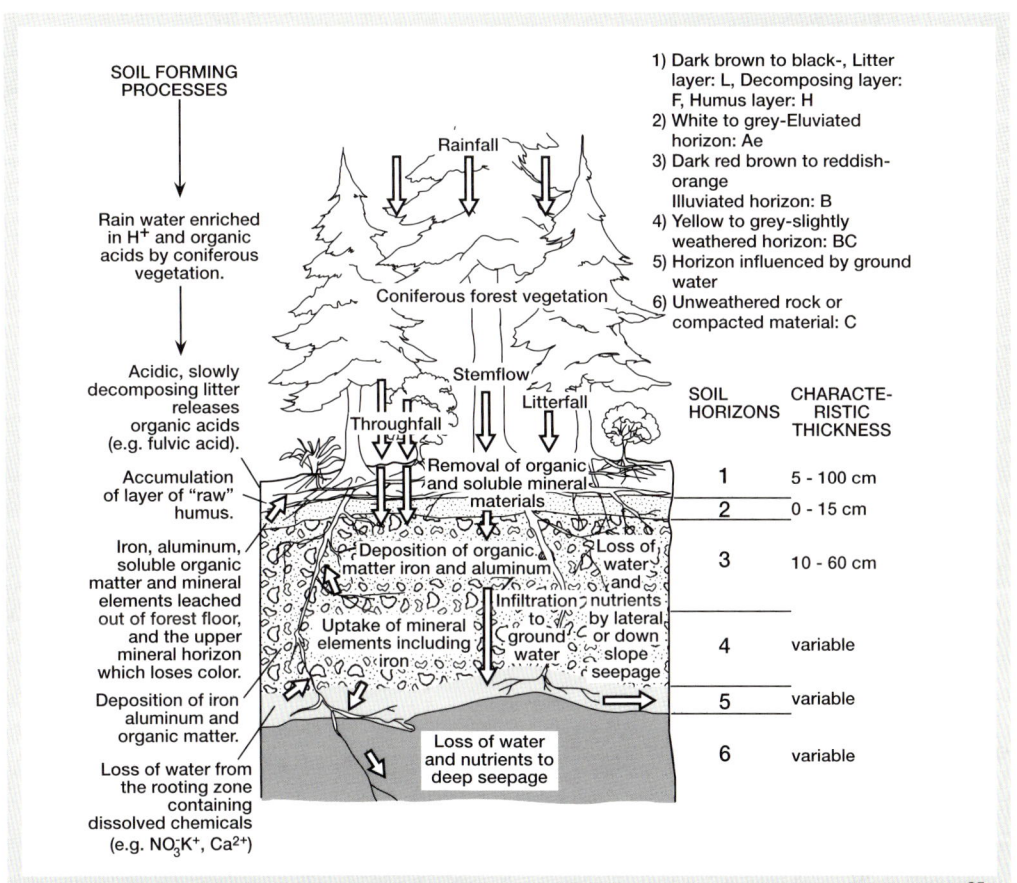

Figure 35. Characteristics of the major horizons and soil forming processes typical of podzolized soils[25].

11.2 Measures to manage nutrient availability

11.2.1 Site preparation

Site preparation is the term used for mechanical measures employed to control the competition between ground vegetation and tree seedlings, the availability of nutrients, the level of groundwater, and physical properties of the soil to enhance regeneration and growth processes. Table 10 provides a listing of the probable impacts of timber harvesting and site preparation on nitrogen cycling. Site preparation is like major disturbances with long-lasting effects. In site preparation, the upper layers of the mineral soil and the humus mechanically mix by ditching, scarifying, or screening. The result is that the amount of competing vegetation (post-harvest pioneer ground vegetation) is less, and more nutrients become available. This causes enhanced restocking and growth of the new tree generation. Planting is also less expensive and technically easier on prepared soils.

Table 10. Probable impacts of timber harvesting and site preparation on nitrogen cycling[26].

Practice	Effects	N-cycle consequences	Probable magnitude
Stem harvest	Remove nutrients, increase soil moisture and temperature, decrease net primary productivity	Decrease uptake Increase immobilization Increase mineralization Increase leaching losses Removal of N in biomass	Moderate Moderate Moderate Small Large
Whole-tree harvest	Remove nutrients, increase soil moisture and temperature, decrease net primary productivity	Decrease uptake Increase immobilization Increase mineralization Increase leaching Removal of N in biomass	Moderate Small Moderate Small Large
Chopping	Crush slash, kill some competing vegetation	Increase immobilization Decrease uptake	Small Small
Burning	Consume slash, kill some competing vegetation, blacken soil surface, reduce acidity	Volatilize N Decrease immobilization Increase mineralization Increase nitrification	Small to large Small Small to moderate Small
Root raking, blading, windrowing	Concentrates slash in rows, moves forest floor and topsoil into rows, exposes mineral soil, controls competition	Redistributes nutrients Decreases immobilization Decreases mineralization Decreases uptake Increases fire loss	Large Small Moderate Small Large
Disking, ploughing	Mixes forest floor and topsoil, reduces compaction, increases aeration, controls competition	Increases erosion Increases mineralization Increases nitrification Decreases uptake	Moderate Moderate Moderate Small
Bedding	Plows soil into raised rows, creating aerobic zone adjacent to anaerobic zone	Increases mineralization Increases nitrification Increases denitrification	Small to moderate Small to moderate Small
Herbicide	Inhibits competing vegetation	Decreases uptake	Small
Thinning	Reduces density, adds slash to the forest floor, increases soil moisture and temperature	Decreases uptake Increases immobilization Increases mineralization Removes N in biomass	Small Small Small Moderate

The impacts of site preparation relate closely to soil temperature. Since site preparation reduces the insulating litter and humus layers, it improves thermal conductivity and the penetration of heat into the soil profile as Fig. 36 indicates. An increase in soil temperature relates to the proportion of the soil surface not covered by organic matter such as litter or humus. Higher soil temperatures accelerate microbial activity and decomposition of the soil organic matter. This then provides more nitrogen in ammonium and nitrate form for plant growth. The availability of mineral nutrients such as

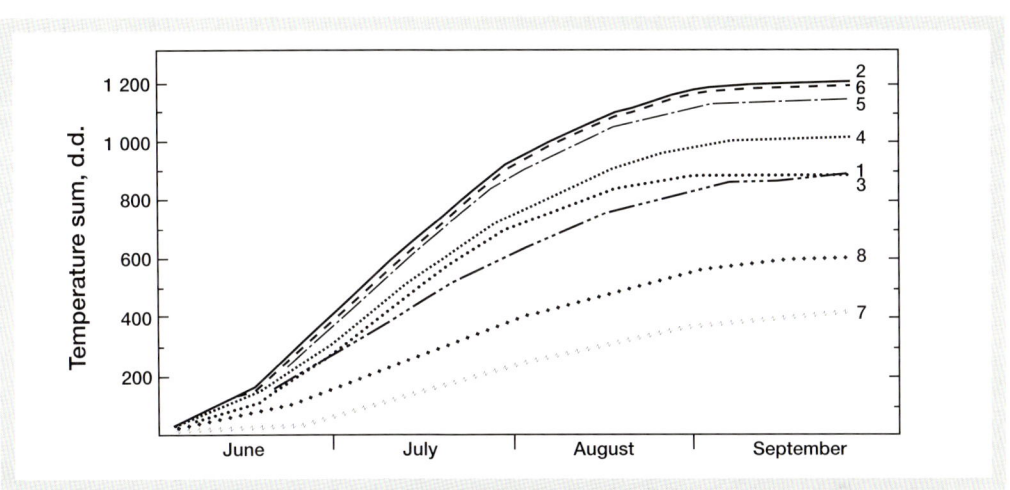

Figure 36. Effect of site preparation on temperature conditions in soil surface in terms of temperature sum, i.e., sum of values of daily mean temperature $\geq +5°C$ from June through September[27]. Explanation: 1=screefing, 2=ridge plowing, 3=shoulder ploughing, 4=gnawing, 5=mounding 40 cm, 6=mounding 50 cm, 7=untreated and unfelled area, 8=untreated open area.

potassium, magnesium, and calcium is also enhanced as part of the upper mineral layer mixes with the humus layer. This makes the parent soil more susceptible to weathering processes. In addition, the site preparation may decrease the soil moisture with a concurrent increase in soil aeration and enhanced root activity and nutrient uptake.

11.2.2 Ditching

Ditching is the measure of controlling water resources in forest soils when excess soil water inhibits tree growth especially on mires – sites covered by peat layer and peat-forming plants such as *Sphagnum* mosses and plant communities that flourish on waterlogged soils with poor aeration. Typically, groundwater table close to the surface is not as great an obstacle to tree growth as poor soil aeration. Many tree species tolerate occasional flooding. Only in extreme cases are mires treeless. Ditching to lower the groundwater table with enhancement of the root activity of trees leads to higher tree survival and growth as Fig. 37 shows. The decomposition of the peat is simultaneously enhanced, and more nutrients become available for plant growth. With time, ditching causes plant communities characteristic of mires to transform to communities typical of upland sites. From the forestry point of view, ditched mires and upland sites have equal worth. Ditching is therefore a means of expanding the area of productive forest land.

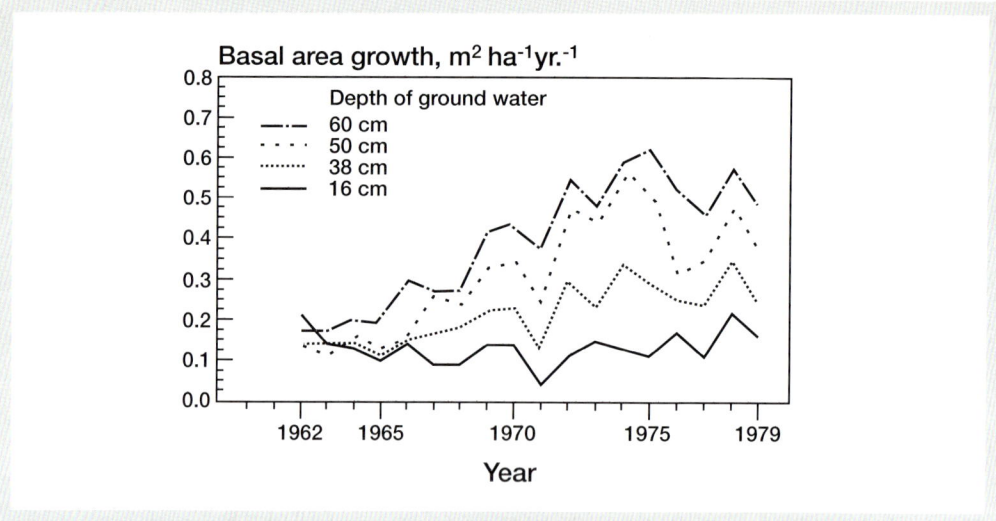

Figure 37. Impact of depth of groundwater table on the growth of Scots pine on a ditched site[28].

In Finland, more than 50% of the total area of 10 million ha of mires were ditched during 1950–1980 to increase the total timber production capacity. Tree growth and the stemwood inventory on these ditched peat land sites have shown substantial increases. In 1989–1994, the annual increment on peat land sites was 17.8 million m^3 (24% of Finland's total forest increment). Of this amount, 14.9 million m^3 occurred on ditched peat lands. In the early 1950s, the annual wood increment on peat land sites was 9.9 million m^3[329]. The impact of ditching on growth depends on site fertility (faster and greater changes have occurred on more fertile sites than on less fertile sites), the intensity of ditching (greater growth on sites with greater drop in groundwater table level than on sites with smaller drop), and on the properties of the tree populations affected (tree species composition and age and size distributions).

11.2.3 Prescribed burning

Prescribed burning is the term for the controlled use of fire to enhance regeneration and tree growth. It makes more nutrients available to the seedlings and reduces competition from communities of ground vegetation. Felling residues, ground vegetation, and part of the litter and humus layers are burned. Prescribed burning has normal use in clear-cut areas with a substantial amount of harvesting residues. Prescribed burning also applies on sites with closed canopies such as those in the southeastern United States to reduce the potential fuel load on the forest floor and risk of wild fires. Prescribed burning increases soil temperatures by decreasing the thickness of the layers of insulating litter and humus, by improving the thermal conductivity of the litter and humus layers, and by improving the absorption of long wave radiation and heat energy in the soil profile. These effects are long-lasting, and they penetrate deeper into the soil profile than similar effects achieved through mechanical site preparation as Fig. 38 shows.

Management of forest ecosystems

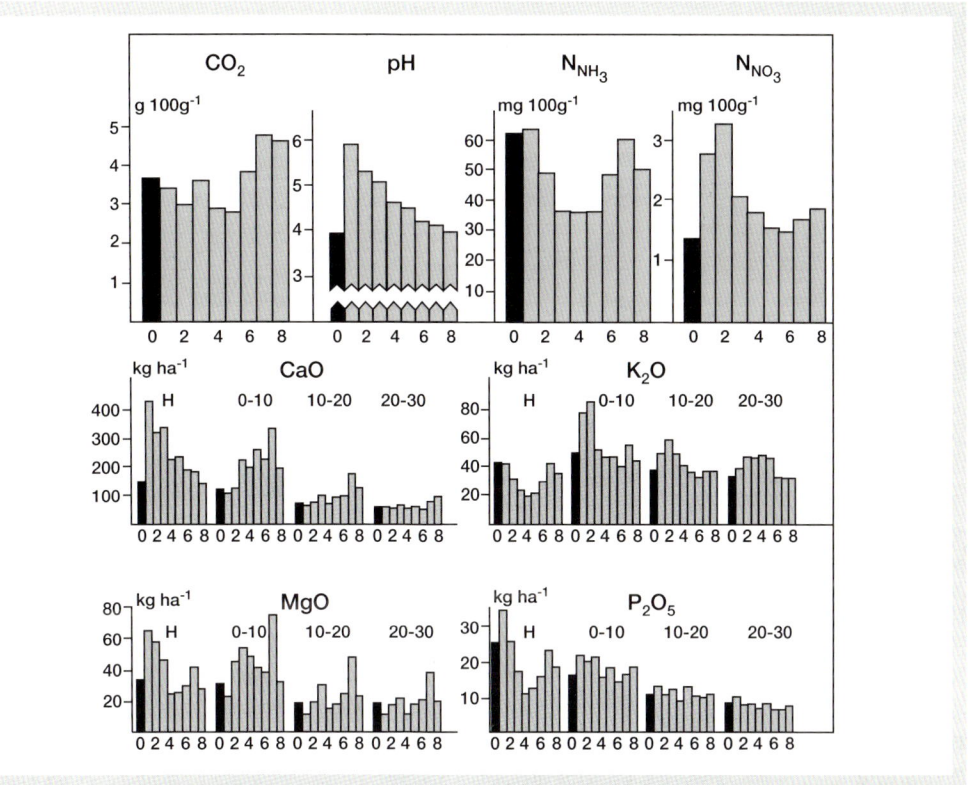

Figure 38. Impact of prescribed burning on the chemical properties of the humus and mineral soil on an upland site[30]. Legend: black bar=area without prescribed burning (before prescribed burning), 1=less than one year passed since prescribed burning, 2=2–3 yr. since prescribed burning, 3=4–6yr. since prescribed burning, 4=7–9 yr. since prescribed burning, 5= 10–12 yr. since prescribed burning, 6=13–19 yr. since prescribed burning, 7=20–32 yr. since prescribed burning, 8=33–50 yr. since prescribed burning, H=humus layer, 0–10=mineral soil layer 0–10 cm, 10–20=mineral soil layer 10–20 cm, and 20–30=mineral soil layer 20–30 cm.

Prescribed burning decreases the acidity of soil and therefore increases the availability of nitrate nitrogen. Some nitrogen also transforms into ammonium nitrogen. The impact of prescribed burning is long lasting on nitrogen and on the availability of mineral nutrients. Calcium and magnesium concentrations in the soil of burned sites can be higher for several decades compared with nonburned sites[30]. This phenomenon is primarily due to enhanced soil temperature and subsequent speeding of the decomposition of soil organic matter. Weathering of the mineral soil may also increase. Many positive factors of prescribed burning on the survival and growth of trees also relate to improved hygiene on forest sites. Many damaging agents die when temperature of the humus layer and upper soil layer exceeds 200°C–300°C during burning. Annosus root-rot (*Heterobasidion annosum*), for example, is less common on burned sites. On the other hand, some forms of damage may be more frequent on burned sites such as Rhizina fungus (*Rhizina inflata*).

CHAPTER 6

11.2.4 Thinnings

More than half the annual nutrient uptake in a forest stand returns to the forest floor as litter[26]. Thinnings can therefore substantially enhance the nutrient cycle. The nutrient ability reflects the quantity and quality of the litter. Since tree species composition affects the composition of foliage litter, the maintenance of an appropriate tree species composition can significantly influence the nutrient cycle. Favoring broadleaved tree species especially improves the within-stand nutrient cycle due to the rapid decomposition of the highly nutritious leaf litter as Table 11 shows. Nitrogen, calcium, potassium, and phosphorus concentrations in the foliage litter of birches (*Betula* sp.) are much higher and the decomposition rate of the litter is much faster than with Norway spruce foliage. In addition, the root system of birches reaches deeper into the soil than that of Norway spruce. This means that in pure birch stands or mixed stands of birches and Norway spruce the nutrient resources and nutrient turnover exceed those of pure stands of Norway spruce.

Table 11. Nutrient concentrations[31], loss of organic matter in leaf and needle litter[32], and depth of root system in stands of birch (stand age 84 yr.)[33], Scots pine (stand age 70 yr.), and Norway spruce (stand age 70 yr.)[34]. Data comes from different sources so results are not directly comparable.

Tree species	Nutrient conc., % of dry weight				Loss of organic matter in 2 yr., %	Depth of horizontal root system, cm
	N	P	K	Ca		
Birch	1.02	0.061	0.526	1.530	46.9	12.9
Scots pine	0.39	0.025	0.085	0.605	55.7	9.3
Norway spruce	0.53	0.067	0.125	1.321	33.3	7.4

Following spacing, thinning allows more solar radiation and precipitation to reach the forest floor. This means improved conditions for the decomposition of the soil organic matter and improved nutrient cycling. In boreal conditions, these changes favor the growth of herbs and grasses. Such growth has a similar impact on the nutrient cycle as does favoring broadleaved tree species. The decomposition of felling residues (foliage, branches, and roots) returns a substantial amount of nutrients to the nutrient cycle for use by the retention stand following a thinning treatment. Thinning can therefore regulate the availability of nutrients and maintain the productivity of forest ecosystems. On the other hand, the applicability of cuttings in nutrient management closely relates to the removal of nutrients with the tree biomass. Whole-tree harvesting in stands of white spruce (*Picea glauca*) can reduce the amount of available nitrogen resources by 70% when compared with conventional stemwood-only harvesting as Table 12 indicates. The impacts of clear-felling are more intense than with intermediate fellings. This introduces the risk that some nutrients made available can leach away from the site.

Table 12. Nutrient content of biomass (kg ha^{-1})[26].

Location	Vegetation	Age	Component	Biomass	N	P	K	Ca
Oregon, USA	Douglas-fir	450 yr.	Leaves	8 900	75	20	70	93
			Branches	48 500	50	10	50	243
			Stem	472 600	190	12	125	284
			Total	530 00	315	42	245	620
Ohio, USA	Poplar	4 yr.	Leaves	750	64	10	50	64
			Branches	1 900	38	4	27	40
			Stem	35 200	119	22	101	159
			Total	37 850	221	36	178	263
Minnesota, USA	Aspen	40 yr.	Leaves	3 600	87	9	47	37
			Branches	16 600	82	11	42	215
			Stem	146 600	200	26	200	600
			Total	166 800	369	46	289	852
	White spruce	39 yr.	Leaves	17 400	153	27	86	256
			Branches	35 800	131	18	83	224
			Stem	100 000	105	14	65	254
			Total	153 200	389	59	234	734
	Red pine	39 yr.	Leaves	13 800	131	19	59	42
			Branches	26 300	63	8	33	67
			Stem	167 000	162	14	88	193
			Total	207 100	356	41	180	302
	Jack pine	39 yr.	Leaves	5 500	65	8	20	20
			Branches	24 000	78	8	26	52
			Stem	121 700	121	10	53	131
			Total	151 200	264	26	99	203

11.2.5 Fertilization

Effects on growth: mechanisms

Fertilization is the measure of adding nutrients to the biogeochemical cycle. In boreal forests, the addition of nitrogen regularly increases tree growth on upland sites. This indicates that the natural cycle of nitrogen is too slow to satisfy the needs of trees. On peatland sites, the inadequate availability of phosphorus and potassium is another example of forest growth limitation from lack of nutrients.

Growth limitation due to nutrient availability relates to the developmental phase of trees, i.e., a close correlation exists between growth and the demand for nutrients (the formation of each unit mass of woody tissue needs a given amount of nutrients) with the concurrent culmination of growth and demand for nutrients. On the other hand, the response to the addition of fertilizers closely relate to the ambient supply of nutrients and water as Fig. 39 shows. The highest increase in growth occurs on nutrient-poor sites with a sufficient supply of soil moisture – sites of intermediate fertility. On the most fertile sites with nonlimiting nutrient supplies and the poorest sites with water as the limiting factor, growth increase may be modest.

CHAPTER 6

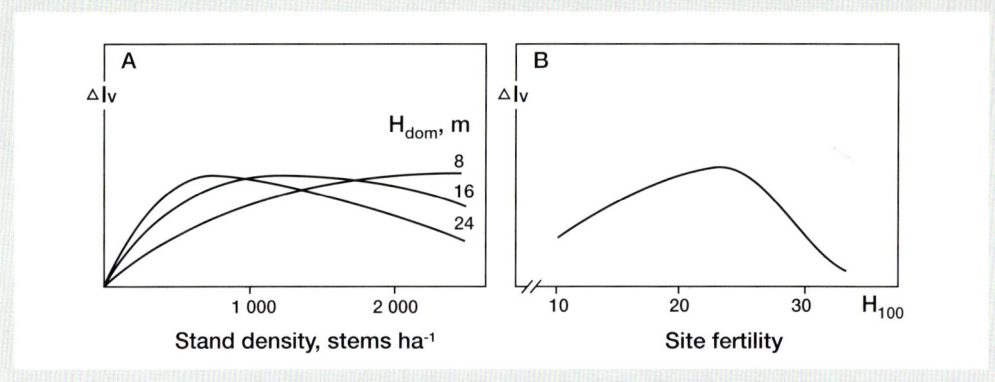

Figure 39. Change in growth (Iv) of Scots pine following nitrogen addition as a function of A: dominant height (H_{dom}) and stand density, and B: site fertility expressed as dominant height over age[35].

The impacts of nitrogen fertilizers on tree growth are relayed through physiological and ecological processes regulated by the biogeochemical nutrient cycle, energy fixation (photosynthesis), and the use of photosynthates for tree growth as Fig. 40 shows. Stem growth and growth of other tree parts improves, but a lag exists between nitrogen addition and the growth response as controlled by the increase of foliage under enhanced nitrogen availability. Furthermore, nitrogen addition enhances the litter fall and therefore the cycling of nutrients within the stand. Thus, the response of trees to nitrogen addition includes short-term physiological and long-term ecological responses. The former relate to the current nitrogen status of the trees, and the latter relate to the biogeochemical cycle of nitrogen.

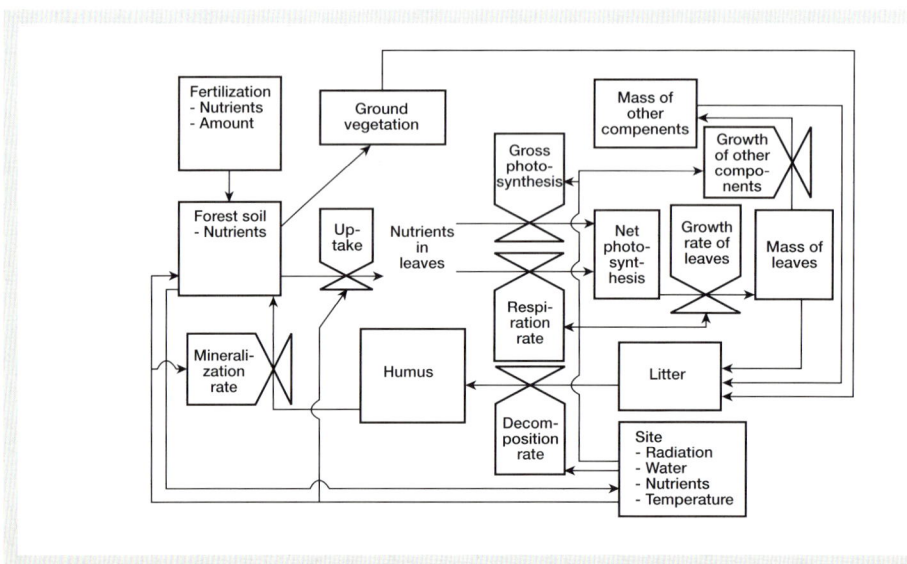

Figure 40. Schematic presentation of some primary effects of nitrogen addition on physiological and ecological processes controlling tree growth in forest ecosystems.

Management of forest ecosystems

Figure 41 shows that the photosynthetic response of Scots pine to nitrogen addition reaches its maximum when the availability of solar radiation and temperature do not limit photosynthesis. On the other hand, the photosynthetic response in stand-grown trees saturates or even decreases with the addition of nitrogen. This implies that excess nitrogen increases the foliage mass and thus reduces the availability of radiation to such an extent that growth decreases with increasing addition of nitrogen. Increased water loss is also probable through enhanced evapotranspiration, followed by the increasing foliage mass with a reduction in the response of Scots pine to further addition of nitrogen. The response of Scots pine to the nitrogen addition relates in several ways to the changes in the foliage mass caused by nitrogen addition.

The responses of trees to nitrogen addition are species-specific as Fig. 42 shows. Many broadleaved species such as poplars and birches need an ample supply

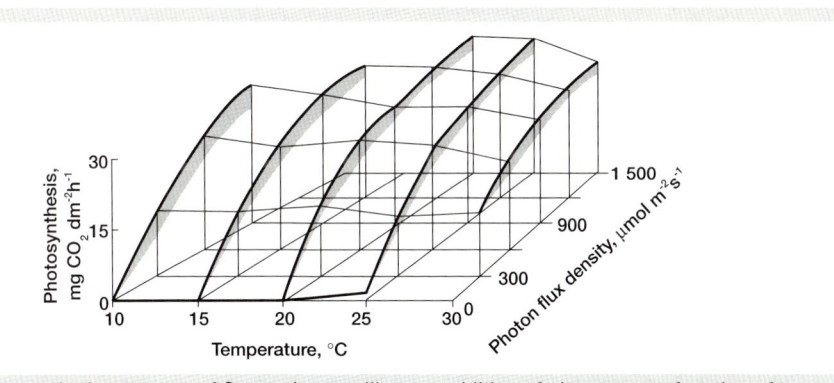

Figure 41. Photosynthetic response of Scots pine seedlings to addition of nitrogen as a function of temperature and photon flux under optimal water supply. Shaded area indicates increase in photosynthesis following nitrogen addition[36].

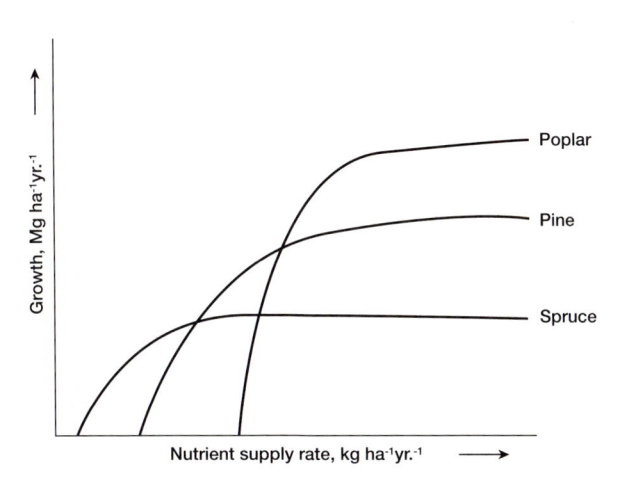

Figure 42. Schematic presentation of the response of different tree species to varying supplies of nutrients[26].

CHAPTER 6

of nitrogen to survive and grow. Consequently, the growth response may be significant following large doses of nitrogen. Many conifers such as spruces and pines flourish even with scarce supplies of nitrogen, although their response to large additions of nitrogen is modest. In forest management, these differences have prime importance for the return of the investment in fertilization and the future development of tree populations treated with nutrient additions. Pioneer tree species are often more responsive to nitrogen addition than climax tree species.

Effects on growth: dynamics of foliage

The differences in response among tree species usually relate to differences in the dynamics of tree foliage, i.e., differences in the renewal rate of the foliage. If the foliage mass compiled by four age classes is $B_i(t_j, t_{j+1})$ ($i = 1,2,3,4$) (typical for Scots pine) and with $N(t_j, t_{j+1})$ as the total amount of nitrogen bound in foliage, then $C(t_j, t_{j+1})$, the nitrogen content in the foliage, is as follows:

$$C(t_j, t_{j+1}) = \frac{N(t_j, t_{j+1})}{\sum_i B_i(t_j, t_{j+1})} \qquad (16)$$

Letting $B_0(t_j)$ indicate the growth of the foliage and $N(t_j)$ indicate the amount of nitrogen concurrently bound in the foliage, then

$$\Delta N(t_j) = p_1 \cdot \Delta B_0(t_j) \qquad (17)$$

where p_1 is the concentration of nitrogen in the foliage.

Assuming that only a part of the free nitrogen in the foliage can be bound in new foliage, i.e., $N(t_j) < N(t_{-j-1}, t_j)$, implies that growth of the foliage is limited to the size

$$\Delta B_0(t_j) = \leq \frac{1}{p_1} \cdot N(t_{t-1}, t_j) \qquad (18)$$

Assuming a linear relationship between growth and nitrogen content of the foliage, the growth of foliage is then

$$\Delta B_0(t_j) = \Delta p_3 \cdot C(t_{j-2}, t_{t-1}) \cdot B_0(t_{j-2}, t_{t-1}) \qquad (19)$$

with the implication that

$$\Delta B_0(t_j) = \min\left(p_3 \cdot C(t_{j-2}, t_{t-1}) \cdot B_0(t_{j-2}, t_{t-1}), \frac{1}{p_1} \cdot N(t_{j-1}, t_j)\right) \qquad (20)$$

The parameter p_3 indicates the growth response related to the change in the free nitrogen in the foliage. The total nitrogen in the foliage is the following

$$N(t_j, t_{j+1}) = N(t_{j-1}, t_1) + I(t_j) - p_1 \cdot B_0(t_j) \tag{21}$$

where $I(t_j)$ is the uptake of nitrogen into the foliage.

The foliage grown in the current year is still part of the crown the following year, i.e.,

$$B_{0i}(t_j, t_{j+1}) = \Delta B_0(t_j) \tag{22}$$

If p_2 is the fraction of foliage that dies each year excluding the needles in the current year, then

$$B_0(t_j, t_{j+1}) = B_{i-1}(t_{j-1}, t_j), \quad i = 1 \tag{23}$$

$$B_0(t_j, t_{j+1}) = (1 - p_2) \cdot B_{i-1}(t_{j-1}, t_j), \quad i = 2, 3 \tag{24}$$

When assuming a linear correlation to govern the relationship between the growth of stemwood ($P(t_j)$) and the amount of the foliage, then

$$P(t_j) = k \cdot \sum_i B_i(t_{j-1}, t_j) \tag{24}$$

where k is a parameter relating the growth of the stemwood to the mass of foliage.

Consequently, one can derive the effect of nitrogen addition on stemwood growth from the total nitrogen bound in the foliage of trees. This also implies that larger aboveground biomasses receive support from smaller amounts of roots due to improved nutrient supply[37].

Growth response of different tree stands

The response to nitrogen addition is maximum in dominant trees regardless of the dosage as Fig. 43 shows. This response pattern indicates that vigorous growth indicates a vigorous response to nitrogen addition. This means that the growth response to nitrogen addition increases with increasing site fertility apart from the most fertile sites. Similarly, very high responses will occur from trees in their state of maximal growth, i.e., growth response diminishes with increasing age and maturity. This trend accelerates with fertilization. Repeated addition of nitrogen will induce increasingly lower responses in growth. Nevertheless, growth response relates to the nitrogen added. Increasing the dosage can only partly compensate for the reduced response

CHAPTER 6

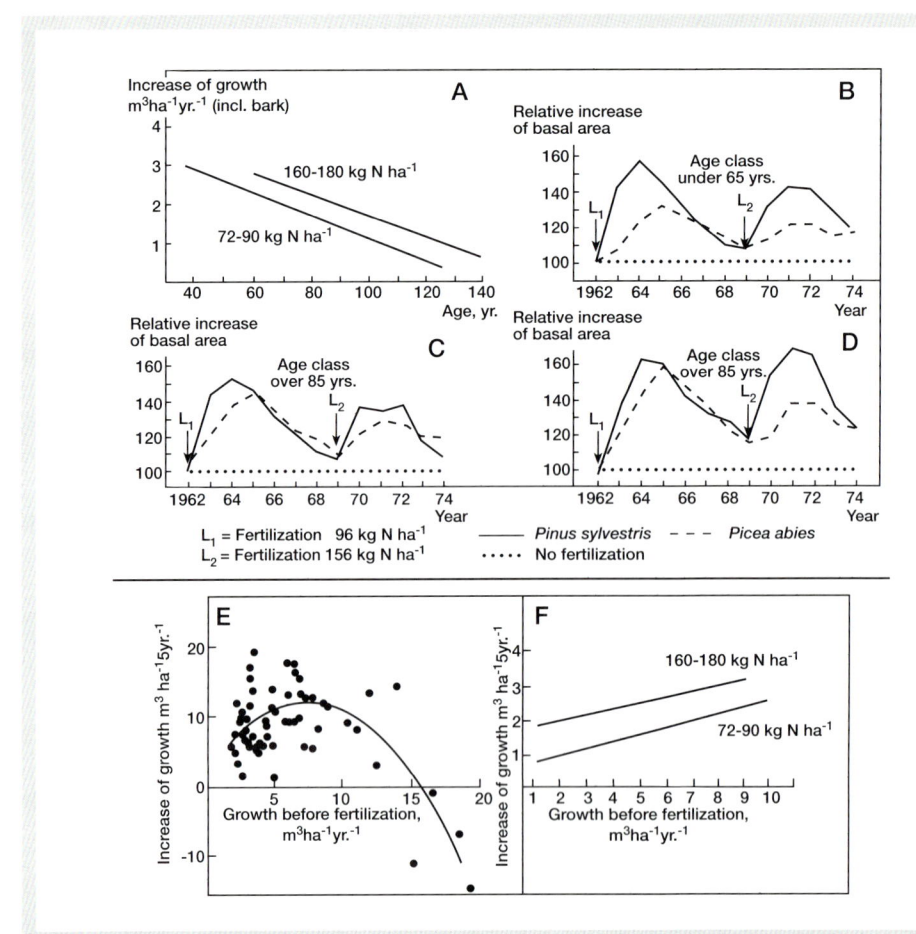

Figure 43. Above: Response of Scots pine to nitrogen addition as a function of stand age and amount of dosage (A). Response of Scots pine and Norway spruce to nitrogen addition as a function of repeated fertilizer dosage (L1 and L 2) and stand age (B–D). Below: Response of Scots pine to fertilizing as a function of growth before fertilizing (E), and growth before fertilizing and the amount of dosage (F) [19].

with time elapsed since nitrogen addition. Small doses of nitrogen applied at short intervals produce more growth than large doses of fertilizer given at longer intervals as Fig. 43 shows.

Large scale fertilization in forestry in Finland began in the early 1960s. During 1965–1990, more than 3.1 million ha were fertilized. This area is approximately 15% of the total forest land of Finland[38]. The annual increment of stemwood increased from 57 million m^3 to 77 million m^3 during this period. Some increase resulted from the effect of fertilizing nitrogen input equaling the total atmospheric nitrogen load over a single rotation[30]. Volume growth increased in Scots pine stands from 4.7 m^3 ha^{-1} yr.$^{-1}$ to 6.3 m^3 ha^{-1} yr.$^{-1}$. In Norway spruce stands, the volume increased from 8.9 m^3 ha^{-1} yr.$^{-1}$ to 9.7 m^3 ha^{-1} yr.$^{-1}$ due to nitrogen fertilizing.

11.3 Impacts of timber harvesting on nutrient resource of forest soils and their implications for forest management

Since tree biomass contains nutrients, the removal of tree biomass from the forest results in a loss of nutrients and may reduce productivity of the forest ecosystem. Nutrient loss relates directly to harvest intensity. This is the rate of harvesting the tree biomass and its various components of foliage, branches, stems, and roots and the stage at which stand development harvesting occurs as Table 13 shows. Nutrient loss increases with increasing proportion of the harvested biomass from that available. In younger stands, nutrient loss is higher than in older stands, since the biomass removed from younger stands includes more living tissue with high nutrient contents than that removed from older stands.

Table 13. Increase in nutrient loss when switching harvesting from stemwood-only harvesting to whole-tree harvesting of Norway spruce and Scots pine[25].

| Species | Age, yr. | Percentage increase in loss of ||||
		N	P	K	Ca
Norway spruce	18	195	233	161	206
	50[a]	114	115	26	40
	85[a]	91	104	42	29
Scots pine	18	188	212	171	129
	28	130	149	97	83
	33	172	150	102	69
	39	164	200	140	88
	44	124	133	108	84
	64	103	114	94	41
	75	77	67	56	59

[a] Average of two stands

Conventional harvesting removes only the stemwood at the rate of 80%–90% of that available. Only 65% the total tree biomass is removed. This means that the harvest residues remaining on the site include those components of tree biomass with the highest nutrient contents (foliage, branches, and roots). In *whole-tree harvesting,* the aboveground tree biomass is harvested so that a substantially greater amount of biomass is removed than with stemwood-only harvesting. In this case, more than 80% of total tree biomass is removed. In *full-tree harvesting*, only the fine roots will remain at the site after removal of approximately 95% of the total tree biomass.

Nutrient losses due to harvesting experience natural compensation if the interval between successive harvests is sufficiently long. The time required by this recovery substantially depends on harvesting intensity and the rate of nutrient replacement as Fig. 44 shows. In boreal forests, the application of conventional timber harvesting and long rotations probably allows sustainable timber harvesting even with slow nutrient inputs. This is contrary to the situation in short rotation forestry accompanied by

CHAPTER 6

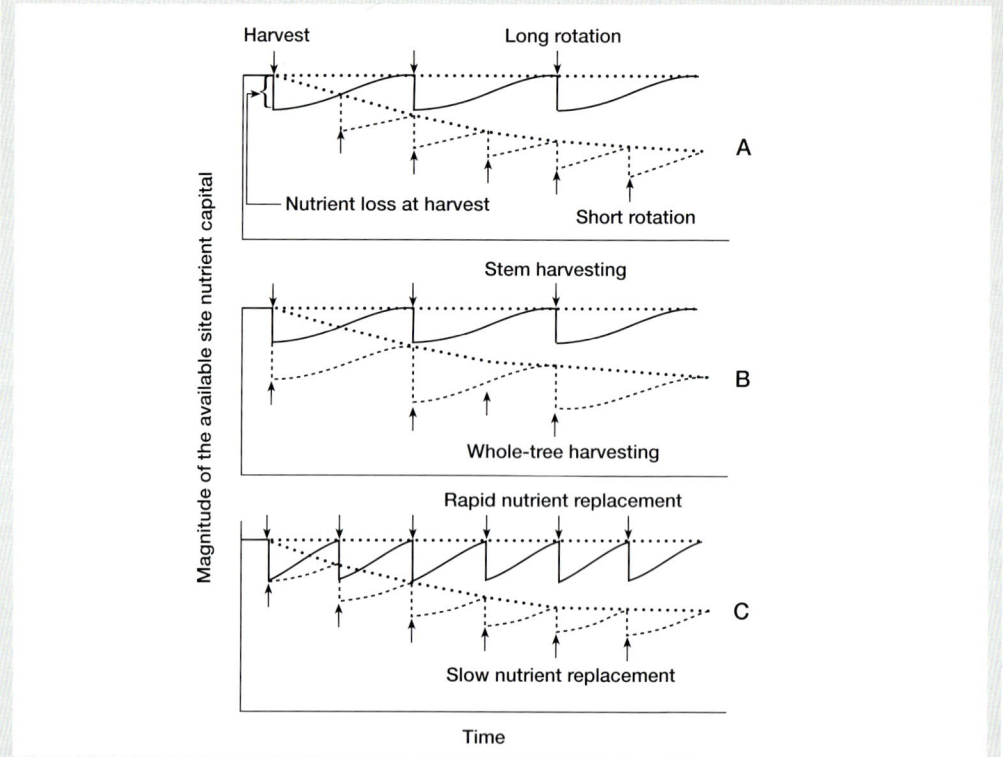

Figure 44. Schematic presentation of how A: rotation length, B: intensity of harvesting, and C: rate of nutrient replacement affect availability of nutrients on the site[25].

intensive harvesting that probably constitutes a detrimental combination unless rapid nutrient replacement occurs. Short-rotation forestry with its intensive biomass harvesting practices requires special care in nutrient management even with rapid nutrient replacement.

11.4 Nutrient management and sustainable forest production

Sustainable management of forest ecosystems means integrating the nutrient cycle, rotation length, and timber harvesting or other commodities to maximize timber production per unit of nutrients consumed by tree growth. Great variability in timber production can occur by combining these factors over the rotation. Applying the management scenarios outlined in Table 14 for Scots pine in the boreal conditions of Finland clearly demonstrate this[40].

The computations used a stand of Scots pine with an initial density of 4 000 seedlings (height 1.3 m) per ha on a site of intermediate fertility. Rotations of 80 and 120 yr. were repeated three times with regular thinnings during the rotation and clear cutting at the end of each rotation. The first four scenarios were exceptions in this respect since they represent natural succession in management without thinnings. The harvesting regime applied was conventional harvesting or whole-tree harvesting. When fertilized, a nitrogen dosage of 150 kg ha^{-1} occurred in years 41 and 61 after the beginning of simulation.

Table 14. Scenarios for management of Scots pine used to demonstrate how management can influence productivity of Scots pine[40].[1)]

No.[1)]	Rotation length, yr.	Length of simulation, yr. (3x rotation length)	Harvesting method	Time, yr. and intensity, % of harvesting [2)]	Time, yr. when fertilized since start of simulation
1	120	360	Cut-to-length logging system		
2	120	360	Whole tree harvesting		
3	80	240	Cut-to-length logging system		
4	80	240	Whole tree harvesting		
5	120	360	Cut-to-length logging system	40–60–80 0.4–0.35–500	
6	120	360	Whole tree harvesting	40–60–80 0.4–0.35–500	
7	80	240	Cut-to-length logging system	30–40–60 0.4–0.35–500	
8	80	240	Whole tree harvesting	30–40–60 0.4–0.35–500	
9	120	360	Cut-to-length logging system	40–60–80 0.4–0.35–500	61–81
10	120	360	Whole tree harvesting	40–60–80 0.4–0.35–500	61–81
11	80	240	Cut-to-length logging system	30–40–60 0.4–0.35–500	41–61
12	80	240	Whole tree harvesting	30–40–60 0.4–0.35–500	41–61

Legend: 1) = scenario, 2) = decimal numbers for intensity of harvesting refer to the share of stem volume removed in thinning and integers to the number of trees retained per hectare in the last thinning.

Longer rotations with regular thinnings and fertilization treatments usually give the highest *total growth* but only slightly more than stands grown without management as Table 15 shows. In managed stands, more than 50% of the total production is harvested as *timber*, but in natural stands this value is less than 30%. Under longer rotations, regular thinnings with fertilization treatment double the timber yield compared with stands grown without management. The differences between stands in terms of total growth are small. This indicates that management may only increase temporarily the productivity of a forest ecosystem, although the timber yield may increase substantially.

CHAPTER 6

Table 15. Simulation results on total growth and timber yield in a stand of Scots pine subjected to different management regimes as described in Table 14.

Scenario	Yield Total, Mg ha^{-1}	Yield Mean, Mg ha^{-1}yr.$^{-1}$	Litter Total, Mg ha^{-1}	Litter Mean, Mg ha^{-1}yr.$^{-1}$	Harvesting Total, Mg ha^{-1}	Harvesting Mean, Mg ha^{-1}yr.$^{-1}$	Share of litter, % of yield
1	1 170	9.8	836	7.0	194	1.62	71
2	1 170	9.8	836	7.0	257	2.15	71
3	813	10.2	541	6.7	167	2.08	66
4	809	10.1	537	6.7	222	2.78	66
5	1 109	9.2	617	5.1	296	2.47	57
6	1 036	8.6	551	4.6	401	3.34	53
7	774	9.6	327	4.1	274	3.43	42
8	657	8.2	282	3.6	322	4.03	43
9	1 290	10.7	756	6.3	312	2.60	59
10	1 219	10.2	694	5.8	429	3.58	57
11	906	11.3	403	5.0	304	3.80	44
12	848	10.6	529	6.6	412	5.15	44

Productivity per unit of nitrogen is 360–400 kg of dry matter produced over the rotation per kg of nitrogen as Table 16 shows. The highest values represent natural stands, and the smallest values indicate the most intensive management. This means that the nitrogen cycle in managed stands is more open with higher loss of nitrogen beyond the ecosystem than in natural stands. These differences closely relate to the amount of nitrogen supplied by the within-tree translocation and the nitrogen cycle through litter fall and decomposition of the soil organic matter. Productivity per unit nitrogen is high whenever the within-tree supply and the supply of nitrogen from the biogeochemical cycle are effective.

Table 16. Consumption of nitrogen in Scots pine stands when applying management scenarios as described in Table 14.

Scenario	Uptake Total, kg ha^{-1}	Uptake Mean, kg ha^{-1}yr.$^{-1}$	Litter return Total, kg ha^{-1}	Litter return Mean, kg ha^{-1}yr.$^{-1}$	Internal cycle Total, kg ha^{-1}	Internal cycle Mean, kg ha^{-1}yr.$^{-1}$	Kg yield per kg N
1	2 970	24.8	2 573	21.4	2 845	23.7	394
2	2 970	24.8	2 573	21.4	2 845	23.7	394
3	2 030	25.4	1 664	20.8	1 852	23.2	400
4	2 035	25.5	1 664	20.8	1 839	23.0	398
5	2 848	23.7	2 183	18.2	2 737	22.8	389
6	2 717	22.6	2 055	17.2	2 457	20.5	381
7	1 984	24.8	1 275	15.9	1 822	22.8	388
8	1 829	22.9	1 182	14.8	1 562	19.5	360
9	3 309	27.6	2 574	21.5	3 450	28.8	390
10	3 123	26.0	2 411	20.1	3 231	26.9	390
11	2 323	29.0	1 504	18.8	2 230	27.9	390
12	2 165	27.1	1 396	17.5	2 077	26.0	392

The balance of nitrogen over the rotation is negative even in the natural stand. This indicates that the harvest of timber could decrease the long-term availability of nitrogen in the forest ecosystem indicated in Table 17. The only exception is the scenario representing a longer rotation and stemwood-only harvesting with regular thinnings and fertilization treatments. The simulation results indicate that nutrient management deserves special attention whenever managing and using forests regularly. Even the loss of nitrogen in timber can substantially reduce the nitrogen resources available to future forest growth. Nutrient loss via the hydrological cycle further increases the loss of nitrogen from managed forests. In summary, it appears that the timber yield increases when using thinning by applying stemwood-only harvesting and fertilization on a regular basis over long rotation. This enables efficient use of nitrogen with no loss beyond the forest ecosystem.

Table 17. Nitrogen balance in Scots pine stands applying management scenarios as described in Table 14.

Scenario	Nitrogen accumulation Total, kg ha^{-1}	Nitrogen accumulation Mean, kg ha^{-1} yr.$^{-1}$	Leaching kg ha^{-1}	Leaching kg ha^{-1} yr.$^{-1}$	Harvesting kg ha^{-1}	Harvesting kg ha^{-1} yr.$^{-1}$	Balance Total, kg ha^{-1}	Balance Mean, kg ha^{-1} yr.$^{-1}$
1	372	3.1	300	2.5	41	1.2	-69	-0.6
2	372	3.1	300	2.5	53	3.0	-282	-2.3
3	249	3.1	298	3.7	130	1.6	-179	-2.2
4	249	3.1	293	3.7	330	4.1	-374	-4.7
5	372	3.1	318	2.7	223	1.9	-169	-1.4
6	372	3.1	320	2.7	595	5.0	-544	-4.5
7	249	3.1	162	2.0	224	2.8	-137	-1.7
8	249	3.1	125	1.6	413	5.2	-291	-3.6
9	643	5.4	325	2.7	232	1.9	+86	+0.7
10	643	5.4	314	2.6	638	5.3	-309	-2.6
11	520	6.5	322	4.0	246	3.1	-47	-0.6
12	520	6.5	319	4.0	699	8.7	-498	-4.2

CHAPTER 6

12 Ecology and management of forest pests

12.1 Introduction

The purpose of this section is to present a concept of *plant-herbivore-systems* and introduce the reader to some ecological concepts that are particularly useful in suggesting ways to develop efficient *pest-management* procedures. The main emphasis is on boreal forests. These are economically the world's most important forest zones. Traditionally, forest pest populations are controlled "from above" by using *natural enemies* like *predators*, *parasites*, and *pathogens*. Another management possibility to control pest populations is "from below" by increasing the *resistance of trees* against pests. This can use different methods like *natural resistance mechanisms, resistant clones*, etc. The theory of plant-herbivore coevolution has developed rapidly since the 1970s[41]. This rapid expansion is the reason why the main interest in forest pest management research in recent years has been "from below" as Fig. 45 shows. Consequently, the first part of this section deals with the factors influencing *herbivorous species abundance* in trees and the resistance mechanisms of trees against pests and pathogens. The second part covers biological and chemical control of forest pests. The final section discusses the ecological and applied aspects of forest pest management.

12.2 Forest pest concept

Current estimates suggest that *phytophagous insects* comprise approximately 25% (or at least 361 000 species) of all living species[42]. In addition, several mammalian orders are phytophagous – lagomorphs, rodents and ruminants, although here the species abundance is lower (2 300 species). The proportion of mainly phytophagous insects varies from order to order. Lepidopterans, aphids, and grasshoppers are almost entirely phytophagous, but only approximately 30% of beetles feed on living plant tissue. Figure 46 shows the amounts of phytophagous insects.

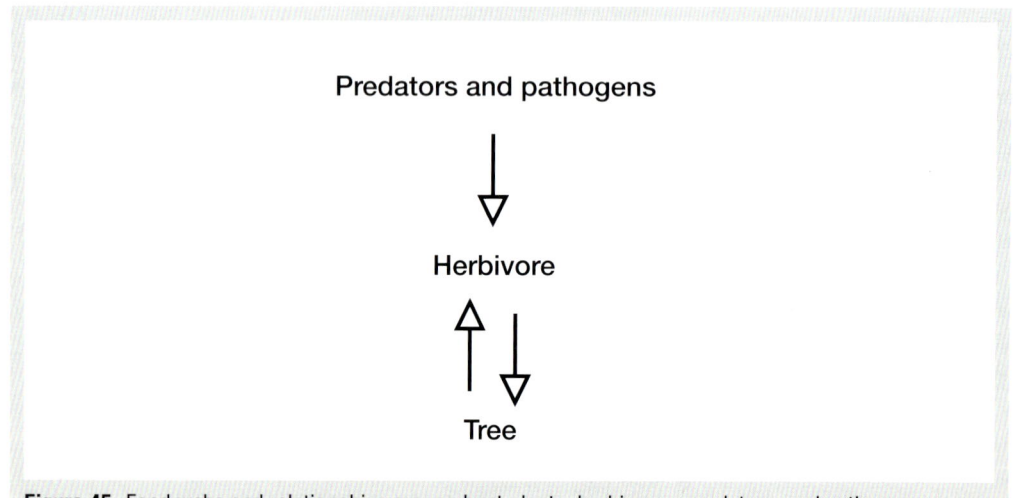

Figure 45. Food webs and relationships among host plants, herbivores, predators, and pathogens.

Management of forest ecosystems

Herbivores exploit their hosts in many ways. Some feed externally by biting and chewing, others exist by sucking the contents of individual cells or the vascular system of a plant. Some species mine into their host while others form galls. The number of *host plants* forming the diet of insects varies considerably. Some species specialize in feeding on one host plant species or genus only. Such species are *monophagous*.

Oligohagous species have a few host plants included in their diet. *Polyphagous species* are generalized feeders that usually have a broad diet consisting of several host plant species or genera. Sawflies and aphids tend to be monophagous or oligophagous. Figure 47 shows that many beetles and lepidopterans are highly polyphagous and sawflies and aphids highly monophagous. All insect groups include major forest pests such as the strictly monophagous diprionid sawflies (*Neodiprion* species, *Diprion* species) and the extremely polyphagous lepidopterans such as the gypsy moth.

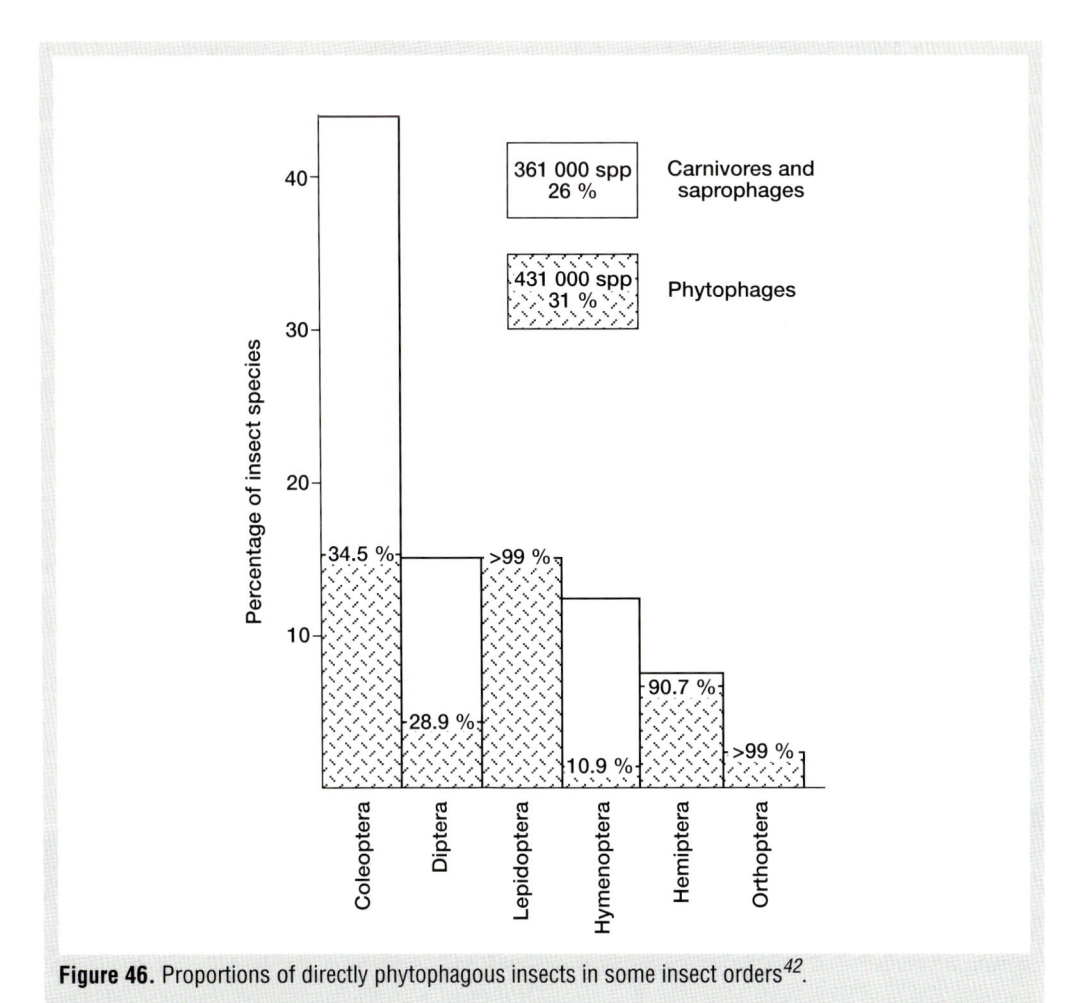

Figure 46. Proportions of directly phytophagous insects in some insect orders[42].

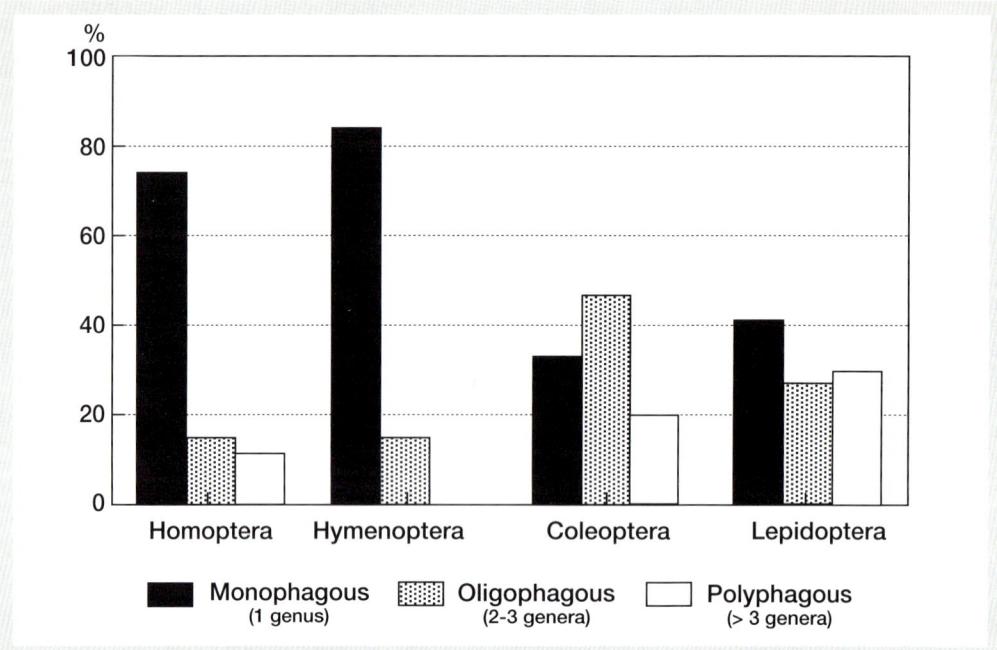

Figure 47. Proportion of mono-, oligo- and polyphagous insect species in four orders of North American immigrant insect fauna[42].

12.3 Characteristics of pest outbreaks

Most *tree-feeding herbivores* cause little damage to growing trees and are therefore not pests. Some mammalian or insect herbivores can cause significant injury and growth loss. They can even kill trees and cause changes in the forest ecosystems if their population densities become high. *Outbreaks* of different kinds of pests vary in their frequency, intensity, duration, and area affected. This has important implications for their management and control[43–45]. Classification of outbreaks falls into the three main categories indicated in Fig. 48: *gradients*, *cycles,* and *eruptions*[43].

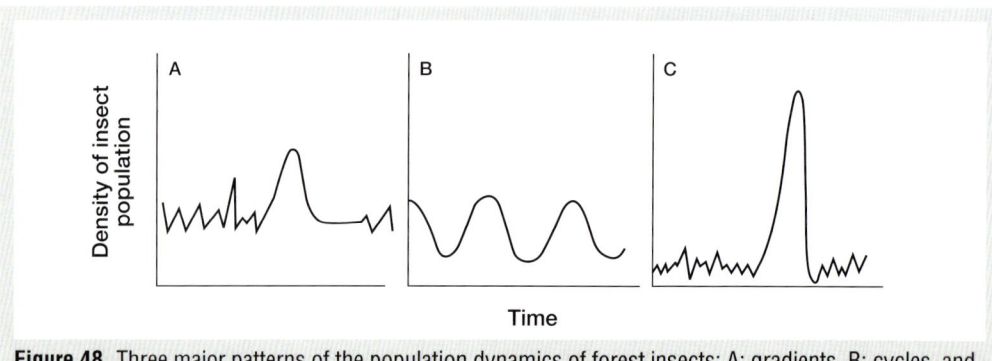

Figure 48. Three major patterns of the population dynamics of forest insects: A: gradients, B: cycles, and C: eruptions[43].

Management of forest ecosystems

Gradients typically do not spread into surroundings of less favorable areas. Gradients occur with those forest herbivore whose abundance depends largely on local environmental factors such as climatic conditions connected with topography, the number of susceptible trees, or the amount of available breeding material that can rapidly increase following forest disturbance. Typical disturbance factors are forest fires, windthrows, or clearcuts. Several species of bark beetles that rely on charred woody material or stumps in clearcut areas demonstrate this gradient pattern (*Hylobius abietis, Tomicus piniperda, T. minor*).

Cyclic outbreaks have captivated ecologists for decades. Cycles occur at regular intervals such as every 8–11 yr. (invertebrates, lagomorphs) or 4–5 yr. intervals (rodents). They often occur with particular site and stand conditions and are frequently the result of host-defensive responses or natural enemies. Good examples of cyclic herbivores are the autumnal moth in Fennoscandian birch forests and rodents (*Microtus* spp., *Clethrionomus* spp.) in the same region. *Eruptive pests* often remain at low population densities for long periods, but outbreaks can develop abruptly. Eruptive outbreaks can spread from local epicenters to cover large areas of forests. They usually last for several years. Examples of eruptive species are diprionid sawflies (*Neodiprion sertifer*) and some bark beetles (*Ips typographus, Dendroctonus* species).

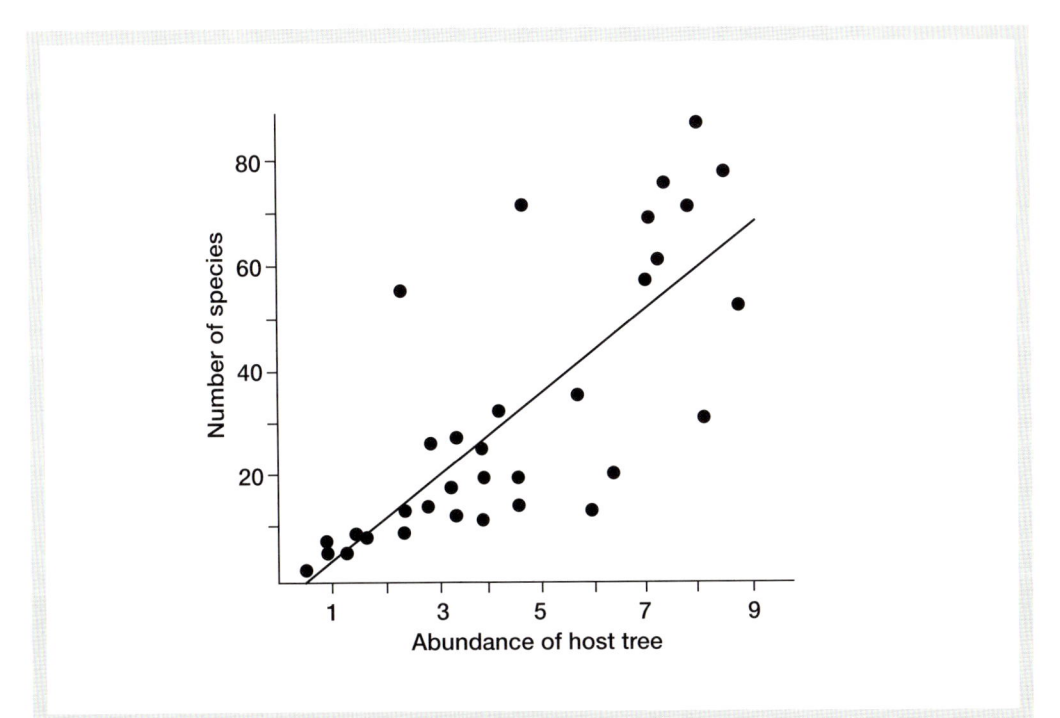

Figure 49. Correlation between number of macrolepidopteran species in Finland and abundance of host trees[47].

12.4 Herbivorous species attacking trees

12.4.1 Tree-distribution related herbivore abundance

Some tree species are afflicted by a large number of herbivores, and some are afflicted by only a few. Several studies have shown that abundance, evolutionary history, and defense tactics of the plants influence the species abundance of herbivores. A positive correlation often exists between species abundance of herbivorous insects and the range, abundance, or both of the host tree species. Plant characteristics such as size, structural complexity, time of occurrence in a particular area, number of habitats occupied, and degree of taxonomic isolation also explain the variation in herbivorous species diversity on trees.

The species abundance of herbivores feeding on trees and shrubs is an increasing function of host plant range and abundance. Widespread tree or shrub species have many associated phytophagous insect species. This is true for lepidopterans, sawflies, and several other herbivore groups as Fig. 49 shows[46]. In most studies, distribution range, abundance, or both have explained 20%–90% of the variation in herbivorous species abundance[46].

12.4.2 Host plant size and growth form

The size of the host plant provides a significant explanation for the abundance of herbivorous species. When comparing this abundance with plants of different growth forms, the following sequence usually occurs[42]: trees > woody shrubs > perennial herbs > weeds and other annuals. At least two factors explain the observed higher herbivorous species abundance afflicting larger plants. Larger plants simply offer more surface area for herbivores. The second factor is that larger plants posses greater architectural diversity that allows greater niche segregation along the tissue-type axis[48]. For example, the larger Finnish tree species are afflicted by more herbivorous species that smaller tree species[47].

12.4.3 Taxonomic affinities of host plant

One set of factors affecting herbivorous species abundance targeting a tree is the degree of its relation to other plants. This is because closely related plants have physical and biochemical similarities that increase the probability of successful colonization by herbivores. This is especially true if these herbivores are not strictly monophagous. Consequently, biochemically isolated trees should have herbivorous species abundances that are lower as the result of reduced immigration rates. Another hypothesis suggests that adaptive radiation of herbivores within a host taxon relates positively with the diversity of hosts[49]. Herbivorous species abundance per host species should increase with greater species abundance of related plants.

Both factors appear valid. Taxonomically isolated species appear to be afflicted by only a few herbivores as the discussion below shows for many introduced tree species. For tree species indigenous to Finland, the number of plant species in the host plant taxon explained 84% of the species abundance of sawflies – extremely special-

Management of forest ecosystems

ized feeders[50]. The relative importance of *tree abundance* and *taxonomic isolation* obviously varies among different herbivore groups. In *Lepidoptera* that are typically polyphagous feeders for example, the factors involved in host plant abundance appear more important. For sawflies that are specialized feeders, the taxonomic diversity of the host taxon is the most important factor affecting species abundance as Fig. 50 shows[51].

12.4.4 Insect recruitment by introduced plants

Plants transferred accidentally or intentionally to other continents or remote areas provide interesting large-scale experiments for the study of insect community dynamics. In many cases, the introduction of tree species to new areas or continents has created long-lasting and large-scale forest management problems.

If an introduced tree species possesses a powerful chemical defense mechanism or is taxonomically (biochemically) isolated from the native flora, it is very probable that type B herbivorous insect recruitment will occur, i.e., recruitment takes place very slowly as Fig. 51 shows. For example, *Eucalyptus* from Australia when introduced to Africa and to North and South America has hardly been colonized by any local polyphagous insects since their introduction a hundred or more years ago[42]. Very slow insect recruitment is also evident in broadleaves transferred from Great Britain to South Africa and vice versa. The insect fauna feeding on introduced species is conspicuously smaller compared with the situation with native trees in both areas[52].

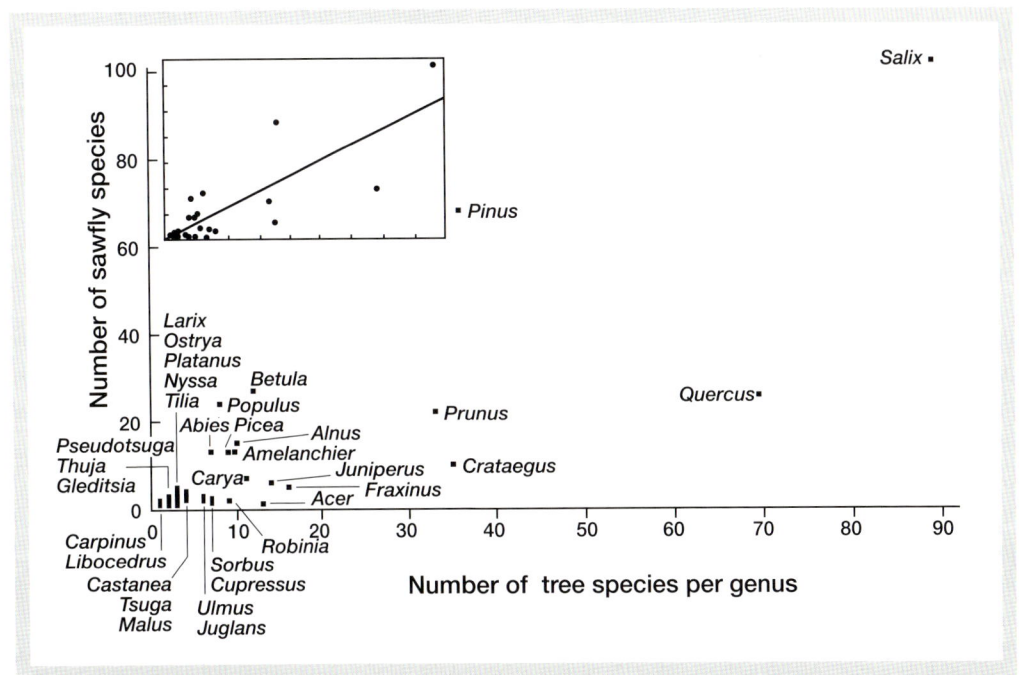

Figure 50. Correlation between the number of North-American sawfly species and the number of tree species in the host plant taxon[51].

Pines introduced to Australia have only a few herbivorous insects, but introduced pine or spruce species in North America or Europe recruit high numbers of herbivorous species within a short time following type A recruitment (*Pinus contorta* and *Picea sitchensis* in Europe) as Fig. 51 shows. Connor et al.[53] clearly demonstrated the role of taxonomic isolation in the rate of insect accrual on introduced plants. The results showed that those introduced *Fagaceous* trees having related species in the locality of introduction accrued a nearly normal number of leaf miners within the one-hundred years since introduction. In contrast, only a few species of leaf miners were recruited by taxonomically more remote tree species.

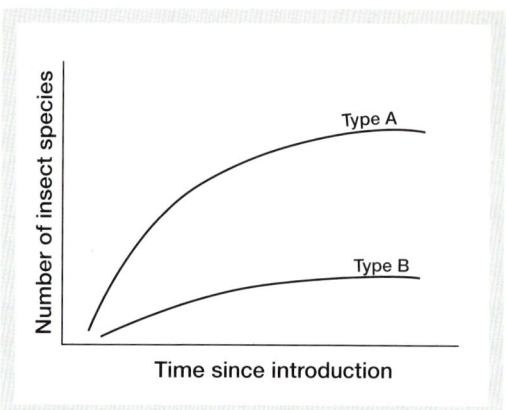

Figure 51. Two extreme types of insect recruitment curves produced by introduced plants. Poorly or generally defended plants approaching type A recruitment dominated with a rapid ecological phase of accrual. Powerfully defended or taxonomically (chemically) remote plants become colonized mainly through evolutionary processes resulting in very slow recruitment type B[46].

12.4.5 Role of introduced herbivores in forest ecosystems

Of the nearly 400 *immigrant insect species* that live on trees and shrubs in North America, 75% are from Europe[54]. Most insects of European origin have become serious forest pests in their new environment, e.g., gypsy moth, winter moth, and European pine sawfly. Kim and McPheron[55] calculated that 40% of the major North American insect pests have exotic origin. Alien, tree-feeding sawflies in North America are more prone to cause outbreaks than are their native brethren[51]. Issues concerning the management of these introduced insect pests have current interest because of the past severe and ever-spreading impacts of introduced insect pests on North American forest ecosystems.

Earlier discussion showed that the total number of phytophagous species attacking a single species of tree increases with the size of the *geographic range* of the plant and the abundance within that range. Additional intrinsic plant characteristics such as size, structural complexity, and biochemical (taxonomic) isolation explain much variation in insect species diversity per tree species.

If one turns this system of knowledge around and examines it from the viewpoint of invading herbivorous insects, the probability of being a successful colonizer is then a function of these same factors: size of the areal distributions of all potential host plants (primarily congeneric and secondarily confamiliar species); the average abundance in their areas of distribution; their taxonomic, biochemical, and ecological similarities with their native hosts; and the invading insect's initiation and concluding feeding on the new host at exactly the right time – host synchrony[56].

12.5 Resistance mechanism of trees against herbivores and pathogens

12.5.1 Constitutive resistance

Secondary metabolites

Herbivores trigger selection pressures in their hosts to evolve defenses or resistance to avoid herbivory. *Plant defense mechanisms* against herbivores (both insects and mammals) fall broadly into one of three categories:

- Physical or mechanical defenses (hooks, spines, tough leaves, etc.).
- Constitutive defense meaning that plants contain noxious phytochemicals that are unpalatable or poisonous or that interfere with insects' digestive systems. These are *defensive allelochemicals*. Low levels of available nitrogen or low water content are also cases of constitutive defense.
- Induced defense is a defense option in resistance activated only when plant tissue is wounded and allelochemicals are produced in the wounded tissues or triggered in other parts of the tree.

Plant *phenolics* are deterrants against herbivores due to their complex formation with proteins. They also have antibiotic properties[57]. Flavonoids play an important role as internal filters by absorbing and reducing the ultraviolet flux reaching certain plant tissues.

Phenolics are very common in broadleaved tree species. Concentrations of phenolics and *tannins* in the leaves may reach levels up to 20% of the dry weight[58]. Phenolics inhibit the growth of wood-decomposing fungi in conifers[59].

Conifer *oleoresin* also simply called resin is a viscous and sticky liquid produced by specialized secretion tissues in the foliage, wood, and roots of a tree. The anatomic organization of secretory structures in conifers varies from simple types containing only resin cells to more complex forms containing highly-evolved resin duct systems. Of the numerous conifer genera, pine *(Pinus)* and spruce *(Picea)* contain the most sophisticated resin-producing structures. Firs *(Abies)*, for example, produce simple resin blisters or cysts[60]. Oleoresin substances give the characteristic aromatic odor to conifers.

Oleoresin primarily contains *monoterpenes, sesquiterpenes,* and *diterpenes*. These are all terpenoids. For most species of pine and spruce, oleoresin primarily contains diterpenes (resin acids) with minor amounts of monoterpenes and sesquiterpenes[61]. Mono- and sesquiterpenes are volatile, but diterpenes are usually not. Without monoterpenes as solvents of oleoresin, resin acids would not flow to or from a wounded spot in a tree.

The relative composition of the individual compounds of oleoresin varies in different conifer species and plant tissues and even between individual trees of the same species[62]. Some monoterpenes and resin acids depend strongly on heredity. This provides a good base for resistance-breeding programs.

CHAPTER 6

Terpenoids act as feeding deterrents to many monophagous and polyphagous insects[63, 64] and mammalian herbivores[65]. Terpenoids provide important chemical resistance against herbivores especially for conifers. For example, the larvae of many Neodiprion sawfly species do not feed on current-year needles but target the older-year classes of needles. The explanation for this feeding behavior is probably due to the diterpenoid podocarpenoid acid that is toxic to the larvae. Current-year needles have high concentrations of this deterrent compound, but it is almost entirely absent from older-year classes of needles[66, 67].

Structurally, *alkaloids* are a very diverse group of compounds containing nitrogen as part of a heterocyclic ring. Alkaloids form as products of amino acid metabolism[68]. Alkaloids are common in herbaceous plants, and many extremely toxic compounds found in plants belong to alkaloids – nicotine, cofeine, morfine, strychnine, belladonna, etc. The alkaloids of conifers have not been throughly researched. *Picea abies* and *Pinus sylvestris* contain alkaloids at levels of approximately 0.03%–0.08% of their fresh weight. The variation depends on species and part of the tree[69].

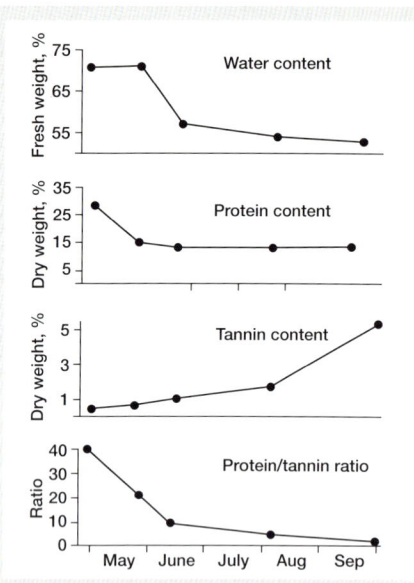

Figure 52. Seasonal variation of water content, protein content, and protein/tannin ratio in oak leaves[70].

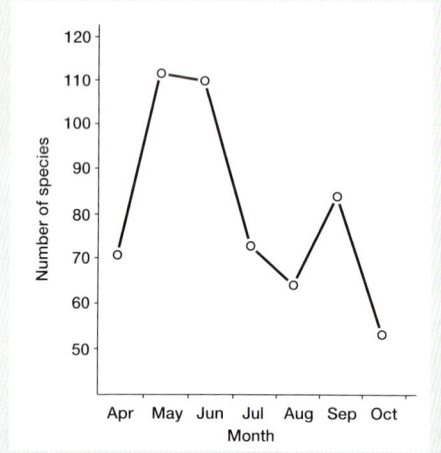

Figure 53. Number of lepidoptrean species feeding as larvae on oak leaves in Britain[70].

Seasonal variation of constitutive resistance

Figure 52 shows the decline in seasonal leaf quality of oak. In most broadleaved trees studied so far, nitrogen and water content are high in young leaves and then decline as the season progresses. Tannin and phenol levels increase with increasing leaf age. Figure 53 shows that many herbivore species feeding on broadleaved trees have timed their feeding period to occur early in the season when leaf quality is optimum – high concentrations of nutrients and low concentration of digestibility-reducing secondary metabolites. Consequently, most outbreaks of species attacking broadleaves occur early in the season. Excellent examples of this are winter moth (*Operotera brumata*) feeding on oak and autumnal moth (*Epirrita autumnata*) feeding on mountain birch. Mountain birch also shows a similar seasonal trend in leaf quality[58].

12.5.2 Induced defense

Delayed induced defense

Delayed induced defense (DIR) or *Long-term induction* (LTI) involves changes in a tree that usually occur one or more years after herbivorous attack. In addition, DIR seldom occurs with minor herbivory. It requires severe herbivory to elicit significant long-term plant responses. Evidence also exists that various experiences can trigger DIR[71].

DIR occurs in at least two lepidopteran defoliators: *Tortrix viridiana* on European larch (*Larix decidua*)[72] and autumnal moth (*E. autumnata*) on mountain birch (*Betula pubescens* ssp. *tortuosa*)[73]. Following defoliation of European larch, the fiber content is high in the needles emerging the next summer. In addition, the protein content of the needles in successive years is very low. Consequently, Fig. 54 shows that such needles are very poor food for the moth larvae and cause high mortality and population collapse.

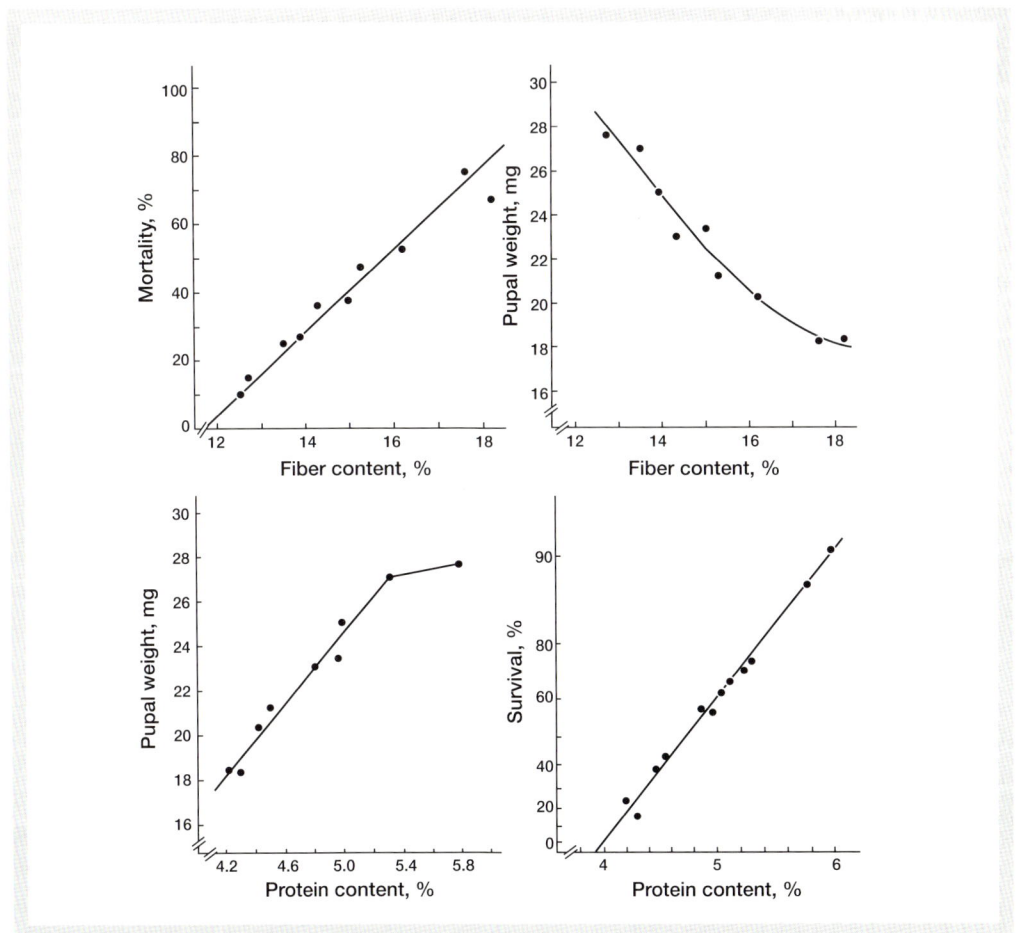

Figure 54. Correlations between needle fiber content, protein content, and survival and pupal weights of larch budmoth on European larch[75].

CHAPTER 6

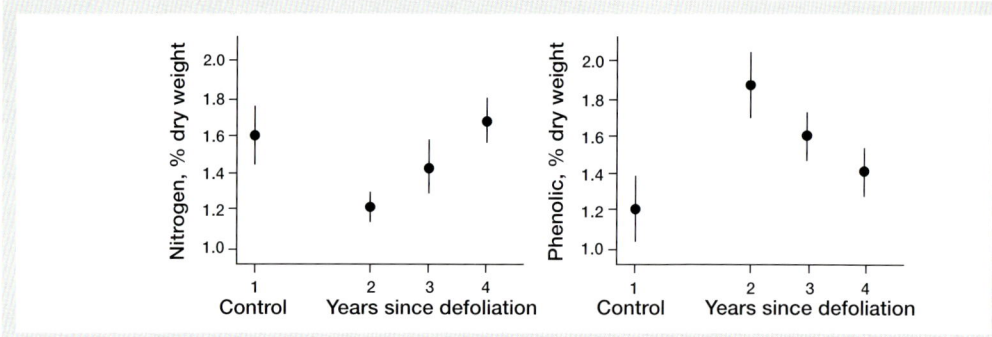

Figure 55. Nitrogen and phenolic content of mountain birch leaves in relation to defoliation history of the trees[76].

In mountain birch, the phenolic content of the leaves is high, and their nitrogen content is low following defoliation as Fig. 55 shows. In both tree species, DIR lasts 3–4 yr. Then leaf quality returns to normal. DIR can last several years and influence several successive insect generations. Due to this time lag, long-term inducible resistance can constitute the driving force behind the cyclic fluctuations in herbivore population dynamics[74]. DIR may well explain the observed 9-yr. cycle of both lepidopteran species shown in Figs. 56 and 57.

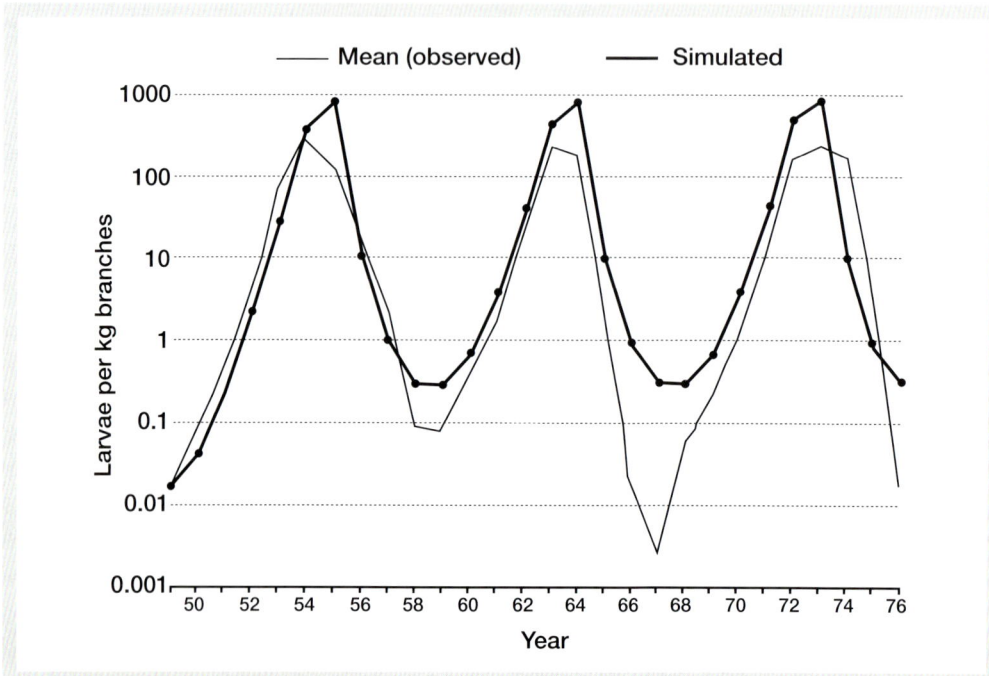

Figure 56. Simulated population fluctuations of larch budmoth based on defoliation-induced changes in foliage quality[72].

Management of forest ecosystems

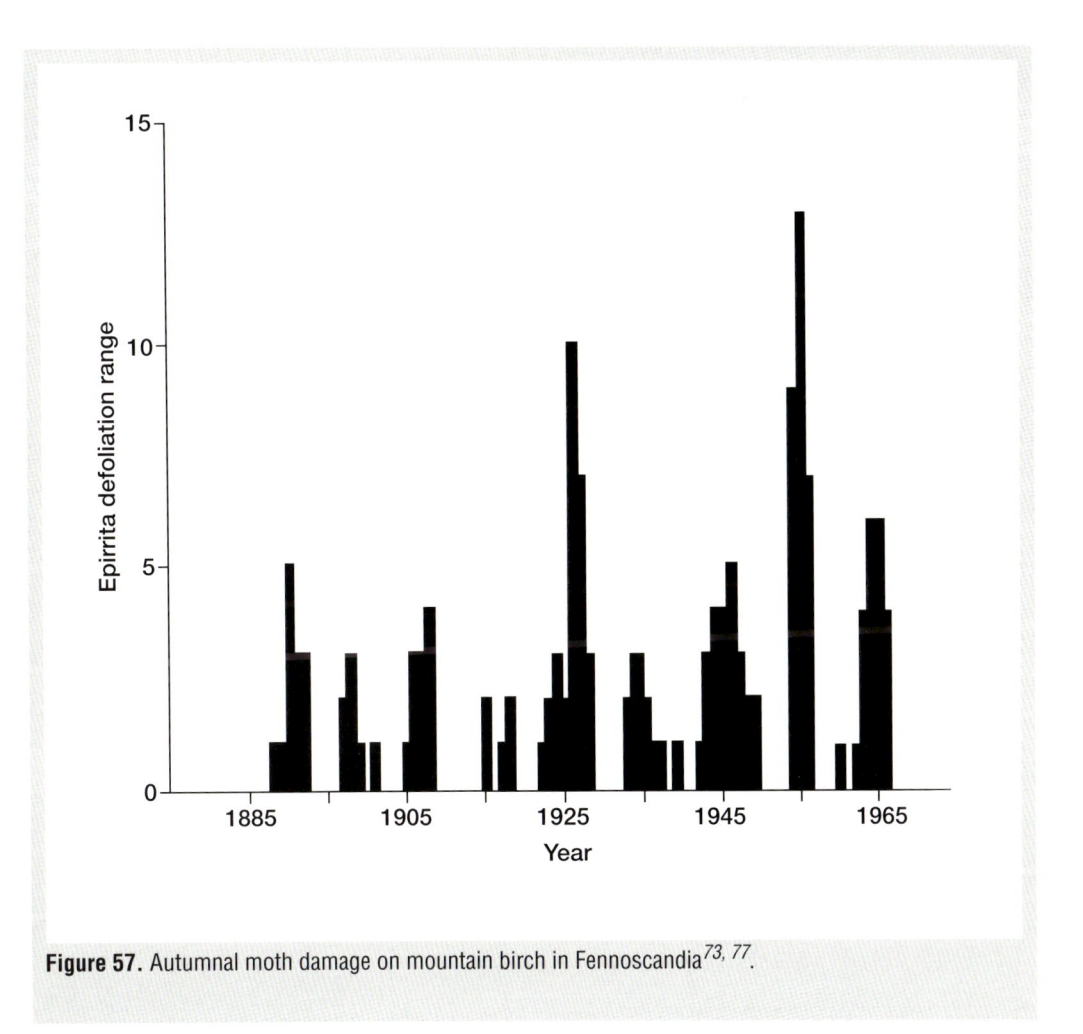

Figure 57. Autumnal moth damage on mountain birch in Fennoscandia[73, 77].

Rapid induced defense

Rapid induced defense (RIR) or *short-term induction* (STI) involves changes in the tree immediately following herbivory or tissue wounding or as many as several days later. While this time frame is too short to cause herbivore population cycles, it may have adverse consequences for the herbivore or pathogen behind the response. In addition, it can influence food preferences and feeding behaviors of herbivores encountering the attacked resource shortly thereafter.

Autumnal moth larvae feeding on mountain birch leaves damaged two days earlier have lower pupal weights and higher mortality than those feeding on nonafflicted control leaves as Fig. 58 shows. Leaves adjacent to damaged leaves also have RIR. This reaction occurs by specific elicitors that activate phenolic production in adjacent undamaged leaves. RIR seems widespread within the plant kingdom. It ranges from herbaceous to woody plants and is effective against many herbivore groups including acarine mites, lepidopterans, and coleopterans[78].

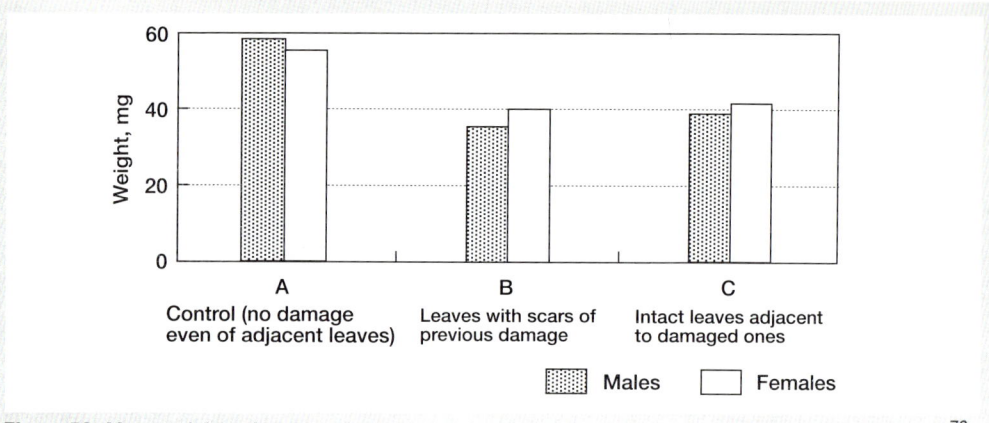

Figure 58. Mean weights of autumnal moth larvae in relation to previous damage to leaves fed to them[79]. A=control, B=leaves with scars of previous damage, and C=intact leaves adjacent to damaged leaves.

RIR is an important mechanism of conifer tree resistance to attack by *bark-beetles* and associated pathogenic *blue-staining fungi*[43]. A tree first responds to bark beetle attacks by producing repellent resin flow. If the beetle attack is successful, the beetles excavate tunnels under the bark as the first phase of gallery construction. This stimulates a wound reaction in the surrounding phloem, and the tissue respond to invasion by fungi carried by the beetles as Fig. 59 shows. Invasion by symbiotic fungi induces rapid necrosis of cells around the wound and stimulation by elicitors of local synthesis of terpenes, phenols, and other allelochemicals through the sugar and protein metabolic pathways. This creates a barrier to further fungal infection and beetle feeding in the phloem and xylem tissues.

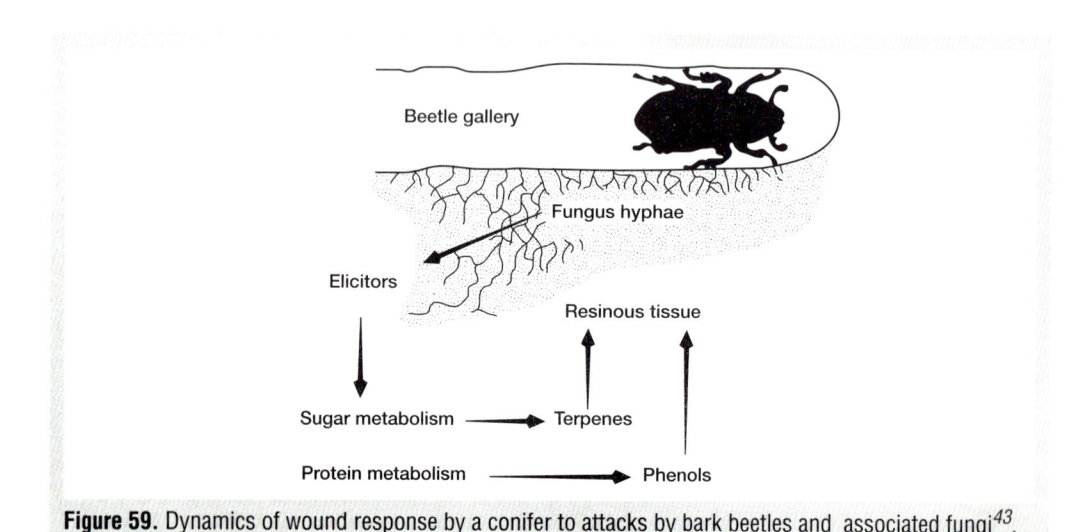

Figure 59. Dynamics of wound response by a conifer to attacks by bark beetles and associated fungi[43].

The ability of trees to prevent beetle and fungus attack depends significantly on the energy resources of the tree. Suppressed trees with low energy resources cannot produce sufficient resins to combat beetles and fungi[80, 81].

12.6 Models explaining variation in chemical defense between plants

Figure 60 shows that several studies indicate that herbivores inhabiting quite diverse ecosystems feed more heavily on tissues of inherently fast-growing species associated with productive environments (fertile soils) than on tissues of inherently slow-growing species inhabiting unproductive environments (infertile soils). A limited capacity to acquire resources and an inherently slow growth rate limit the ability of a plant to compensate for herbivory through growth. For example, studies on the effect of winter-browsing on the growth of boreal woody plants have consistently demonstrated that inherently slow-growing species such as spruce and alder can compensate less to herbivory by growth than with more rapidly growing species such as willows and aspen as Fig. 60 shows. Trees and shrubs that cannot easily compensate herbivory damage by adding extra growth must deter herbivory by defensive mechanisms. Usually, low-resource adapted plants have high concentrations of terpenes, phenolics, and other carbon-based allelochemicals in their tissues.

Trees and shrub that have evolved in productive environments have been selected for rapid growth to win over their neighbors and therefore dominate the available light and nutrient resources. In this way, inherently rapid growing woody plants have been favored by selection in the allocation of resources to growth at the expense of *chemical defenses*[82]. High-resource adapted trees have also evolved below-ground storage organs that contain large nutrient and carbohydrate reserves for use in vegetative regeneration following herbivory.

Herms and Mattson[41] concluded that plants have a dilemma to solve – to grow or to defend. A trade-off exists between growth and defense, and plants fall into classification of growth-dominated and differentiation (defense) dominated plants. Growth-dominated plants have physiological traits characterized by fast growth rates, phenotypic plasticity, high compensative ability, and high induced defense. Differentiation (defense) dominated plants have slow growth rates, low phenotypic plasticity, low compensatory growth, and high constitutive resistance. Table 18 compares the two types.

CHAPTER 6

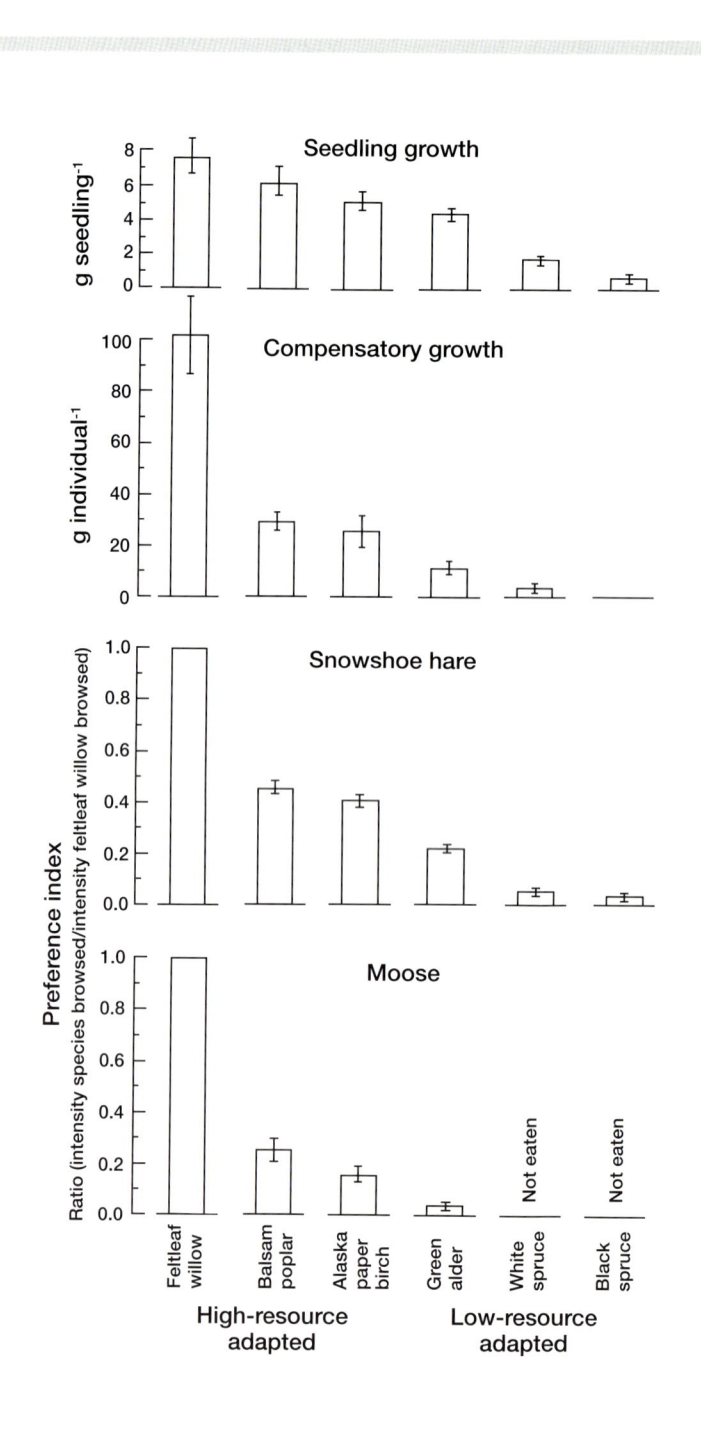

Figure 60. Seedling growth, compensatory growth, and palatability of some Alaskan woody plants to snowshoe hare and moose[82].

Table 18. Some traits predicted to be characteristics of growth and differentiation-dominated perennials[41].

Plant trait	Growth-dominated plants	Differentiation-dominated plants
Relative growth rate	High	Low
Resource-acquisition rate	High	Low
Resource-use efficiency	Low	High
Allocation to leaf area	High	Low
Storage reserves	Low	High
Secondary metabolism	Low	High
Respiration rate	High	Low
Phenotypic plasticity	High	Low
Constitutive resistance to herbivores	High	Low
Induced resistance to herbivores	High	Low
Competitive ability		
-Resource-rich environments	High	Low
-Resource-limited environments	Low	High

12.7 Management of forest pests

12.7.1 Tree resistance

The above discussion has indicated that trees posses a wide variety of resistance mechanisms against herbivores that have evolved as the result of natural selection. One successful forest management method to decrease damage is to select tree provenances that have high "natural" resistance against pests. For example, resistance of birches or birch clones against mammalian herbivores, hares, and voles correlates with the number of resin droplets in the twigs as Fig. 61 shows. Resin droplets contain high amounts of papyripheric acid. This is toxic to many mammalian herbivores. Some birch species such as *Betula resinosa* and *B. platyphylla* have high amounts of resinous glands. They are consequently very resistant against mammalian browsing[83].

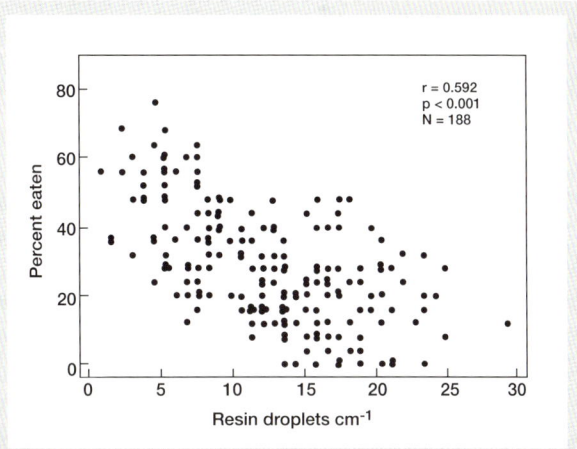

Figure 61. Relationship between mean number of resin droplets in *Betula pendula* seedlings and feeding intensity by mountain hare[83].

White pine weevil, *Pissodes strobi,* is a major cause of failures with plantations of Sitka spruce, *Picea sitchensis,* in coastal British Columbia, Washington, and Oregon. This pest is also becoming a menace to spruce plantations causing considerable losses in yield and quality of wood[84]. Results from various trials provide strong evidence of genetic variation in the susceptibility of tree species to weevil damage. White spruce is especially very resistant to white pine weevil. The damage level in white spruce was 5.5% while native provenances suffered a damage level of 23%. This resistance is probably of chemical origin from terpenes[84].

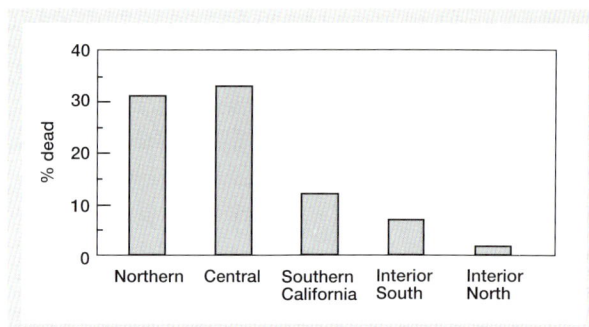

Figure 62. Mortality of white fir provenances caused by fir engraver, *Scolytus ventralis*[85].

Fir engraver, *Scolytus ventralis*, and associated pathogenic brown-stain fungus, *Trichosporium symbioticum*, attack white fir *Abies concolor* and other true firs causing considerable damage. Results obtained by Ferrell and Otrosina[85] indicate that certain provenances are very susceptible to the fir engraver and its fungal symbiotic, but other provenances are very resistant as Fig. 62 shows. Resistant firs react to invasion attempts from the insect by forming a resinous necrotic wound in the phloem and outer sapwood to prevent the spread of the fungus and kill or repel the beetles.

12.7.2 Biological control

Concepts

Discussion above mentioned management of pest populations "from below" by increasing resistance of trees against the pests or by reducing the availability of their food resource. Another control possibility is "from above" by manipulating natural enemies such as predators, parasites, and pathogens. Vertebrate predators such as shrews and rodents are most effective at limiting the growth of insect pest populations when pest populations are relatively sparse as Fig. 63 show. Invertebrate predators and parasites such as

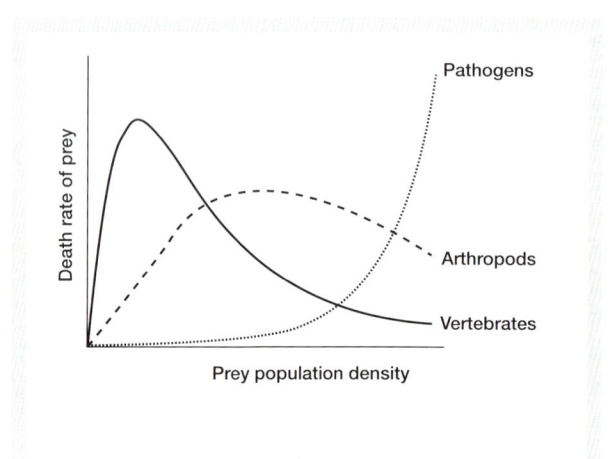

Figure 63. Relative impact of various natural enemies on mortality of gypsy moth larvae in relation to prey population density[43].

Management of forest ecosystems

spiders, spidermites, beetles, and ichneumonids are most effective when pests occur in intermediate densities. Pathogens such as bacteria, viruses, and fungi generally regulate populations when pests occur at very high densities. Many biological control efforts against forest insect pests have used several predatory or parasitoid species. The proportion of successful cases of control has remained modest at 17%–37% as Table 19 shows.

Table 19. Summary of biological control efforts taken in the United states and Canada against insects pests[86].

Success of biological control	United States — Forest insect pests and agricultural pests	Canada — Forest insect pests	Canada — Agricultural pests
Pest species	91	36	27
Pest species effectively controlled	18	6	10
Percentage controlled	20	17	37
Species of parasites and predators released	485	104	85
Species of parasites and predators established	95	36	28
Percentage established	20	35	33

12.7.3 Examples of biological control

Viruses

European pine sawfly (*Neodiprion sertifer*) is a wide-spread pest afflicting Scots pine (*Pinus sylvestris*) in Europe and North America. A *nuclear polyhedrosis virus* (NPV) frequently causes considerable larval mortality during the declining phase of outbreaks. The virus reproduces rapidly in the nuclei of the midgut's epithelial cells of the infected larvae and transmits readily within and among larval colonies. Because of its high and rapidly induced larval mortality shown in Fig. 64, NPV and other baculoviruses have found frequent use in biological control as Table 20 shows.

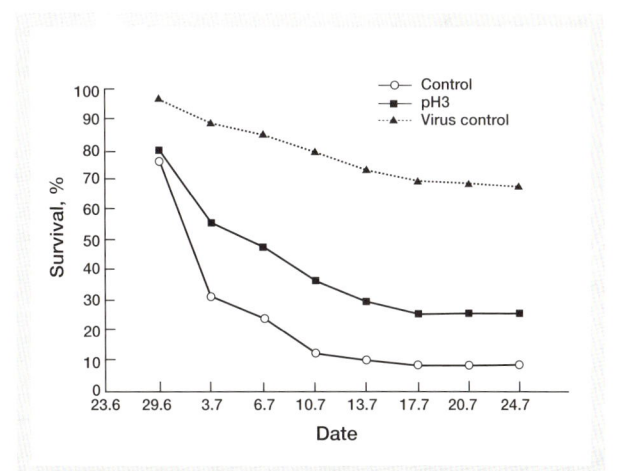

Figure 64. Survival of *Neodiprion sertifer* larvae treated with a nuclear polyhedrosis virus suspension and fed with needles subjected to acid-rain treatment (pH 3). Control is a treatment with irrigation only. Virus control indicates survival of larvae not exposed to virus[87].

The interaction between sawfly larvae and NPV depends on the quality of the host's foliage. The efficacy of NPV on young larvae decreased when feeding the larvae on pine foliage treated with acid rain[87]. Consequently, the efficacy of biological control can depend on environmental factors such as air pollution.

Table 20. Examples of forest insects potentially controllable by means of baculoviruses[45] (NPV=nuclear polyhedrosis virus, GV= granulosis virus).

Insect pests	Baculovirus	Host tree
Lepidoptera		
Choristoneura fumiferana	NPV	Spruce (*Picea* spp.)
Choristoneura muriana	GV	Spruce (*Picea* spp.)
Hyphantria cunea	NPV	Broadleaves
Lymantria dispar	NPV	Broadleaves
Lymantria fumida	NPV	Japanese fir
Lymantria monacha	NPV	Conifers
Malacosoma disstria	NPV	Aspen (*Populus* spp.)
Orgyia pseudotsugata	NPV	Douglas fir (*Pseudotsuga menziesii*)
Hymenoptera		
Gilpinia hercyniae	NPV	Spruce (*Picea* spp.)
Neodiprion lecontei	NPV	Pines (*Pinus* spp.)
Neodiprion pratti pratti	NPV	Pines (*Pinus* spp.)
Neodiprion sertifer	NPV	Pines (*Pinus* spp.)
Neodiprion swainei	NPV	Jack pine (*Pinus banksiana*)

Bacteria

A common biological control method uses the spore-forming bacterium, *Bacillus thuringensis* (Bt). Bacteria crystals containing toxic proteins can be artificially produced in fermentation chambers. Once insects ingest these cells, the toxin releases in the gut of the insect causing disruption of the epithelial cells and subsequent death. The bacteria therefore act like a highly efficient narrow-spectrum insecticide. Because environmental and socio-political constraints have restricted or totally stopped the use of synthetic insecticides in North America and Europe, the use of *Bacillus thuringensis* has rapidly increased in pest control over large forest areas.

Several forest lepidopterans are susceptible to Bt. This has led to the isolation and development of several strains of bacteria that are effective against particular insects. The variability among and within different strains of Bt can result in failure to control the target insect when using the wrong strain[45].

12.7.4 Chemical control

Some insecticides originate from plants and are allelochemicals used by plants in defense against herbivores. Pyrethrins and nicotine are good examples of these. Earlier widely used dichlorodiphenyltrichloroethane (DDT) and related insecticides like aldrin and dieldrin face bans in many countries today. Experience has shown that the insecticides affected the natural enemies of the pests more than the target species. Consequently, pest populations could increase rapidly after spraying in the absence of natural control. Because the insecticides are relatively stable compounds and fat-soluble, they became enriched in food-chains killing top-predators. In addition, many pests rapidly evolved a resistance against the widely used insecticides.

For these reasons, the use of insecticides in forest pest control has decreased dramatically during the past two decades. The organophosphates such as malathion and fenitrothion that are insect growth regulators are the common products used today. These materials disturb insect development by inhibiting chitin synthesis and disrupting the moulting process. Organophosphates find efficient use in controlling many lepidopteran forest pests such as *Bupalus piniarius, Choristoneura fumiferana,* and diprionid sawflies like *Diprion pini*.

Other insecticides still used are the synthetic pyrethroids bioresmethrin and permethrin. These are effective on a wide range of pests and are nonhazardous to mammals.

12.7.5 Behavior-modifying chemicals

Concepts

Secondary metabolites that trees produce primarily for defensive purposes are often attractive to specialized insects. These insects use secondary metabolites in locating and identifying host trees. Bark beetles are the most common forest pests. Most examples are from this group of insects. Because many bark beetles feed on conifers, attractants are mainly the volatile terpenes that conifers produce in high quantities indicated in Table 21.

Table 21. Some examples of host trees' secondary metabolites that serve as attractants for forest insects[45].

Insect	Host	Chemical
Scolytus scolytus	Elm	α-cubenone
Ips grandicollis	Pine	geraniol limonene methyl chavicol myrcene
Dendroctonus frontalis	Pine	3-carene α-pinene
Tomicus piniperda	Pine	α-pinene α-terpineol
Hylobius abietis	Pine	α-pinene α-terpineol 3-carene
Pissodes strobi	Pine	limonene

Many secondary chemicals are also involved in host acceptance after contact is made and oviposition or feeding begins. Secondary metabolites may stimulate feeding (phagostimulants) or ovioposition or deter it (inhibitors or deterrents) depending on the insect species involved. Sawflies are especially very cautious in selecting the tree or branch for ovipositioning. Alkaloids and terpenoids are particularly good feeding deterrents. The concentration and within-tree distribution of deterrents can influence the degree and distribution of insect damage on trees. As mentioned earlier, the larvae of European pine sawfly feed only on older year classes of needles of pines and avoid the current year foliage that contains toxic diterpene.

Management of forest ecosystems

Pheromones

Many different chemicals transmit information between organisms. These are semiochemicals. The best-known semiochemicals are *pheromones*. They mediate in interactions between individuals of the same species. Pheromones posses great potential in forest insect control programs. The most important are sex pheromones of lepidoptera emitted by the female to attract the male and the aggregation pheromones of many bark beetle species. Either sex can release aggregation pheromones. They unite the sexes on the host tree for mating and synchronize a simultaneous attack on the tree to destroy its defensive systems.

Many pheromones are of dietary origin. Insects may use them directly. Sometimes an insect itself or a microorganism associated with it will modify them. Some species of bark beetles can metabolize α-pinene to trans-verbenol and verbenone. These are part of their aggregation pheromone.

Figure 65 depicts the principles of chemical communication in the attack system of the spruce bark beetle *Ips typographus* involving aggregation and antiaggregation pheromones[81]. The main components of the aggregation pheromone are methybutenol and cis-verbenol in the initial attack and mass aggregation phase. Two components (verbenone and ipsenol) act as antiaggregation pheromones. They are released after the females have entered the gallery, and they appear to regulate gallery density and cause the shift of attack to fresh areas of bark or neighboring trees[81].

Pheromone traps have wide use in surveying, detecting, and monitoring many lepidopteran (gypsy moth, pine beauty moth, nun moth, etc.) and bark beetle species. Recently, some diprionid sawfly pheromones have been identified by a pheromone-trapping system customized for this forest insect. Pheromone traps can monitor population levels in areas susceptible to outbreaks. The main aim in monitoring is to detect increases in pest numbers to plan appropriate control methods in advance and restrict treatments to areas of high pest density. When the insect numbers in traps approach

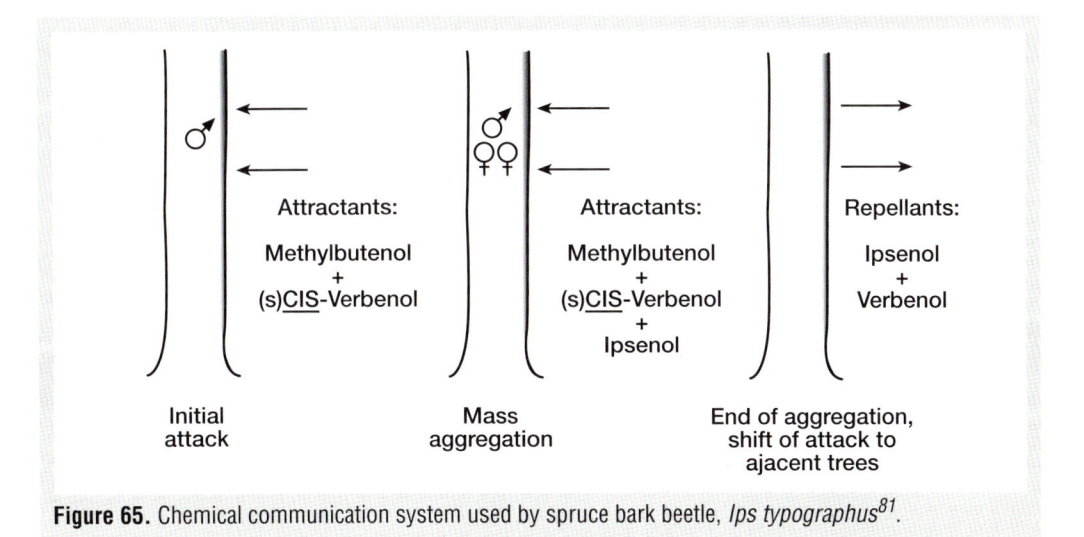

Initial attack	Mass aggregation	End of aggregation, shift of attack to ajacent trees
Attractants: Methylbutenol + (s)CIS-Verbenol	Attractants: Methylbutenol + (s)CIS-Verbenol + Ipsenol	Repellants: Ipsenol + Verbenol

Figure 65. Chemical communication system used by spruce bark beetle, *Ips typographus*[81].

CHAPTER 6

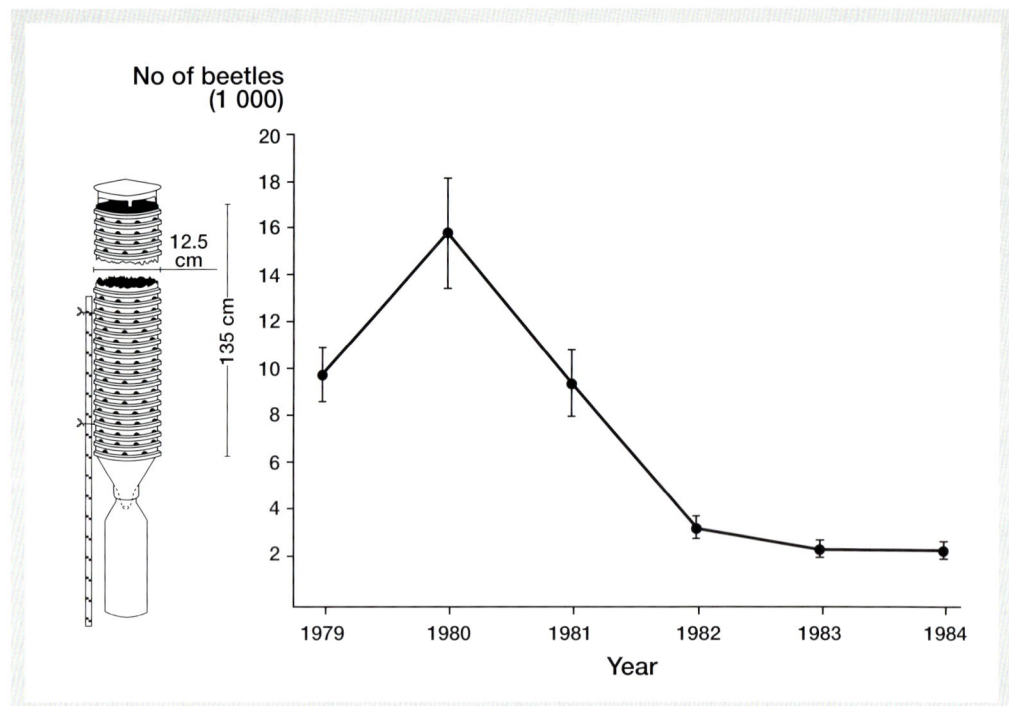

Figure 66. Mean catches of *Ips typographus* per pheromone trap in southeast Norway during a severe outbreak of beetle damage[81].

some empirically determined threshold value, intensive sampling can occur in localized high-risk areas.

Mass trapping is a method of directly controlling pest populations. It has found use against lepidopteran and coleopteran pests. Many trials have used mass trapping with the gypsy moth in North America and Europe, but the results have been inconclusive[88]. A large-scale mass trapping occurred during a large-scale outbreak of spruce bark beetle *Ips typographus* in central Sweden and southern Norway in 1978–1982. The epidemic started in drought-stricken areas in 1977 and killed approximately one million cubic meters of spruce each year. Figure 66 shows that the epidemic declined in 1983–1985. The primarly factors contributing to the decline of the epidemic were probably the extensive control campaign that included salvage logging and mass pheromone trapping, an increase in host resistance following the elimination of the most susceptible stands, and abundant rainfalls that increased the tree resistance.

CHAPTER 6

References

1. Daniel, T. W., Helms., J. A., and Baker, F. S., Principles of Silviculture, McGraw-Hill, New York, 1979, pp. 6–14.

2. Guittet, J. and Laberche, J. -C., Ecol. Plant. 9:111(1974).

3. Kellomäki, S., Hänninen, H., Kolström, T., et al., Silva Fennica 21(1):1(1987).

4. Pukkala, T., Silva Fennica 21(1):37(1987).

5. Parviainen, J. The Finnish Forest Research Institute, Research Papers 43:1(1982).

6. Remröd, J., Val av tallprovenienser i Norra Sverige - analys av överlevnad, tillväxt och kvalitet i 1951 års tallproveniensförsök, Rapp. Uppsats. Instn. Skogsgenet. Skogshögsk. 19, Stockholm, 1976, 132 p.

7. Uotila, A., Folia Forestalia 639(1):1(1985).

8. Heikinheimo, O., Commun. Inst. For. Fenn. 37(2):1(1948).

9. Anon., Metsä 2000-ohjelma, tarkastustoimikunnan mietintö, Komiteamietintö 1992:5. Maa-ja metsätalousministeriö, Helsinki, 1992, 112 p.

10. Aaltonen, V. T., Commun. Inst. For. Fenn. 9(3):1(1925).

11. Shinozaki, K. and Kira, T., Jour. Inst. Polytech. D7(1):35(1956).

12. Kira, T., Ogawa, H., and Sakazaki, N., Jour. Inst. Polytech. D4(1):1(1953).

13. Yoda, K., Kira, T., Ogawa, H., et al., Journ. of Biol. 14(1):107(1963).

14. White, J., in Demography and evolution in plant populations (O.T. Solbrig, Ed.), Blackwell Scientific, Oxford, 1980, pp. 21–48.

15. Braathe, P., Medd. Norske Skogsforsoksv.11:425(1952).

16. Jack, W., Irish For. 28:13(1971).

17. Oker-Blom, P. and Kellomäki, S., Folia Forestalia. 509(1):1(1982).

18. Kimmins, J. P. and Scoullar, K. A., Forcyte-10. A user's manual, University of British Columbia, Vancouver, 1984, pp. 27–29.

19. Vuokila, Y., Folia Forestalia. 448(1):1(1980).

20. Vuokila, Y., Acta For. Fenn. 110(1):1(1970).

21. Nyyssönen, A., Acta For. Fenn. 60(4):1(1954).

22. Savill, P. S. and Sandels, A. J., Forestry 56(2):109(1983).

CHAPTER 6

23. Möller, C., The influence of thinning on volume increment, World For. Series 1:1(1954).

24. Vuokila, Y., Folia Forestalia 247(1):1(1975).

25. Kimmins, J. P., Forest Ecology, Macmillan Publishing Company, New York, 1987, pp. 245– 255.

26. Binkley, D., Forest Nutrition Management, John Wiley & Sons Inc., New York, 1986, p. 163.

27. Kauppila, A. and Lähde, E., Folia Forestalia 230(1):1(1975).

28. Miina, J., Management of Scotch pine stands on drained peatland: a model approach, Research Notes 43, University of Joensuu, Joensuu, 1996, pp. 1–29.

29. Tomppo, E. and Henttonen, H., Metsätilastotiedote, The Finnish Forest Research Institute, Helsinki, 1996.

30. Viro, P. J., Commun. Inst. For. Fenn. 67(7):1(1969).

31. Johansson, M. -B., Forestry 68(1):49(1995).

32. Mikola, P., Commun. Inst. For. Fenn. 43(1):1(1955).

33. Laitakari, E., Acta For. Fenn. 40(1):1(1934).

34. Kalela, E. K., Acta For. Fenn. 57(2):1(1950).

35. Kukkola, M. and Saramäki, J., Commun. Inst. For. Fenn. 114(1):1(1983).

36. Linder, S. and Rook, D. A., in Nutrition of plantation forets (G. D. Bowen and E. K. S Nambiar, Ed.), Academic Press, London, 1984, pp. 211–236.

37. Axelsson, B., in IUFRO Symposium on forest site and continuous productivity (B. Ballard and S. Gessel, Ed.), USDA Forest Service General Technical Report PVW-163, USDA, Portland, 1983, pp. 305–311.

38. Anon., in Statistical Yearbook of Forestry 1995 (M. Aarne, Ed.), The Finnish Forest Research Institute, Helsinki, 1995, pp. 1–351.

39. Mälkönen, E., Derome, J., and Kukkola, M., in Acidification in Finland (Kauppi, P., Ed.), Springer-Verlag, Heidelberg, 1990, pp. 325–347.

40. Kellomäki, S., Silva Carelica 8, Joensuun yliopisto, Joensuu, 1991, pp. 443–459.

41. Herms, D. A. and Mattson, W. J., Quaterly Review of Biology 67:283(1992).

42. Strong, D. R., Lawton, J. H., and Southwood, T. R. E., Insects on Plants, Community Patterns and Mechanisms, Blackwell Scientific, Oxford, 1984, pp. 1–4.

43. Berryman, A. A., Forest Insects, Principles and Practice of Population Management, Plenum Press, New York, 1986, pp. 37–80.

44. Berryman, A. A., Dynamics of Forest Insect Populations, Patterns, Causes, Implications, Plenum Press, New York, 1988, pp. 1–596.

45. Speight, M. R. and Wainhause, D., *Ecology and Management of Forest Insects*, Clarendon Press, Oxford, 1989, pp. 10–15.

46. Tahvanainen, J. and Niemelä, P., *Annales Zoologici Fennici* 24(3):239(1987).

47. Neuvonen, S. and Niemelä, P., *Oecologia* 51(3):364(1981).

48. Strong, D. R. and Levin, D. A., *American Naturalist* 114(1):1(1979).

49. Price, P. W., *Evolution* 31(2):405(1977).

50. Neuvonen, S. and Niemelä, P., *Oikos* 40(4):452(1983).

51. Haack, R. A. and Mattson, W. J., in *Sawfly Life History Adaptations to Woody Plants* (M.R. Wagner and K.F. Raffa, Ed.), Academic Press, Boca Raton, 1993, pp. 515–519.

52. Southwood, T. R. E., Moran, V. C., and Kennedy, C. E. J., *Journal of Animal Ecology* 51(2):635(1982).

53. Connor, E. F., Faeth, S. H., Simberloff, D., and Opler, P. A., *Ecological Entomology* 5:205(1980).

54. Mattson, W. J., Niemelä, P., Millers, I., and Inguanzo, Y., *Immigrant phytophagous insects on woody plants in the United States and Canada: an annotated list*, General Technical Report nr NC-169, US Department of Agriculture, St. Paul, 1994, pp. 1–24.

55. Kim, K. C. and McPheron, B. A., in *Evolution and Insect Pests: Patterns of Variation* (K. C. Kim and B. A. McPheron, Ed.), John Wiley & Sons, New York, 1993, pp. 3–25.

56. Niemelä, P. and Mattson, W. J., *BioScience* 46(10):741(1996).

57. Bernays, E. A., Cooper-Driver, G., and Bilgener, M., *Advances in Ecological Research* 19(1):263(1989).

58. Haukioja, E., Niemelä, P., Iso-Iivari, L., et al., *Reports of the Kevo Subarctic Research Station* 14:5(1978).

59. Lindberg, M., Lundgren, L., Gref, R., and Johansson, M., *European Journal of Forest Pathology* 22(2):95(1992).

60. Fahn, A., *Secretory Tissues in Plants*, Academic Press, New York, 1979, pp. 176–218.

61. Bernard-Dagan, C., in *Mechanisms of Woody Plant Defenses against Insects* (W.J. Mattson, J. Levieux, and C. Bernard-Dagan, Ed.), Springer-Verlag, New York, 1988, pp. 93–116.

62. Kainulainen, P., *Kuopio University Publications C., Natural and Environmental Sciences* 58:1(1997).

63. Gershenzon, J. and Groteau, R., in *Herbivores. Their Interactions with Secondary Plant Metabolites. Volume I. The Chemical Participants* (G.A. Rosenthal and M.R. Berenbaum, Ed.), Academic Press, San Diego, 1991, pp. 165–219.

CHAPTER 6

64. Langenheim, J. H., Journal of Chemical Ecology 20(6):1223(1994).

65. Duncan, A. J., Hartley, S. E., and Lason, G. R., Canadian Journal of Zoology 72(10):1715 (1994).

66. Ikeda, T., Matsumura, F., and Benjamin, D. M., Journal of Chemical Ecology 3:677(1977).

67. Niemelä, P., Mannila, R., and Mäntsälä, P., Annales Entomologici Fennici 48(2):57(1982).

68. Hartmann, T., in Herbivores. Their Interactions with Secondary Plant Metabolites. Volume I. The Chemical Participants (G. A. Rosenthal and M. R. Berenbaum, Ed.), Academic Press, San Diego, 1991, pp. 79–121.

69. Stermitz, F. R., Tawara, J. N., Boeckl, M., et al., Phytochemistry 35(4):951(1994).

70. Feeny, P., Ecology 51(4):565(1970).

71. Haukioja, E., Suomela, J., and Neuvonen, S., Oecologia 65(3):363(1985).

72. Baltensweiler, W. and Fischlin, A. in Dynamics of Forest Insect Populations. Patterns, Causes, Implications (A. A. Berryman, Ed.), Plenum Press, New York, 1988, pp. 331–351.

73. Haukioja, E., Neuvonen, S. and Niemelä, P., in Dynamics of Forest Insect Populations. Patterns, Causes, Implications (A.A. Berryman, Ed.), Plenum Press, New York, 1988, pp. 163– 178.

74. Haukioja, E., Oikos 35(2):202(1980).

75. Benz, G., Zeitschrifte fur Angewandte Entomologie 76:31(1974).

76. Tuomi, J., Niemelä, P., Haukioja, E., Siren, S., and Neuvonen, S., Oecologia 61(2):208 (1984).

77. Tenow, O., Zoologiska Bidrag från Uppsla. Suppl. 2:1(1972).

78. Tallamy, D. W. and Raupp, M. J. (Ed.), Phytochemical Induction by Herbivores, John Wiley & Sons, New York, 1991, pp. 1–431.

79. Haukioja, E., Inducible defences of white birch to a geometrid defoliator, Epurrita autumnata, 1982 Proceedings of the Fifth International Symposium on Insect Plant Relationships, Pudoc, Wageningen, pp. 199–203.

80. Berryman, A. A. and Ferrell, G. T., in Dynamics of Forest Insect Populations. Patterns, Causes, Implications (A. A. Berryman, Ed.), Plenum Press, New York, 1988, pp. 555–577.

81. Christiansen, E. and Bakke, A., in Dynamics of Forest Insect Populations. Patterns, Causes, Implications (A. A. Berryman, Ed.), Plenum Press, New York, 1988, pp. 479–503.

82. Bryant, J. P., Tuomi, J., and Niemelä, P., in Chemical Mediation of Coevolution (K. S. Spencer, Ed.), Academic Press, San Diego, 1988, pp. 367–389.

83. Tahvanainen, J., Julkunen-Tiitto, R., Rousi, M., and Reichardt, P., Chemoecology 2(1):49 (1991).

84. Kiss, G. K., Yanchuk, A. D., Alfaro, R. I., et al., in Dynamics of Forest Herbivory: Guest for Pattern and Principle (W. J. Mattson, P. Niemelä and M. Rousi, Ed.), USDA Forest Service, General Technical Report NC-183, St Paul, MN, 1996, pp. 150–158.

85. Ferrell, G. T. and Otrosina, W. J., in Dynamics of Forest Herbivory: Guest for Pattern and Principle (W. J. Mattson, P. Niemelä and M. Rousi, Ed.), USDA Forest Service, General Technical Report NC-183, St Paul, MN, 1996, pp. 187–191.

86. Krebs, C. J., Ecology. The Experimental Analysis of Distribution and Abundance, Harber & Row, New York, 1972, pp. 372–373.

87. Saikkonen, K. and Neuvonen, S., Oecologia 95(1):134(1993).

88. Webb, R. E., in Insect suppression with controlled release pheromone systems (A. F. Kydonieus and M. Beroza, Ed.), Vol II, CRC Press, Boca Raton, 1982, pp. 27–56.

CHAPTER 7

Timber procurement

1	**Introduction**	**311**
1.1	Timber assortments	313
2	**Harvesting and timber haulage**	**316**
2.1	Harvesting methods and systems	316
2.2	Logging machinery	319
2.3	Storing timber	333
3	**Organizing timber harvesting**	**334**
4	**Planning and monitoring of timber procurement**	**335**
5	**Damage to timber**	**342**
5.1	Causes of damage	342
5.2	Abiotic damage	342
5.3	Fungal decay	343
	5.3.1 Introduction	343
	5.3.2 Brown rot	344
	5.3.3 White rot	344
	5.3.4 Soft rot	344
	5.3.5 Comparison of rot types and prevention of rot	344
5.4	Attacks by insects and aquatic organisms	345
5.5	Decay caused by bacteria and biological reactions	346
5.6	Influence of storage on mill processes	346
5.7	Decay in chip storage	348
5.8	Storing of bark	352
6	**Environmental impacts of timber harvesting**	**352**
6.1	Common conceptions	352
6.2	Energy consumption and emissions	354
6.3	Impacts on forest site	355
6.4	Impacts on biodiversity, multiple-use, and landscape	356
7	**Timber trade**	**357**
7.1	Purchasing timber on domestic markets	357
7.2	International timber markets	358
	References	**360**

CHAPTER 7

Pertti Harstela

Timber procurement

1 Introduction

Timber procurement covers wood harvesting and all the activities necessary to produce timber from the forest for mills or other consuming activities. It is a primary function in forestry. Procurement activities involve purchasing and measurement of timber in the round or roundwood, planning and supervision of operations, and business management. *Wood harvesting* covers *logging*. This includes the cutting and off-road hauling of timber from the stump to a roadside storage point or landing and *transportation of timber* from the landing to the mill yard. Procurement activities also require establishing an enterprise organization, work organization, payment systems, safety and heath systems, risk-control systems, and other process-management systems. The history of wood procurement includes the steps presented in Table 1 as viewed from the world perspective with Finland serving as an example.

Table 1. Turning points in the history of wood procurement in Finland.

Period	Development step
1500s onward	Large-scale industrial use of wood; international markets for tar and wood-based products
1700s	First commercial sawmills in Finland; seasonal, untrained labor; axes, handsaws, horse-based hauling and river driving (floating) as main technological solutions
1940s	Development of tools, maintenance of tools, and work techniques; improvement of workers' living conditions
1950s	Introduction of farm-tractors in off-road hauling
1960s	Introduction of chainsaws and forestry tractors; permanent and skilled cutters and machine operators
1970s	Large-scale adoption of multi function machines (processors and harvesters)

Permanent and skilled cutters and machine operators began to replace seasonal labor beginning in the 1970s. Simultaneously, entrepreneurs working as contractors under the supervision of large forest-industry companies entered the harvesting scene. In some countries, major problems arose following the large-scale employment of unskilled or irresponsible contractors working without proper supervision on logging operations. In most cases, the current practice is for forest-industry companies to buy their timber on the stump from private forest owners and do the logging themselves. Simultaneously with logging by companies, Finnish farmers continue the practice of *delivery sales* of timber. This means the landowner attends to timber cutting and off-

CHAPTER 7

road hauling to the landing by the roadside. Today, the proportion of delivery sales is approximately 25% of the total annual cuttings.

The common endeavor to increase productivity and decrease costs in timber procurement led to mechanization of forest operations. Introduction of chainsaws in the 1960s made a significant impact. The mechanization of off-road hauling took its first steps even earlier in the 1950s when farm tractors began hauling timber from the woods. Bulldozers found use especially in tropical high forests in the extraction of large-diameter and heavy stems and logs. The first tractors purposely built for forestry were the skidders developed in North America in the 1960s. In the middle 1960s, the first wheeled forwarders were introduced in the Nordic countries.

A major step forward in applying the cut-to-length cutting method was the innovation of long-reach grapple loaders. This eased the manual input in bunching of timber. Direct logging on forwarders replaced mechanized pre-skidding. Similar development occurred in tree-length harvesting with the introduction of clam-bunk and grapple skidders. This meant that skidding operations no longer required a helper for the skidder operator. The development of felling and cutting machines in the form of feller-bunchers, delimbers, slashers, processors, and eventually harvesters began almost simultaneously with the development of forestry tractors. More development work and time was necessary before their use became profitable especially in thinnings. In mountainous conditions, the development of cable-skidding and cable ways made large advances. Lighter versions have been introduced for use in thinnings.

Work studies have played major roles in the technical development of harvesting. Work productivity as an objective has recently been accompanied with efforts focusing on ergonomic and safety factors, timber quality, and environmental issues. The state of health of the forests has also become an important issue. The focus of development has shifted from technical aspects to the control of the process by advanced planning systems employing the principles of modern logistics. Figure 1 illustrates that timber procurement is a complicated management system.

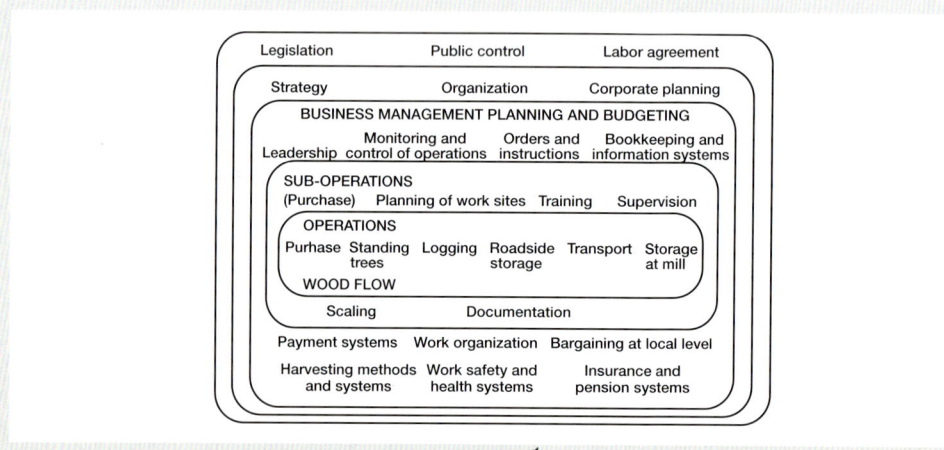

Figure 1. Timber procurement as a management system [1].

Timber procurement

1.1 Timber assortments

The primary commercial timber assortments in Finland are softwood *saw logs*, hardwood saw logs, and soft- and hardwood *pulpwood*. Smaller amounts of *special assortments* and *fuelwood* are harvested for specialized production, construction projects, and energy generation. Bucking (crosscutting) of stems into timber assortments has a major impact on the value of industrial timber. Bucking is the first stage of milling (conversion).

Planning for bucking of stems into logs in the woods can use simulation programs so the distribution of saw logs will comply with the sawmill's requirements as Fig. 2 indicates[2]. The simulation program allows the user to examine the degree to which the stems from a stand of trees can be crosscut to provide the log distribution necessary at the sawmill. The model enables the user to select appropriate lots of standing for a specific market situation. The results can also select an appropriate value-and-requirement matrix for application in semi-automated bucking. The bucking programs installed in state-of-the-art harvesters use the value-and-requirement matrixes for different dimensions and grades. At sawmills, logs are graded by dimension and quality class based on visible knots and defects and even invisible defects following x-ray tomography – a recently introduced practice.

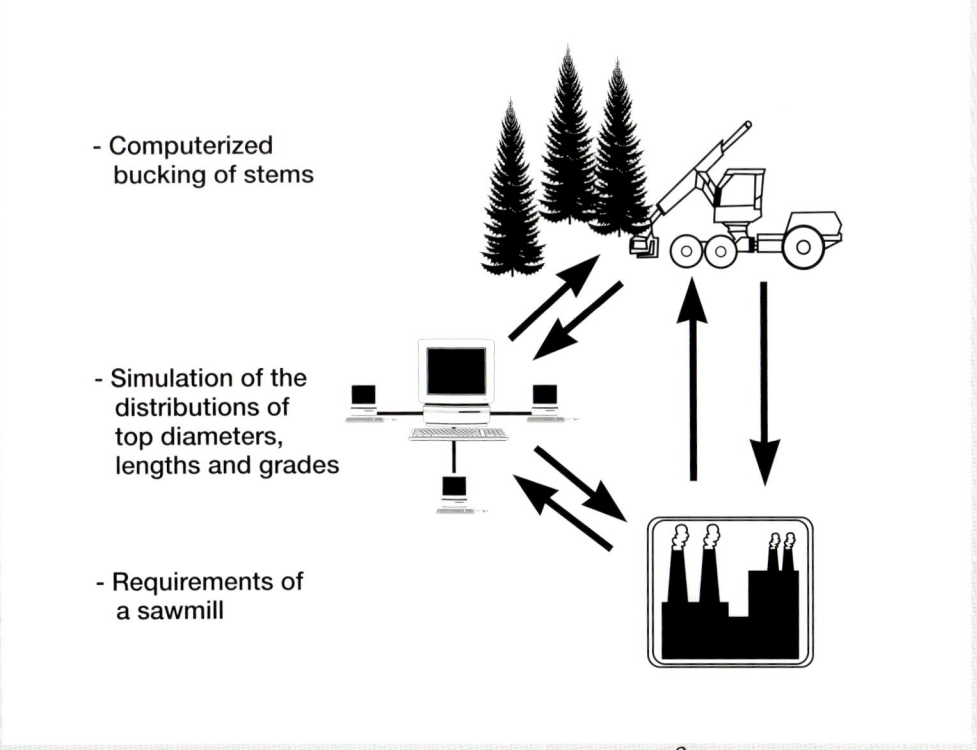

Figure 2. Selecting a bucking (crosscutting) program for a marked stand[2].

CHAPTER 7

The most important timber assortments in order of descending value are the following:

- special forms of wood such as curly-grained hardwood
- special-quality logs and telephone poles
- saw logs and peeler logs according to tree species
- spars and beams
- pulpwood by tree species (Spruce pulpwood has wide use in groundwood pulp mills, and its quality requirements are more stringent than those of pine, birch, and other hardwood species used in sulfate pulp mills.)
- fuel wood , firewood, etc.
- chips: stemwood (with or without bark), whole-tree or logging residue chips
- crushed logging residues.

The *quality specifications* of timber assortments differ from country to country and even from company to company according to the mill process and trade policies. The following text therefore only presents a general description and some examples of current Finnish practice. All timber assortments have minimum requirements for dimensions and quality[3]. Timber assortments may also have specific quality classification systems. As an example, pine saw logs are classified according to their value when sawed. The quality classes for saw logs have the purpose of optimizing the recovery rate and quality of sawed goods. Classes may be designed by experimental sawing or computer simulations. The minimum top diameter is typically determined and grading rules are designed for dimensions and defects such as knots of various types, sweep, crookedness, and presence of rot[4]. Table 2 illustrates the influence of dimensions and quality classes on pine saw logs in the value of customary sawed goods[5].

Table 2. Value of Scots pine saw logs (trade name redwood logs) as a function of log's top (small-end) diameter and quality class[6].

Log's top (small-end) diameter mm, under bark	Quality class I	II	III
	Relative value		
150	92	84	83
250	126	104	93
350	142	111	98

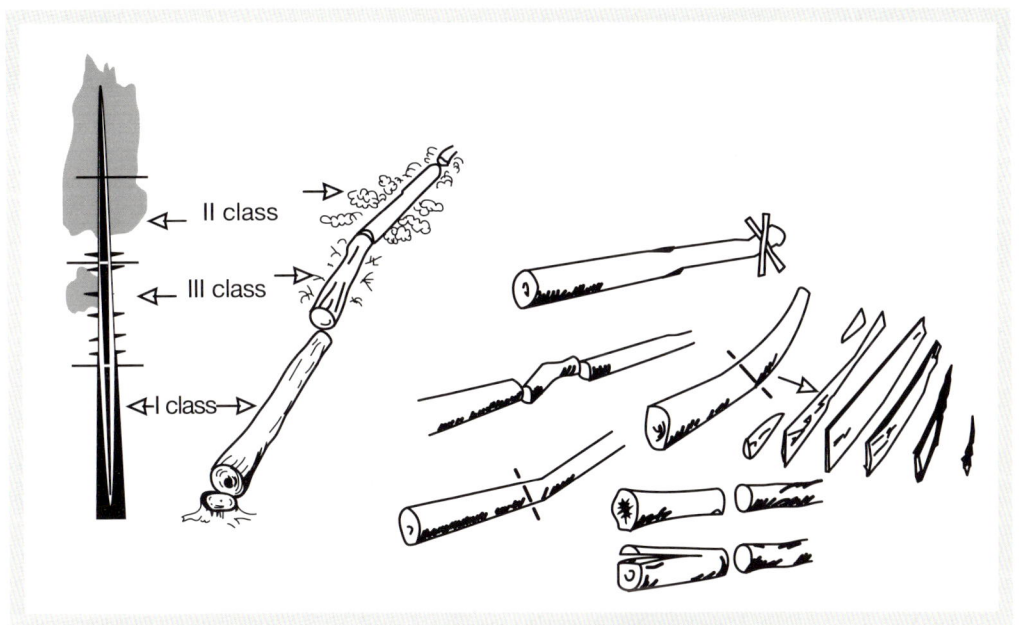

Figure 3. Some basic principles applying to bucking (crosscutting) sawlogs.

Another grading system considers the end use of the sawed goods produced. Class I is for butt logs containing only a few knots. These logs find main use in producing high-quality sawed goods. Class II is for logs from the top part of the bole containing live knots. These sawed goods find typical use in the production of paneling and furniture. Class III logs come from the middle of the stem and contain numerous dry knots. The sawed goods from them have use mainly for construction purposes, the building of log-houses, etc. Figure 3 shows the three classes. In some countries, saw logs are cut to one particular length. A better recovery rate and quality control result when using several alternative lengths. In Finland, log lengths increase in modules of 30 cm starting from 310 cm and extending to 610 cm.

Spruce pulpwood used in the manufacturing of groundwood pulp has stringent quality requirements. The wood must be fresh or "green" and free from pathogenic infections and rot. If these requirements are not met, the practice in Finland is to use spruce pulpwood with pine pulpwood. The best results in groundwood pulping result when using straight bolts of pulpwood. Pine and hardwood pulpwood in Finland have primary use in sulfate pulp mills, but other locations use pine in groundwood pulp mills if spruce is not available.

The quality requirements when using wood in chemical pulping are not particularly stringent, but freshly-felled wood free of defects ensures better debarking results, recovery rate, and pulp quality in the sulfate process[3, 5]. Mixing freshly-felled and dry pulpwood also causes problems in pulping. When considering the total costs, the optimum length of pulpwood bolts for drum debarking and chipping is 4–5 m for softwood and 3–4 m for birch pulpwood that is often crooked.

Recently, even pulpwood assortments have been subdivided according to the pulping process and end-product qualities[7]. Pulpwood from early thinnings differs in processing behavior and fiber properties from pulpwood logs obtained from older trees. Logs with defects such as rot can be separated from sound logs, and different diameter classes can be separated. The age of the tree or cambium are a good basis for sorting pulpwood. For groundwood pulping, spruce heartwood and latewood percentages are additional factors significantly influencing energy consumption. Separation of small and large pulpwood bolts before processing has diminished wood loss in drum debarking.

Other bases for the sorting of pulpwood are tree age (first thinnings, thinnings, and final cuttings) and location of the piece of pulpwood along the stem (butt log, top log, or sawmill chips). The growing site and geographical region may also provide bases for sorting. These sub-assortments can be processed as separate batches or mixed with the remainder in a suitable proportion. Stumpage prices can also function as the basis for distinguishing sub-assortments. One must consider that additional sub-assortments mean additional costs in harvesting. Harvesting costs and benefits gained at the mill processing stage must be optimum, and the site and level of assorting determined accordingly.

2 Harvesting and timber haulage

2.1 Harvesting methods and systems

Harvesting methods are established ways of harvesting timber and include conducting operations and the format of the product as Table 3 shows.

Table 3. Harvesting methods.

Method	Features of the method
Cut-to-length or shortwood method	Delimbed stems are cut into logs at the stump and then hauled to roadside landing and from there to the mill
Tree-length or stem method	Delimbed stems or stems with tops cut off are transported to roadside landing or to the mill
Whole-tree method	Stems with branches are transported
Tree-section method	Tree sections with branches intact are transported to the mill
Chipping method	Stems or trees are converted into chips in the woods or at roadside landing

Timber procurement

Tree-length and *cut-to-length methods* are the most commonly used methods. The *tree-length method* with load-pulling tractors (skidders) has global use and is the most common in industrial operations. The *cut-to-length method* is enjoying increasing usage, especially because of the growing interest in thinnings. The cut-to-length method and load-carrying tractors (forwarders) have proven their superiority in Nordic conditions. Several factors influence the profitability of these methods as the following comparison shows:

- The cut-to-length method is relatively more profitable for smaller mean stem volume and vice versa as Fig. 4 shows. The load size in forwarding is rather independent of the size of the stems. In skidding, the size of the load decreases with decreasing stem size.

- A shorter forest hauling distance is relatively more advantageous when using the tree-length method because the importance of load size decreases. Loading and unloading stems consume less time than when dealing with assortments as Fig. 4 shows.

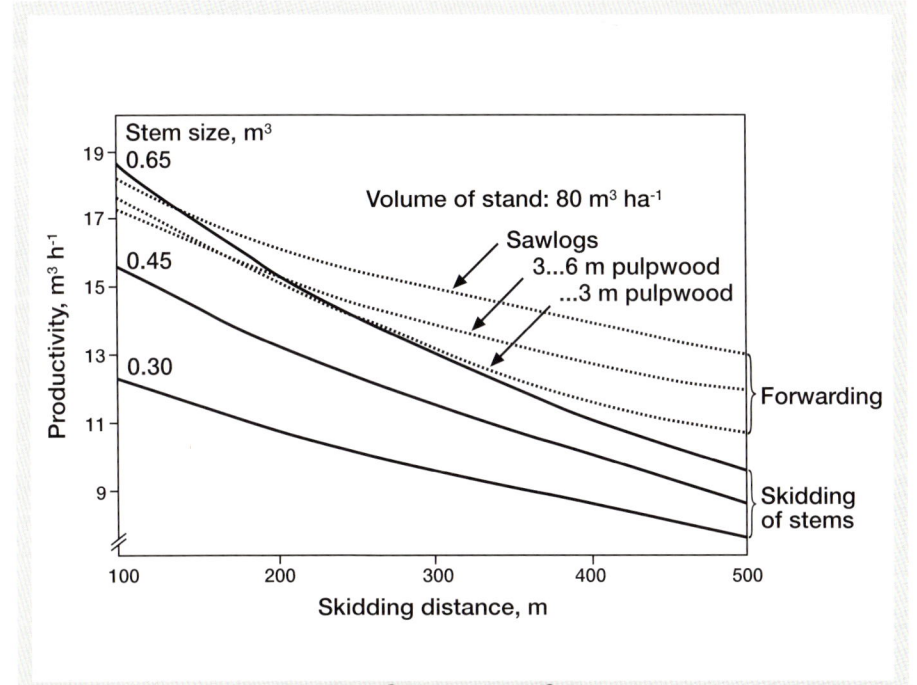

Figure 4. The productivity of skidding[8] and forwarding[9] as functions of stem size and forest haulage distance. Although the studies referred to here are not fully comparable to the levels of productivity, the trend in the influence of condition factors is shown.

- The cut-to-length method allows minimization of transportation distance to the mill and liberty in scheduling transportation of the different assortments. If cross-cutting occurs at the upper landing, the situation is the same as with the tree-length method.
- Sometimes it is difficult and even expensive to set up the large landings needed by the tree-length method. This is especially true in small-scale forest ownership.
- In thinnings, the cut-to-length method causes less damage to the retention stand than the tree-length method. Small stem size favors the use of the cut-to-length method.
- Soil compaction, rutting, and the risk of erosion decrease when using forwarders instead of skidders. Whole-body vibration – a health hazard of forestry machine operators – is less with forwarders because of their slower driving speed.
- Load-carrying forwarders keep the timber clean. Contaminants such as grit causes undue wear in mill equipment and other problems.
- The load volume of stems in truck transportation along public roads is smaller than that of cut-to-length timber unless the stems are loaded butt-ends and tops mixed. This is possible only when using heavy-duty loading equipment. Floating of stems is often not possible because of technical and economic reasons.
- Bucking of stems is faster and easier to combine with other elements of cutting in the forest than at landings. A centralized conversion plant is an expensive and inflexible investment.
- Centralized bucking allows the use of modern automation and optimization technology. X-ray tomography may be feasible in the future to determine the internal quality of logs. A common belief is that the quality of bucking is better when done in centralized conversion plants rather than in the woods[6]. A conversion plant can also allow one to minimize the inventory of sawed goods with resultant savings in interest costs. The option of bucking stems precisely according to the sawmill's production program is also possible then. Bucking in the woods by using harvesters is also under intensive development and meets the production requirements of sawmills better and better. It is also semi-automated. Development work in this area is continuing.
- Using the tree-length method involves the simultaneous adoption of two basic methods with the cut-to-length method because the tree-length method is not suitable for application in thinning stands and in some other conditions. This would increase distances for transferring machines between operations and cause other expenses.
- In mountainous conditions the productivity of cable-way systems may be higher in hauling stems than when dealing with cut-to-length timber.

Harvesting systems are differently organized man-machine/man-equipment combinations forming a sequence of harvesting operations. As an example, the cut-to-length method can be organized to form the following harvesting system: felling and cutting by harvester, off-road transport by forwarder, intermediate storage at landing, and transportation to mill by truck. The tree-length method can be organized as follows: felling and delimbing by feller/delimber-machine, off-road transport by skidder, intermediate storage at landing and cross-cutting there by a "slasher" and transportation to mill by truck. Numerous harvesting systems are in use nationally and internationally. The first of the above examples is the most common one in the Nordic countries. The second often finds use in North America for instance.

One can also depict harvesting systems as "hot" or "cold" chains of the operation. The former means interdependence between the various links of operations with the output of the next link. This depends on the output of the preceding one and vice versa. Because of the limited area at landings, a skidder cannot haul much timber to the landing unless the cross-cutting machine is working at the same rate. The cross-cutting machine has nothing to process unless the skidder is operating. This is a typical "hot" logging system. An example of a "cold" system is cut-to-length cutting by harvester followed by forwarding. A long interval can exist between these operations. If the productivities of these two machines differ, the number of machines or operating hours necessary for an area will be larger than one operation.

2.2 Logging machinery

A variety of equipment and machinery is available for logging. This ranges from *basic tools* and *animal skidding equipment* via intermediate technology such as *farm tractors* and *bulldozers* to advanced special technology such as *harvesters* and sophisticated *cableways*. The concept of appropriate technology depends on the stage of development in a particular country, the scale of operations, and the logging object and conditions. For example, farm tractors can have a wide variety of auxillary devices. They can have use in the forest owners' own logging operations or in plantation forestry in developing countries as Figs. 5 and 6 show. The text below only discusses machinery in common use in industrialized countries.

In *motor-manual cutting*, chainsaws fell, delimb, and crosscut trees. When applying the cut-to-length method, the usual practice is for the worker to bunch those pieces of timber that are not too heavy for manual movement. Directional felling is an important technique. It orients the felled timber to ease the job of forest hauling, to avoid damaging retention trees, to have the felled stems at an appropriate height for delimbing, and to reduce the distance for moving pieces of timber manually when bunching them.

CHAPTER 7

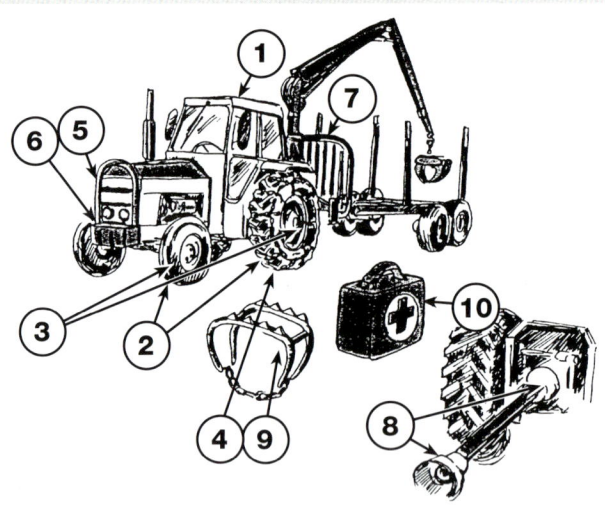

1. Security cabin
2. Tires
3. Shelter of valves
4. Friction chains
5. Shelter of the bottom and nose
6. Front weights
7. Shelter of the load
8. Shelter of the power transmission shaft
9. (see 4)
10. First aid kit

Figure 5. Equipping a farm tractor for forest hauling of timber in rocky terrain as recommended by TTS-institute, a Finnish research organization specializing in rationalization of forestry and agricultural work.

Figure 6. A farm-tractor mounted with forwarding equipment[10].

Timber procurement

Chainsaws are the most common motorized tool used in forestry. They have evolved from heavy and unpractical tools to their present-day light and efficient qualities. The 1950s, 1960s, and 1970s were the golden years of the chainsaw. The advances in light metals and engine technology were decisive factors in the development of modern chainsaws. The most important technical innovations have been the following:

- replacement of conventional saw blades by chains and saw teeth by grooved cutters comprising a side and top plate
- centrifugal clutch that couples the chain automatically into motion with engine acceleration
- membrane carburetor that allows the engine to run regardless of the position of the chainsaw
- reduced weight and size were important especially in delimbing and when working with small-diameter trees
- progress in reliability
- vibration damping
- improvements in safety features especially against kickback
- numerous other technical improvements.

The development of chainsaws continues despite their long development period and partial replacement by cutting machines.

The first step in the mechanization of timber cutting was the development of mobile *debarkers*. Although these machines working at landings still have use in some countries, debarking in the industrialized world usually occurs at sawmills with rotator-debarkers or at pulp mills with drum debarkers.

Mechanized cutting means the use of *cutting machines, i.e.,* fellers, feller-bunchers, feller- skidders, delimber-bunchers, slashers (bucking-piling machines), processors, and harvesters. The last two find use in cut-to-length systems. *Harvesters* execute felling, delimbing, bucking and bunching operations and volume scaling, color marking for grading, debarking by feeder rolls (used in Eucalyptus stands) and treatment of stumps against fungal attack. They are the most advanced cutting machines and therefore deserve coverage in this otherwise brief presentation of logging machinery. The two types of harvesters are the following:

- two-grip harvesters with a felling head mounted on a hydraulically operated boom with the delimbing-slashing-bunching device located on the prime mover
- one-grip harvesters with a light and compact felling-delimbing- bucking head mounted on the end of a hydraulically operated boom.

One-grip harvesters are appropriate for use in thinnings, other selective cuttings, and clear-cuttings if the mean stem size is not too large. If such a harvester is used only in clear-cuttings and the tree size is big, a two-grip harvester may be a better choice.

Different sizes of harvesters range from very small used only in thinnings such as Fig. 7 shows to heavy units. A medium-sized harvester having a mass of 10–11 Mg is a good all-purpose machine for thinnings and clear-cuttings unless the trees are very large. Fig. 8 shows an example.

A state-of-the-art harvester has the ergonomically designed cab, seat, and control devices shown in Fig. 9. Good visibility from tree base to the crown and efficient lights are essential, because these costly machines often operate in two shifts so that some hours are dark. The unit has good ground clearance, and its soil compaction properties are acceptable for the reasons discussed when introducing forestry tractors. The power transmission uses hydraulics, and an electro-hydraulic system controls the machine. Automation controls the power transmission, measurement of timber, and crosscutting. The technical reliability and maintainability have improved remarkably during the past ten years including the central lubrication systems.

Automated measurement systems are sufficiently accurate and reliable for commercial scaling and practical operations with proper calibration. A mechanical roll produces the electrical impulses indicating measured length and feeder rolls or delimbing knives serve as diameter sensors. Prediction of the taper curve of the stem uses the first measurements, and some manufacturers have developed systems that can learn the features of the trees of a stand. *Bucking* has been partly automated so that a computer suggests the crosscutting points according to value-and-demand matrixes for logs of different dimensions and quality classes. Definition of log quality is still the responsibility of the harvester operator, since that person finally selects the crosscutting points.

Figure 7. A light rubber-tracked one-grip harvester[11].

Timber procurement

Figure 8. A medium-sized one-grip harvester[12].

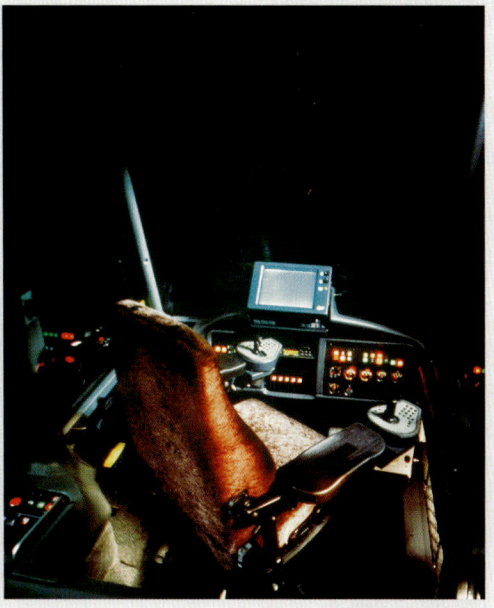

Figure 9. A view into the cab of a state-of-the-art logging machine[12].

Central conversion plants (lower landings and processing terminals) are used in some countries when applying tree-length or whole-tree systems. Massive cranes and conveyers are often necessary to handle large stems. A central conversion plant may be at the mill or at a special terminal enabling optimization of transportation of the various assortments to mills.

Special *forestry tractors such as skidders* and *forwarders* are available for the forest hauling of timber. The following technical features that improved the maneuverability are typical: large and wide low-profile tires, tracks, bogies, or crawlers; frame steering; hydraulic power transmission systems; high ground clearance; all-wheel drive; and hydraulic grapple loaders.

Big wheels or bogies allow easier clearance of obstacles and reduce the surface pressure imposed by the machine on the soil. Frame steering is more accurate in terrain driving with heavy loads than wheel steering. It also prevents the tractor from becoming stuck and can even remove snow by using the front axle's pendulum-like movement.

The first forestry tractors were *skidders* that dragged logs, stems, or trees partly on the ground. A conventional skidder has a winch for moving stems near the tractor, lifting the end of a load off the ground, and binding the load onto the tractor. A worker on foot is necessary to fasten the load onto the wire of the winch. In more developed skidders, hydraulic grapples or grapple-loader and clam-bunks eliminate the need for the choker man and increase productivity. Figure 10 is a photograph of a clam-bunk skidder.

CHAPTER 7

Figure 10. A clam-bunk skidder[13].

Forwarders find use with the cut-to-length method. They carry the entire load in the wheeled unit. Another criterion for classifying forestry tractors is their size – the mass or load-bearing capacity of a tractor, the engine capacity, or both. The classification presented in Table 4 is common.

Table 4. Classification of forwarders according to their load-bearing capacity and engine capacity.

Forwarder	Load-bearing capacity, Mg	Engine capacity, kW
Mini-tractors	<5	<40
Small-sized tractors	5–7	40–60
Medium-size tractors	8–11	50–70
Heavy-duty tractors	12–15	60–100
Very large tractors	>15	>100

The productivity and cost figures in Table 5 are the result of various size classes of forwarders in average conditions. Studies of various machines have used clear-cutting conditions. Because the study referenced here is several years old, the level of productivity achieved today may be higher. The results nevertheless indicate the relative productivities of the size classes.

Table 5. Productivity and relative unit costs of forwarders[9].

Forwarder	Productivity, $m^3 h^{-1}$	Relative unit costs
Very large forwarders	13.9	118
Heavy-duty forwarders	11.7	100
Medium-sized forwarders	10.6	100
Small-sized forwarders	<7.5	107

The medium-sized forwarder of Fig. 11 is suitable for clear-cuttings and thinnings. Because work sites in Nordic countries are often small and consist of clear-cutting and thinning, medium-sized forestry machines have common use. This minimizes the cost of moving the machine from one stand to the next. In an area dominated by clear-cuttings, heavy forwarders may be the most economical choice. If a leading objective is to reduce damage to retention stands and soil disturbances, the small-sized tracked forwarder of Fig. 12 may be the best choice.

Figure 11. A medium-sized forwarder[12].

CHAPTER 7

Figure 12. A small rubber-tracked forwarder[11].

In thinnings, medium-sized or small forwarders are very common. With these machines, the width of the unit determines the width of the strip roads – typically approximately four meters. Forwarder widths are usually less than 270 mm. The forwarder in front of the tractor is narrow compared with its height. This and its high ground clearance oppose the requirements of sidewardstability. Positioning the engine, main frame, and other heavy elements as low as possible can influence the center of gravity. In any case, strip roads for a forwarder require that the sideward slope is as small as possible. If strip roads run up and down the hillside to avoid sideward slopes, forwarders are satisfactory even in mountainous conditions on slopes as great as 40%.

Mountainous conditions require cableway skidding if the gradients or slopes exceed 45%–50% (24°–27°)[14, 15]. This is especially true if the micro-terrain is rough. Cable systems including ballooning may find use even in flat terrain in tropical rainforests to avoid soil disturbance and erosion[16]. The use of cableways means greater logging costs compared with tractor-based forest hauling.

Winch and high-lead systems are relatively inexpensive and simple to use, but soil disturbance they cause can lead to erosion. Plastic tubes may find use for small pulpwood logs primarily in thinnings. Skyline systems lift the load completely off the ground and therefore avoid soil disturbance. Running skyline with an interlocked winch system is an important innovation in cable logging. Because no braking of the return line is necessary, cable tensions are much smaller than in conventional running-skyline systems[15]. Planning cable logging is an art in itself, but the development of computerized systems using terrain models offer substantial improvements.

Figure 13. A new mobile whole-tree chipper for harvesting and transporting whole-trees and logging residues (Photo by Antti Asikainen).

Harvesting low-quality forest biomass for *energy generation* is a new challenge. In industrial energy generation, process residues, bark, and wood have found traditional use as fuel, but interest in using small-diameter trees and slash from clear-cuttings for energy generation is increasing. The harvesting costs of slash are smaller than those of small trees from thinnings. Some logging technologies for thinnings such as multi stem processing by harvesters, flail-delimbing-debarking-chipping/drum debarking, and harvesting of tree sections with branches allow recovery of more biomass for energy.

Integration of energy wood harvesting with industrial wood harvesting to reduce costs is an interesting option. This integration can use the same machines for harvesting both types of wood. An option is to separate the industrial raw material component from the energy component on the same machine – chain delimbing-debarking-chipping machine. Other options are more complicated systems – upgrading plant for segregation of bark from chips. Energy wood is often converted to chips or crushed material. Several kinds of *chippers* are available ranging from small farm-tractor-mounted models to large units mounted on trucks. Some such as shown in Fig. 13 have use on strip roads. Some have use at upper or lower landings. Several working principles are also used from disk chipper to crushers (hogs). The following techniques are under intensive development in Finland:

- accumulating feller-forwarder for small trees
- terrain (mobile) whole-tree chipper (energy wood harvester)

CHAPTER 7

- combined flail-delimbing-debarking-chipping technologies for upper landings or flail-delimbing-debarking and drum debarking technologies for central conversion plants (terminals)
- one-grip harvester capable of multi-stem processing.

Foliage of trees contains much nutrient bound by trees. At least some crown mass should therefore remain on the logging site in nutrient-poor forest stands.

Road, rail, or *water* techniques can transport the timber to the mill. The most economical method depends on the route network, infrastructure, stage of development of the country, transportation distance, and timber assortment. In Finland, *road transport by truck* is cost effective if the distance is less than 100–150 km. *Bundle floating* is the most profitable way over longer distances if a suitable floating route exists. In addition, contact with water must not harm the timber. This method is slower than other transportation. *Rail* and *barge* transport are also economical alternatives over long distances.

A typical timber truck for carrying cut-to-length timber in Finland has the following features: truck and trailer with gross weight of 60 Mg, load capacity of 40–42 Mg, engine output 300–400 kW, gear boxes to 18 speeds, and a variable number of axles, drift-in axles, lift-bar axles, etc., according to the specifications of the purchaser. Figure 14 is a photograph of a timber truck. Trucks are often self-loading, but the crane is detachable.

Figure 14. A timber truck for transporting cut-to-length timber assortments[10].

Figure 15. A log stacker with a telescopic boom[10].

Leaving it at the landing site will increase the load carrying capacity of the truck. Some trucks and trailers have releasable bolsters to drop timber bundles easily from the truck into water for floating. These also facilitate storing at the edge of a body of water by tilting the vehicle. At the mill terminal, unloading of timber trucks or railway cars typically uses an overhead traversing crane or log-stacking forklift trucks (log stackers).
Figure 15 shows a photograph of a unit with a telescopic boom. The optimum size and model of log stackers at a mill terminal depends on the bundle size coming in on trucks or wagons, the amount of timber handled annually, and the shifting distances at the terminal. The room available at the terminal is another factor because the height of piles depends on the log stacker's reach[17].

CHAPTER 7

Development of equipment, work methods, working techniques, and even mechanization have provided huge increases in the productivity of forest work. Moving the debarking stage from the woods to the mill and constructing permanent forest truck roads have further strengthened this development. Shortening the off-road hauling distance and organizing and planning forest operations more efficiently have also offered further improvements. Table 6 provides a review for certain countries.

Table 6. A comparison of the increase in labor productivity in logging[18].

Years	Bulgaria	British Columbia, Canada	Baden, W. Germany	Finland
	\multicolumn{4}{c}{Annual increase in productivity, %}
1953–1960	-	-	6.0	4.3
1960–1964	4.8	7.8	5.7	6.7
1966–1970	0.6	3.8	8.5	9.4
1970–1974	3.0	-0.9	3.9	12.2
1974–1980	2.6	1.7	7.1	7.6

Before the 1960s, the increase in labor productivity in logging for Finland was low. It has since increased faster than in the other countries included in the comparison. Figure 16 shows a continuing increase during the past few years.

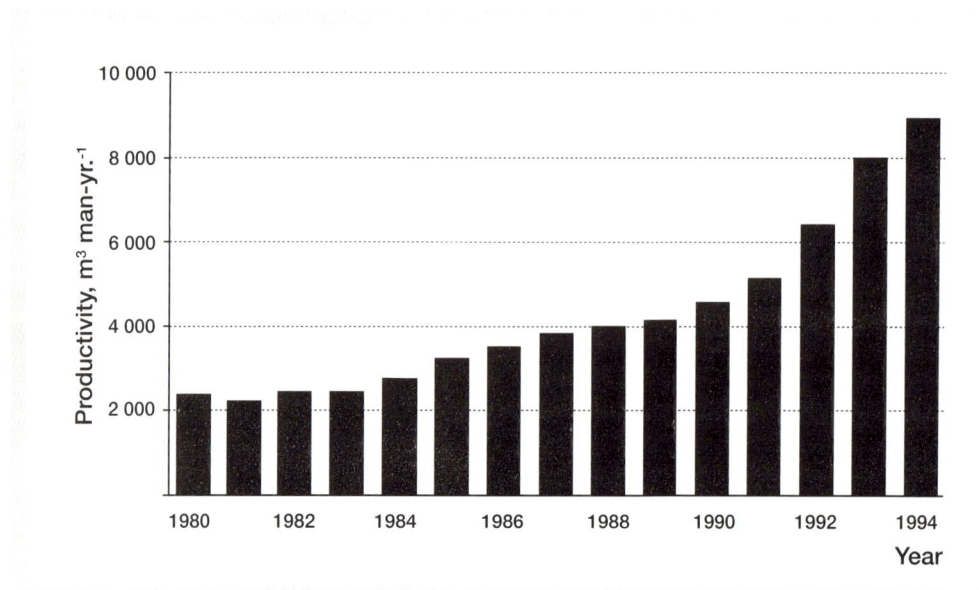

Figure 16. Labor productivity in logging in Finland from stump to roadside according to results obtained by Metsäteho[19].

Timber procurement

Opening forest areas by constructing roads

Constructing *roads*, railways, or cableways can open a forest area. Using natural lake and river systems with canals is also possible. Forest truck roads often complement a public road network. In Finland, three categories of forest truck roads are entitled to government support on private lands: main forest truck roads, secondary forest truck roads, and branch roads.

After planning the *road system (network)*, the next stage is to design and locate individual roads. Road location is particularly difficult in steep terrain[14]. Figure 17 shows some basic models for road networks. Although road construction may be easier along ridges, the location of roads should allow forest hauling of timber down slopes as much as possible. With some cable logging systems, uphill hauling is preferable. In addition to forestry use, road construction must also consider other uses.

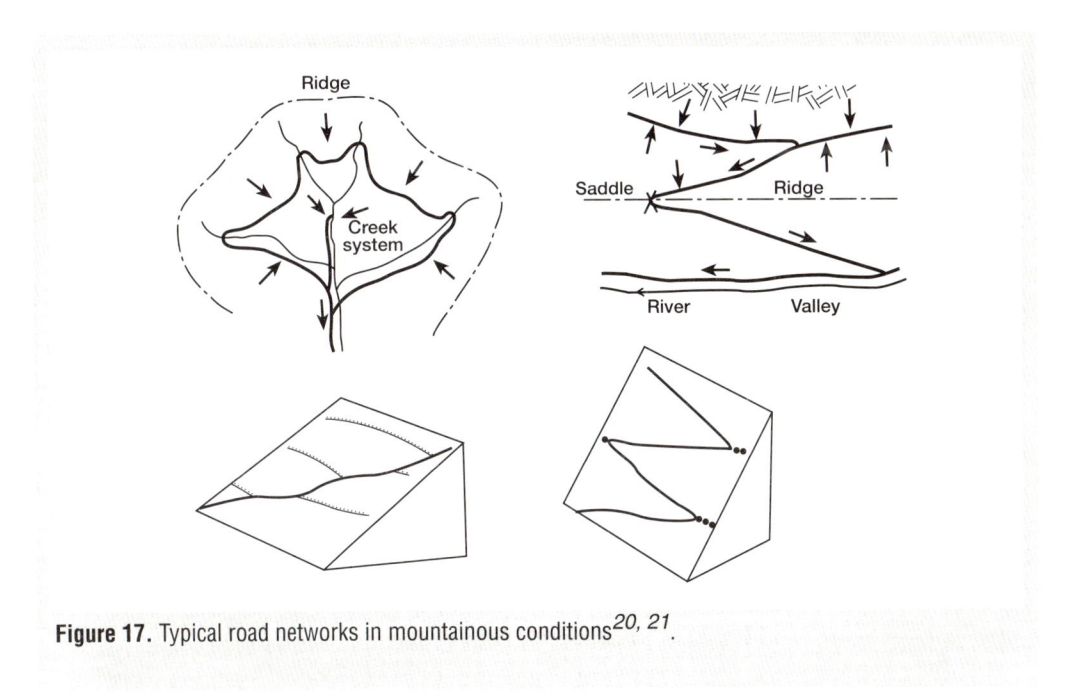

Figure 17. Typical road networks in mountainous conditions[20, 21].

Although straight, parallel roads can theoretically cover an area most effectively, many factors determine the ideal road network and the location of an individual road. Figure 18 illustrates locating a road to avoid wetlands and environmentally valuable areas while retaining a fairly straight line.

Optimization functions can estimate the best *road spacing* in larger forest areas, but these seldom apply in determining the profitability or length of an individual road. Actual *location of the road* is outside the scope of these functions. Several *computerized methods* are available to find the best possible road location. In mountainous conditions, *digital terrain models (DGM)* are particularly useful tools because the steepness of the terrain is a problem. Models that produce *road layouts* take topographical and

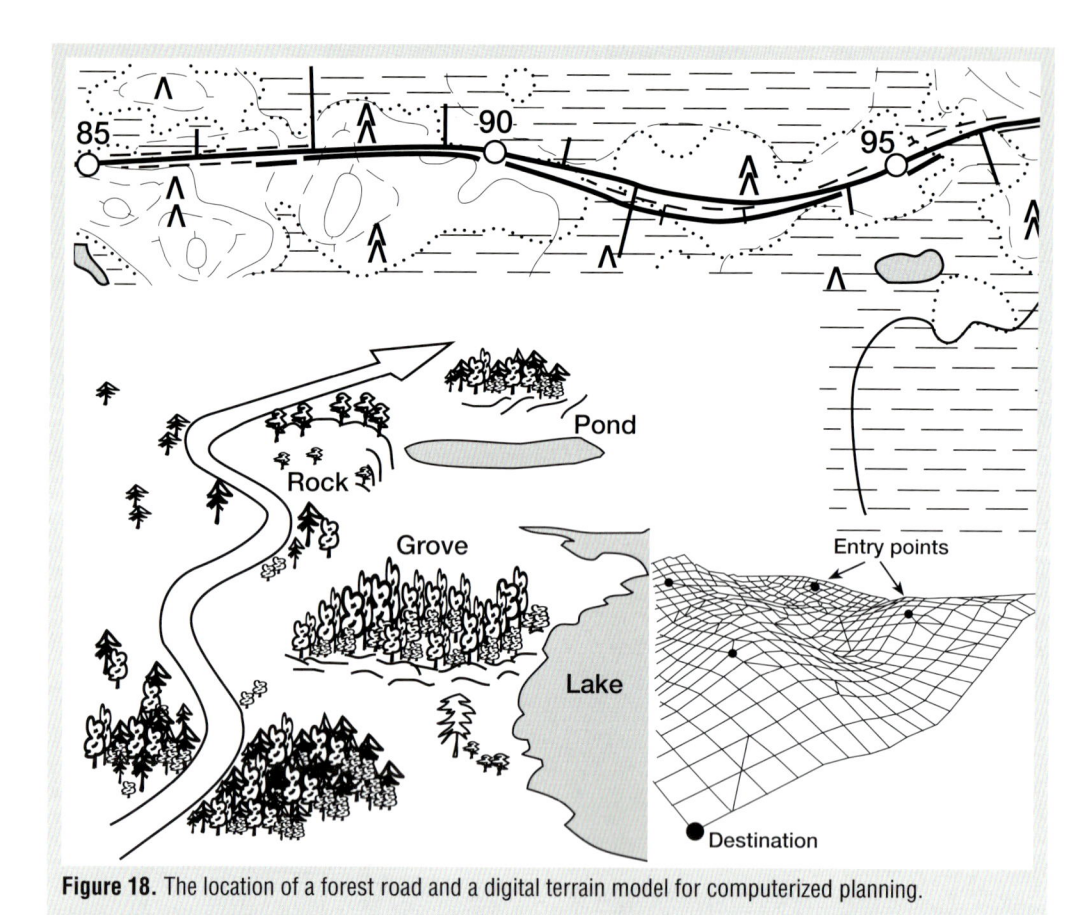

CHAPTER 7

Figure 18. The location of a forest road and a digital terrain model for computerized planning.

geological characteristics, required earthworks, building and maintenance costs, environmental aspects, and harvesting costs into consideration.

Planning of forest roads includes terrain reconnaissance, divider location; road location line including curves; pegging; measurements for earthworks; planning of culverts, outlets, drifts, and bridges; junctions; profile drawings; earth mass computations, and selection of methods, materials and machines. Documentation includes road maps and earthwork computations. *Computerized models* for planning a road profile, calculations of earth masses, and landscape planning area also available.

Many kinds of forest truck roads depend on the conditions and needs to transport timber seasonally or all year. A good forest *road embankment* consists of three main layers:

- surface course (wearing layer): fine particles of different sizes and low porosity of earth material are necessary to form an even material with good bearing and wearing capacity
- base course (pressure dividing layer): coarse material with fine particles and good binding such as moraine with clay are necessary
- sub-base (isolating layer) preventing capillary movement of water: even and fairly coarse structure.

To keep the road dry, the embankment itself must be higher than the surrounding terrain. Drains and surface layout must direct the water from the road.

Construction of forest truck roads consists of planning cuts and profiles, sub-grade preparations and building embankments, culverts, bridges, river controlling, quarrying and transportation of fill and surface material, rock quarries (removal), crushing of stones and surfacing, and preparation of associated areas including stabilization of steep slopes. The two principal methods used in making cuts and embankments are the bulldozer and excavator methods.

A bulldozer is especially efficient in moving soil in the direction of the road but also sidewards. Bulldozers have main use in hilly conditions on soils with good carrying capacities – sandy and gravel soils. They also find use in mountainous conditions because of the lower costs compared with those of excavators.

The excavator method is especially useful in even terrain on soils with low bearing capacities because an excavator can easily lift soil from side drains and ditches onto the main body of the road to achieve the high road structure. This is also a good method for small operations where it is not economical to use a variety of machines. The method is environmentally satisfactory because it can bury large stones and stumps. On slopes, laying a wall of stones can prevent erosion.

2.3 Storing timber

Timber may be stored in the woods, at the roadside, or beside some other transport route at a terminal and ultimately in the mill yard. When using *whole-tree* or *tree-length methods, upper* or *lower landings* or *conversion plants* where bucking occurs also serve as timber storage points. In these cases, enough room must exist for processing the stems and storing the resultant timber assortments. For safety reasons, these activities must avoid crossing machine routes and allow units to work sufficiently far apart. If the carrying capacity of the soil is insufficient, stems easily become contaminated with grit and similar impurities.

Using the *cut-to-length method* decreases the space requirement. Forming piles of timber assortments is even possible over roadside ditches. A *good storage* place has the following features:

- sufficient area to turn a truck and trailer around
- sufficient room for the piles of different timber assortments to transport the assortments separately and independently
- adequate soil carrying capacity and road surface for trucks
- piles within reach of the truck's loader
- no power lines or other obstacles causing hazards for the operation of the truck or loader
- easy access to a public road without endangering other road users
- piles sufficiently large to provide full load without repeated moving from one pickup place to the next.

One should place pieces of timber under the *piles of commercial timber* to prevent the logs from becoming soiled and collecting stones during loading. The section covering damage to wood discusses other aspects of storing timber.

3 Organizing timber harvesting

Owners of large-scale harvesting operations, an association of forest owners, large contractors with concessions to forest areas, enterprises purchasing the timber from forest owners and selling it to industry, and the wood departments of forest industry companies can organize large forest areas. These organizations can use the services of independent *entrepreneurs* (*contractors*) in their operations. Commercial roundwood exchanges resembling stock markets can also sell timber.

In many countries, the *organization* of wood departments uses regional responsibility while also having some functional features. The development of organizational structures has followed the common trend toward a flatter hierarchy and an "open" and elastic structure including projects and autonomous teams. External services and outsourcing are also possible. Small companies especially have started to form enterprise networks.

Procurement enterprises use private forest machine *entrepreneurs* or *contractors* who previously worked under the strict supervision of the wood department foremen. Entrepreneurs today share increasing responsibility by belonging to company *work teams*. Quality management systems monitor the quality of work. In the late 1970s when the productivity and mechanical availability of company-owned forestry machines were compared with those owned by private contractors, the availability and frequently the productivity of the former were worse. This accelerated the change from company-owned machines to use of contractors. When work-site planning and scaling of timber (automated measurement by on-board computers on harvesters) shift from company staff to entrepreneurs, overhead costs decrease. Shifting of work-site planning and similar responsibilities presupposes good training of contractors and operators and random supervision of work quality.

Earnings may use time rate, bonus time rate, daily task, or piece rate. The commonly held concept is that productivity is best when using the piece-rate *payment system* and worst with the time-rate system. The strain on workers, the quality of work, and work safety can suffer as a result. Some forest managers believe that piece rates should not be used in silvicultural works because of the poorer work quality that results. *Quality management systems* provide a method of continuously enhancing the quality of activities and products concurrently with productivity, cost-effectiveness, and environmental and work safety aspects.

When time rates in chainsaw cutting replaced piece rates in Sweden in the 1970s, productivity declined by 10%–17%, and direct costs rose by approximately 30%. The earnings levels decreased by 10%. Work accidents decreased by 9%, and absence due to the accidents decreased by 48%. Due to the higher costs, the time rate system has probably speeded mechanization in forestry.

4 Planning and monitoring of timber procurement

The basic problem in timber procurement like other forms of production is to make the goods (timber assortments) and services available in the right form, at the right place, and at the right time. Optimizing the value of timber in conversion and procurement costs is also a primary objective. To achieve these objectives, every organization must invest in *planning, monitoring,* and *controlling of activities*. The first step is to define the objectives of the organization. Business enterprises base the profitability of their activities on a *business idea* and *strategic planning*. Many timber procurement organizations today emphasize *customer-oriented production, quality politics* and *rapid response* to unexpected changes in the enterprise environment in their business strategy.

Although one year is a common time span for *tactical planning*, resource planning requires longer planning perspectives because training and recruitment of people, investments, and financial arrangements usually take longer. A typical tactical plan is an annual timber procurement plan including the amounts of different assortments by district and a procurement *budget*. This normally has monthly, weekly, or even daily targets that undergo regular updating. Figure 19 illustrates the principle involved. Optimization models minimize procurement costs.

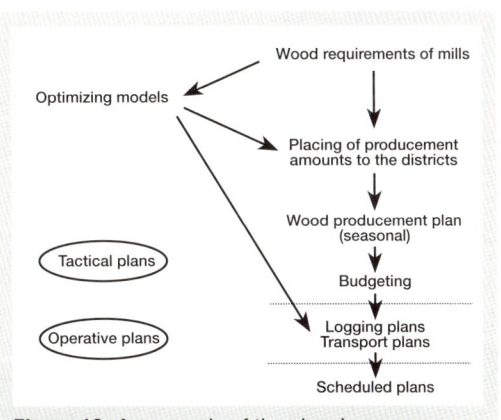

Figure 19. An example of the planning process involved in organizing timber procurement for one year.

CHAPTER 7

A detailed timber procurement plan can include the following items:

- timber requirements
- storage points in the beginning (woods storage points, roadside storage points, and mill-yard storage)
- reserves
- standing timber (purchased)
- standing timber (marked stands on company's own land – delivery purchases)
- purchasing (standing timber – marked stands and delivery-sales purchases)
- entry of timber to roadside storage points
- logging of purchased marked stands (logging from own forests and deliveries)
- exchange of timber between companies
- transportation to mills (truck, rail, water, and deliveries by other enterprises)
- terminal storage points.

In practice, compiling such a plan requires good knowledge of the prevailing conditions and experience with uncertainty factors. Realization and updating of plans are like a game. Nature, competitors, and the unexpected incidents in one's own operations form the opposite team.

In the past, timber-procurement planning goals were to distribute the procurement volume among districts, to meet the minimum quality requirements of timber, and to minimize or at least maintain reasonable costs. Increased competition and the awareness and demands of customers today force enterprises to maximize their benefit/cost ratio. The goals of procurement include maximizing *timber quality*, minimizing *storage time* and invested capital (interest costs), etc. Customer-oriented business ideas involve *real-time steering* of procurement according to the market situation.

All this has resulted in the application of the principles of modern logistics and new management techniques. The efficacy of the management system enables the fast flow of timber from the stump to the mill. The basic prerequisites for this are the following:

- simple organization
- short pass-through time
- ability to handle different products (assortments, qualities, and dimensions)
- ability to set preferences.

In this context, *operational planning* refers to the programming of real physical operations. Weekly or even daily work programs for individual machines and persons are therefore an essential component of operational plans. Operational plans use the *work-site plans* described later. A schedule of projected stands or work sites allows one to choose suitable compartments for a certain period of operations.

A prerequisite of modern, customer-oriented planning and control of operations is a *real-time information system*. Harvesters and forwarders have mobile telephone modems and microcomputers for reporting their daily outputs of timber and for receiving instructions on required log lengths and dimensions complying with the market situation. *Computer-aided bucking* uses the value matrixes of different dimension and quality class combinations of logs and the desired distribution of different dimension and quality class combinations. Choosing appropriate work sites and compartments and the optimum bucking programs for different work sites can use computer simulations shown in Fig. 20.

Timber trucks have similar communication equipment as forestry machines such as satellite-assisted navigation called a *Geographical Positioning System (GPS)*. The systems also use *Geographical Information System (GIS)* containing stored data on stands and roadside inventories linked to a road map. Timber truck drivers receive instructions concerning the inventories, transport volumes, timetables, most profitable routes, and even return loads via a data communication networks. GIS today is a tool for all planning activities from large optimization models and *Decision Support Systems (DSS)* to field surveys aided by portable personal computers.

This kind of an information and control system helps shorten the travel time for timber from the stump to the mill yard, optimize the distribution of raw material to different destinations, improve the quality of timber, and avoid unnecessary overlapping in the transport of timber. For big companies with several mills to supply and big quantities of timber, such an advanced planning system is a prerequisite for efficient operations.

CHAPTER 7

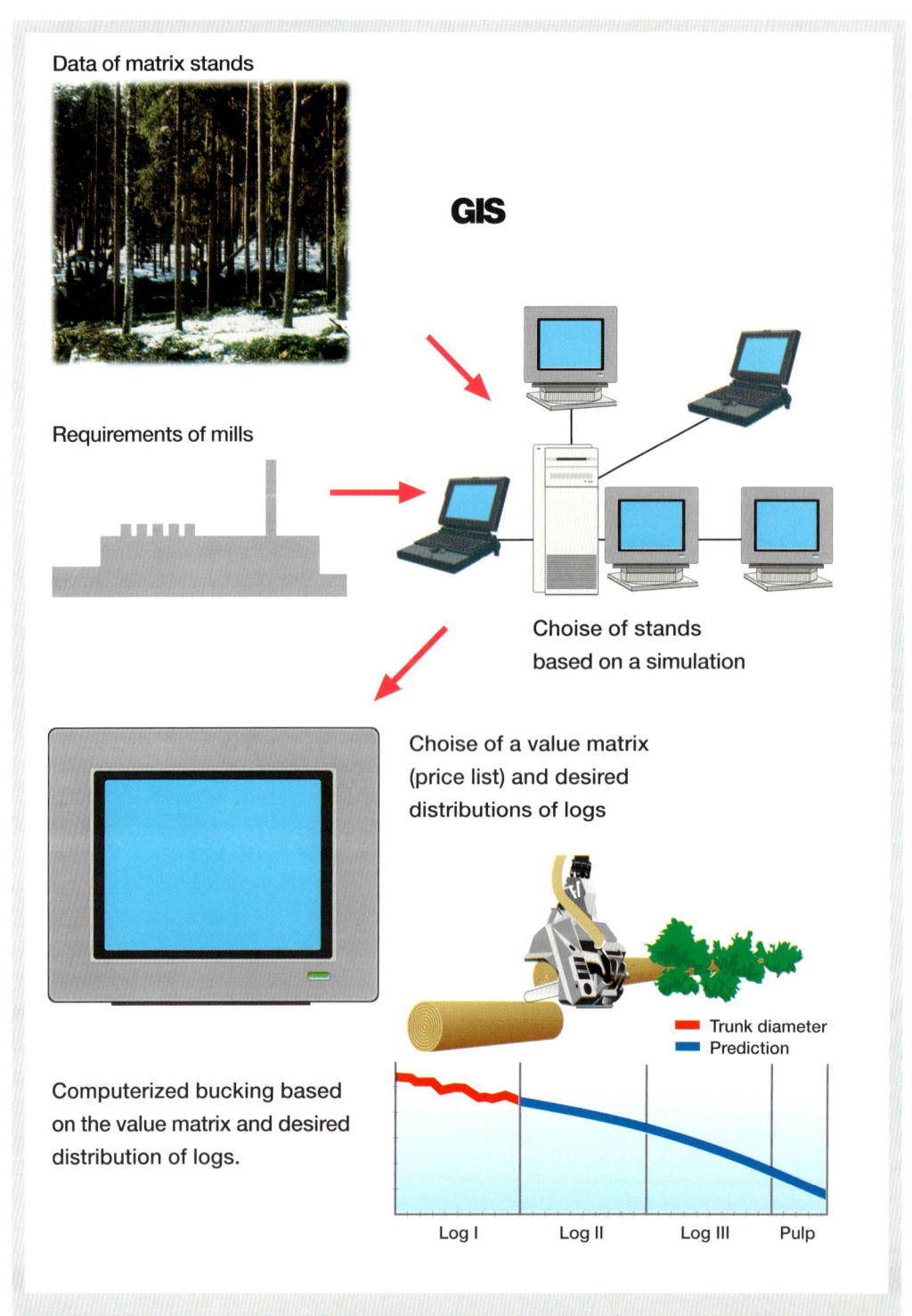

Data of matrix stands

GIS

Requirements of mills

Choise of stands based on a simulation

Choise of a value matrix (price list) and desired distributions of logs

Computerized bucking based on the value matrix and desired distribution of logs.

Trunk diameter
Prediction

Log I   Log II   Log III   Pulp

Figure 20. A logistic system for monitoring and controlling harvesting operations on a real-time basis.

Timber procurement

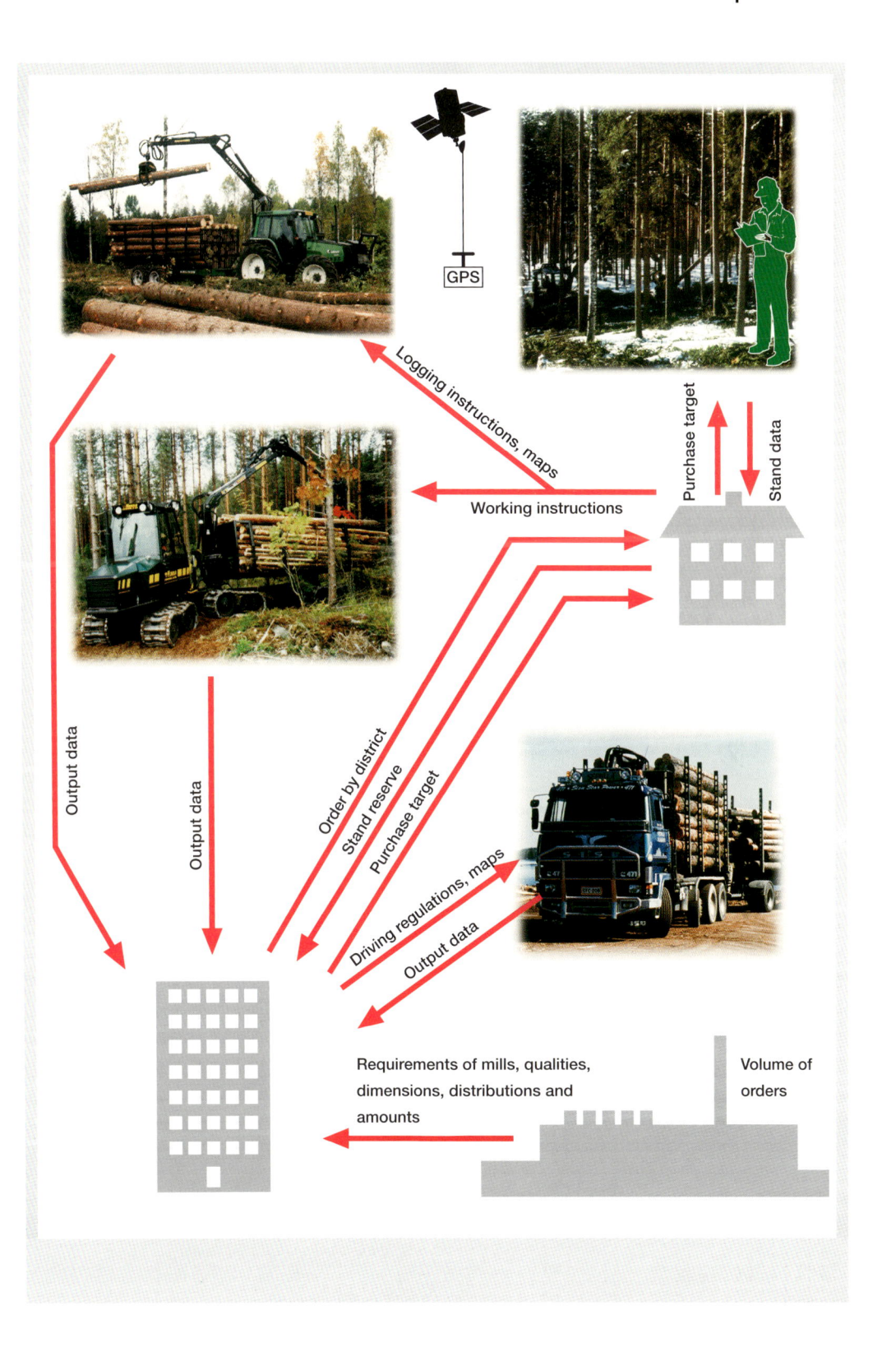

339

CHAPTER 7

In operational planning, *Operation Research* (*OR*) methods assist in compiling work programs for different machines, machine systems, and work teams. *Heuristic models* and applications of *Artificial Intelligence* (*AI*) have use in planning routes and return-load programs for timber trucks. Planning and control systems are usually hierarchical and serve the control of production teams and upper echelons in organizations. Systems serve common management systems such as "*management by results*" and "*quality management*." *Quality management systems* can include "*environmental systems*", "*safety-and-risk control systems*," and certification of production. Development is progressing from fixed and structural models to interactive and flexible *DSS* making *ad hoc* queries possible that use different kinds of data, models, and AI.

Work-site plans provide information for operational planning, and they guarantee productive, safe, and environmentally friendly harvesting operations as Fig. 21 shows. The need for planning depends on the prevailing conditions, but it is especially important in small-scale, private forest holdings where conditions vary considerably. After choosing the sites for cutting and marking the boundaries of the stands, the next step is to decide whether the road network needs expanding. Actual work site planning can include the following:

- suggestions for silvicultural treatments by compartments and sub-compartments to meet forestry, multiple-use, and environmental and landscape goals
- boundaries of sub-compartments for different harvesting methods, seasons, machines, and labor
- estimation of timber volumes according to timber assortments and quality classes
- estimation of condition factors for productivity and cost computations and piece rates
- planning of timber storage to meet the requirements of good storage (see *Storing of timber*) and to minimize off-road hauling distances
- marking of environmentally valuable places needing special treatment or to remain outside the cutting operation
- marking of trees to be removed in selective cutting unless it is done by cutters or machine operators (setting the density of the retention stand)
- planning of main strip roads and strip roads unless this is done by operators
- productivity, cost, and resource need calculations
- compiling work-site map including index, roads, compartments and sub-compartments, landings, strip road network, environmentally important places, places needing special treatment, safety aspects (power lines, paths, etc.).

Timber procurement

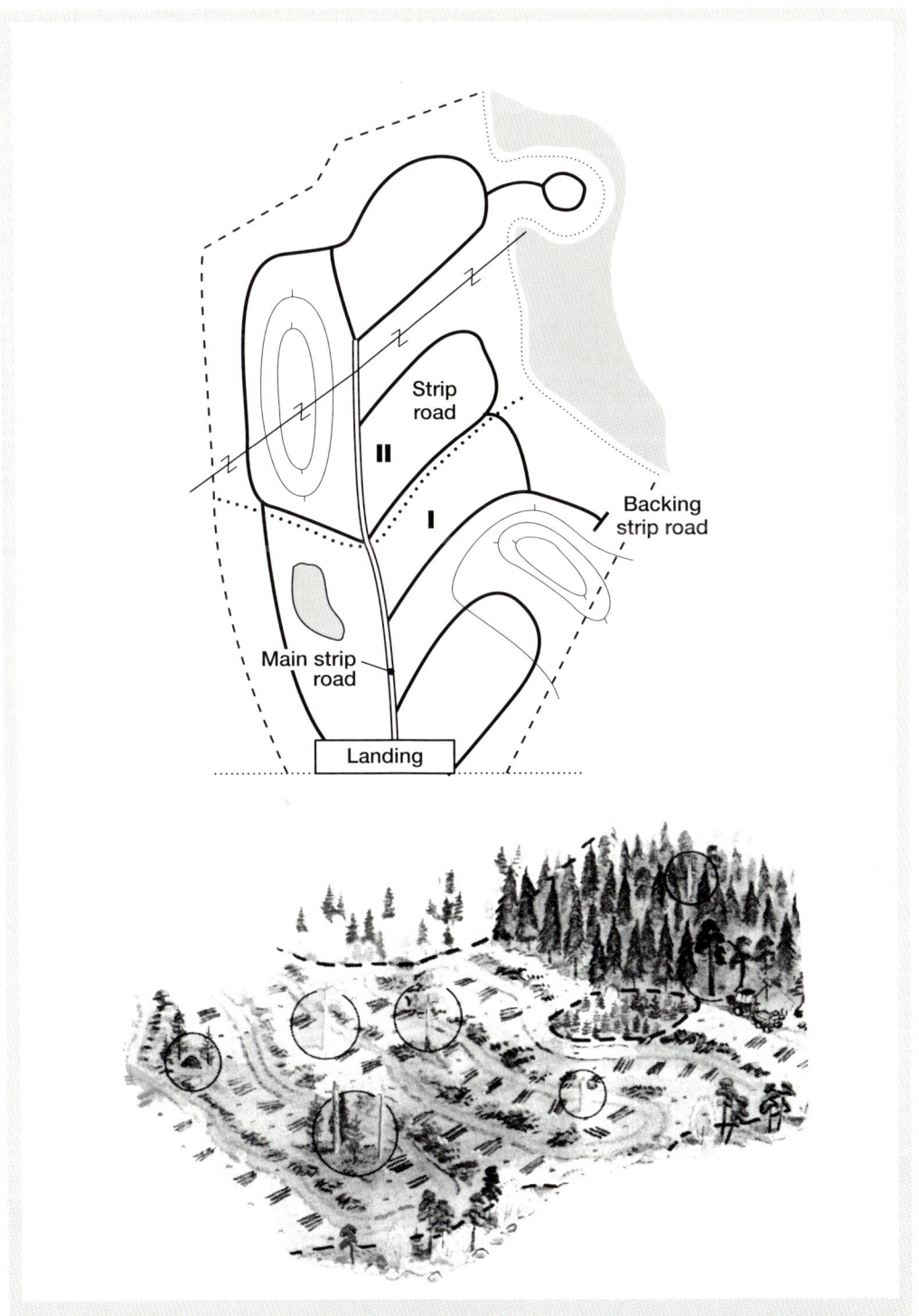

Figure 21. Work site map[20] and some environmental aspects of promoting biodiversity (Skogsarbeten).

CHAPTER 7

Environmental aspects are an important part of work-site planning. Key biotypes, riverine forests, and other buffer zones may be protected or cut by applying a special method, tree groups may be retained, ancient trees and decaying wood material may be left to promote biodiversity, etc.

A well planned *strip road network* increases the productivity of off-road hauling and prevents soil disturbance and damage to the retention stand in thinnings. Many factors require consideration when planning strip roads and skid trails. Information on these matters is available in relevant textbooks[20].

5 Damage to timber

5.1 Causes of damage

The woods of different tree species vary in their *resistance to decay.* Many species of trees have an outer zone under the bark – sapwood – that is more susceptible to decay than the heartwood. Moisture content is a factor significantly influencing the risk of decay in wood. The following items can cause damage and destruction of timber: mechanical blows, wear or contamination, weathering, and other physical alterations; exposure to fire, dampness, drying, or chemicals; chemical reactions, evaporation, and other modifications; fungal decay; insects, and aquatic organisms; attacks by other animals; decay or discoloration caused by bacteria, viruses or other microorganisms, and impacts of higher plants.

Climatic conditions and biotic organisms have such global variation that the following presentation is a generalization. The species named only indicate examples.

5.2 Abiotic damage

Timber can become damaged or *soiled* during the logging and hauling stages. Typical elements causing *mechanical breakage* are felling, delimbing, crosscutting, and skidding. Incorrect felling technique without adequate felling notches or undercuts on the side opposite the felling cut may cause splitting of the tree butt. Big trees can easily break on steep slopes if felled down the slope. Very "aggressive" feeding rolls on harvesters can damage the surface of logs especially at the beginning of the growing season. Delimbing knives can cut into stemwood when delimbing crooked stem – especially broadleaved tree species – unless the feeding and delimbing devices have proper design and adjustment. When working with a one-grip harvester, supporting the butt-end of the log on the ground can avoid splitting of logs during crosscutting.

Charred wood and soot are a nuisance in the pulping process, but they are not always visible on the trees exposed to forest fire a long time ago. Dirt and pieces of rock or plastic are further problems in mill processes. Pieces of timber placed below the piles and care in loading are important in eliminating these problems. The public should know not to put plastic material in piles of timber.

Adequate *drying* after debarking protects timber from many microbial agents of decay during extended storage and reduces transportation costs due to the loss in weight of the timber. Inadequate drying when storing unbarked timber makes wood more susceptible to rot. Drying itself is harmful for most industrial processes. Groundwood pulp mills and sawmills especially require freshly felled timber. Even the sulfate process is easier to manage when using freshly felled timber. The partial breaking of the bark sheath during logging is therefore disadvantageous because it promotes the drying of the wood in summer.

Free water in wood occurs in the cell lumens and cavities, and bound water occurs in cell walls. After removal of the free water and other fluids by capillary and static pressure and diffusion, the bound-water begins to remove by diffusion. Then the structure of wood can change, and its permeability can diminish. Pit aspiration significantly reduces permeability in softwoods especially in the heartwood. One result of drying is checking. This can spoil saw logs and peeler logs. Shrinkage is maximum in the tangential direction and causes radial cracks. Dry wood is harder, and the bark attaches more tightly to the wood. This causes raw material loss or higher bark content in chips following drum debarking.

Chemical reactions, respiration, and evaporation alter the chemical properties of wood. Storing of timber causes dark coloring of the wood due to *chemical reactions* such as oxidation of extracts and fungal decay and the *excretion of chemicals*. As an example, tannin easily excretes in water string from the bark of spruce into the wood.

Ambient temperature, moisture, wind, precipitation, pile size, log length, bark damage on log surfaces, and wood properties influence drying. Debarking or splitting of the logs and storing the timber in direct sunlight with exposure to wind can intensify drying. The seasoning of whole trees after felling and cross-wise piling of logs are additional means of accelerating drying. Covering the top of timber piles to prevent rainwater from entering the piles is useful when storing firewood. Storing the timber in water or irrigating it can hinder drying. The above-water parts of bundles of timber can remain dry when floating timber unless the bundles are turned occasionally or irrigated.

5.3 Fungal decay

5.3.1 Introduction

Biotic decay occurs mostly during the storing and seasoning of timber. The durability of timber exposed to the weather or in contact with the soil is largely the result of its *resistance to fungal decay*. Fungal attack can cause *discoloring* of wood. *Blue-stain fungi* do not impair the mechanical strength of wood, the hyphae stain the wood using the cells' soluble contents as their nutrition. *Wood-rotting fungi* also affect the color of wood, although their main impact is to cause rot. The optimum moisture content for the growth of most wood-rotting fungi is 35%–75% on a dry-weight basis–the minimum is approximately 22%. Fungal growth can start at the temperature of 65°C, although the optimum in the boreal zone is 22°C– 30°C.

Fungi *decompose wood* by secreting acids and enzymes that dissolve the wood's cellulose, hemicelluloses, lignin, and pectin. The decay process requires oxygen and moisture, and it releases carbon dioxide and water. Acidity increases in the wood, and this increases the consumption of NaOH in the pulping process. Decayed wood is soft and crumbles easily.

5.3.2 Brown rot

Fungi that initially use holocellulose but leave most lignin untouched cause brown rot. These fungi usually occur in coniferous trees, but approximately 27% of the species specialize in causing decay in broadleaves. Rotten wood is brown or even black and forms small cubes on drying.

5.3.3 White rot

White rot is typical for broadleaved trees, but 18% of species attack only conifers. The fungi that cause white rot consume lignin 2–3 times faster than holocellulose especially during the first stage of decay. The fiber yield in chemical pulping will therefore be reasonable unless the wood has bad decay. White-rot fungi are interesting alternates in the development of biological pulping, but the problem is that these fungi decompose lignin faster when they have access to carbohydrates. This means the simultaneous decomposition of lignin and cellulose.

One type of white rot has the term *corrosion rot* or *white pocket rot* because the afflicted wood often has uneven decomposition. In the first stage of decay, both cellulose and lignin decompose evenly, but lignin is used later in the light colored pockets leaving pure cellulose. The decomposition of lignin advances from the cell membrane toward the cell wall. After consumption of the lignin, a large proportion of cellulose also decomposes.

5.3.4 Soft rot

The third type of rot is *soft rot* that forms several hard surfaces perpendicular to the wood fibers. The wood retains its form and hardness for a long period, but it can break at right angles to the fibers. Because the fungi causing soft rot need less oxygen, soft rot is typical in moist conditions as in poles contacting soil. In chip storage, these fungi are also an important agent of decay since they thrive in the high temperatures often occurring in chip piles.

5.3.5 Comparison of rot types and prevention of rot

In the first stage of decay, the fastest loss of dry matter occurs when brown rot is present. Decomposition of cellulose is fastest in soft rot and brown rot[22, 23]. Various works have studied the effects of rot types on beech wood in detail[22, 23].

In the Nordic conditions, the loss of dry matter in roundwood stored in piles is typically 2%–5% during the first summer of storage. Table 7 presents the relationships between weight loss and impact-bending strength. The impact of rot on the strength properties of wood is dramatic even in the early stages of the decay process.

Table 7. The relationships between loss of mass and impact-bending strength of wood[24].

Loss of mass, %	Brown rot Hardwood	Brown rot Softwood	White rot Hardwood
	\multicolumn{3}{c}{Value of sound wood, %}	
1	6–27	28–38	21
4	60–70	25–55	26
6	80	62–72	50
10	70–92	85	60

A short *storage period* and retention of a high moisture content in the wood by avoiding breaking the bark, irrigating, or using water storage are practical means of avoiding fungal decay. A common goal is fast flow of timber from the woods to the mill. Some companies consider two months as the maximum storage time in summer for pulpwood. If timber requires storage for a very long period, the moisture content should be as low as possible – preferably below 27%. This usually requires debarking and seasoning of the timber in a well organized storage facility.

5.4 Attacks by insects and aquatic organisms

The *nutritional requirements* of timber-boring insects vary greatly. Some of them or their newly-hatched larvae feed only on the inner bark and in some cases merely on the fungi in their galleries without destroying the actual wood. These insects can carry with them blue-stain or wood-rotting fungi. Other insects penetrate into the wood causing damage to the extent that collapse of wood structures occurs. Examples are certain termites.

Like fungi, insects can also have classifications:

- by the natural order of insects - beetles *(Coleoptera)* and termites *(Isoptera)* – do the most damage to wood)
- according to the stage of their attack on wood or bark (living trees, logs, storage piles, buildings, and other wood structures)
- by their source of nutrition (cellulose, starch, fungus, etc.)
- by the state of wood suitable for them (dry wood or fresh wood).

Living conditions favorable to insects also vary. Some favor shade and moist conditions, some prefer sunlight and warm places. Timber can often become infested only if the female insects can deposit fertile eggs into the bark or wood. One way to *prevent their attack* is to avoid the *storage* of timber in the woods during the *swarming time* of the insects. *Other means* include chemical treatment, storing in water, covering piles, debarking timber, and short storage times. One must know the living habits of the insects in question to find appropriate prevention measure.

In *tropical* and *subtropical zones,* insects pose more serious threats. The approximately 2 000 species of *termites* are especially noteworthy. From the practical point of view, classification of termites uses their habits and mode of living:

- subterranean termites: underground-nesting termites, mound- building termites, and carton-nest termites
- wood-dwelling termites: dry-wood termites and moist-wood termites.

Attacks by ground-dwelling termites can be *prevented* by blocking their access to a wooden construction or pile of wood by mechanical shields. This is expensive for timber piles. Other preventive measures include chemical treatment and rapid transfer of timber from the stump to the mill. Biological agents are also under consideration. Termite-resistant natural timbers do exist; although they are not totally termite-proof. The heartwood of some species is very resistant.

Aquatic organisms can attack wooden structures and timber during floating or storage. This is especially true in tropical sea water and the brackish waters of lagoons or deltas. If logs are only stored for a short period before shipping, the risk is small. For instance, the microscopic larva of the shipworm (*Teredos*) can settle on the wood surface and bore minute entrance holes. On developing into adults, they form bigger galleries deeper and deeper into the wood. Almost no signs of infestation are detectable on the log surface, but severe internal damage is a reality.

5.5 Decay caused by bacteria and biological reactions

For other wood-damaging microorganisms, *bacteria* are worth mentioning. Bacterial degradation can occur during storage or floating of timber for long periods. This is more common in softwoods, but it also occurs in hardwoods that resist soft rot. Bacteria can reduce the amount of pectin, sugar, and starch in wood, but the loss of cellulose or lignin is minor. In the conditions of the Nordic countries, notable effects of bacteria can be visible in pine saw logs and birch peeler logs stored in water for several weeks. Bacteria can destroy cell pits and pith ray cells and make the wood more permeable to liquids. This can hamper painting and other surface coating of sawed goods.

Other biological and chemical functions play significant roles especially in the storage of chips. A specific section deals with this topic.

5.6 Influence of storage on mill processes

To give an idea of the impacts of storing wood on industrial processes, some examples of the results of research conducted mainly in the Nordic countries follow. For instance, *storing of logs* can cause the following costs in *sawmilling*[25]:

- Sprinkling, extra handling, and local transportation can increase costs.
- Handling damage, sunken logs, and eroding of bark can lead to loss of raw material.
- Difficulties in sorting by quality due to discoloration or other biochemical or biological defects can occur; loss of bark may hamper sorting by diameter.

- In a frame-saw mill, capacity could decrease because of dry logs.
- Dry logs or soiled logs can result in increased frequency of sharpening and maintenance of saws and blades.
- Very wet or dry or frozen bark can hamper debarking.
- Energy requirement for kilns increases when drying sawed goods made from wet-stored logs.
- Drying time may increase when drying timber sawed from wet- stored logs.
- Capital invested in stored timber means interest expenses.
- Energy consumption of kilns decreases when drying sawed goods made from land-stored logs and the value of bark as fuel increases.

Different ways of storing and transportation influence the properties of spruce *mechanical pulps* in summer due to drying and chemical reactions. In winter when temperatures fall very low in the Nordic countries, no noticeable changes happen. Shortly after the end of winter in approximately June, the difference in the brightness of pulp made from pulpwood with different storage histories is minor or negligible. Two months' storage in water in summer means a decrease of approximately two units in pulp brightness. This then means significant extra costs in bleaching. In August, the brightness of pulp made from floated and irrigated timber is 2–4 units lower than that of pulp made from green pulpwood. When the dry matter content was 42%–60%, no distinct impact on the strength, linting, or printing properties of the pulp or newsprint made from it occurred[26]. If floated bundles of timber are not irrigated, drying occurs in the upper parts of the bundles. When this happens, the strength of pulp decreases, and difficulties arise in the groundwood process itself. New processes are not so sensitive to dry pulpwood[27].

In another experiment, the land-storage of pulpwood for 1–2 months caused a fall of one unit in brightness and no distinct reduction in the strength properties of the resultant groundwood pulp. A storage time of 3–4 months clearly influences the rip and tensile strength of pulp. Even longer storage times decrease brightness and tensile strength[28].

The impacts of storing wood on *chemical pulp* are smaller than those on groundwood pulp. The wood loss is approximately 3%–5% per year of storage. Drying, fungal decay, and chemical reactions can cause additional wood loss in debarking and increase cooking time and consumption of chemicals[29, 30]. Although turpentine recovery does not decrease significantly during the storing of roundwood, that of pine oil does decrease[31]. If pine oil is not used, the loss of resin during the storing is beneficial to the cooking process. Besides the losses incurred during the storage of timber, reduction of interest costs due to invested capital is another reason to favor short storage times.

CHAPTER 7

5.7 Decay in chip storage

Often the *turnover time* in the big chip storage piles at pulp mills is only a few days and no biological deterioration occurs. Prolonged storing and transport times can be involved because of the need to keep buffer stock in a mill, breaks in production, long transportation distances, or using chips for energy generation.

Wood and especially bark and foliage among chips are susceptible to *biological* and *chemical reactions* and *fungal decay* during prolonged storage. The following text uses articles and textbooks by Assarsson[32], Bergman[33, 34], and Hakkila[35]. At some small consumption points, chips used for energy generation can be shielded from rain, and artificial drying can be applied. The moisture content of chips is usually rather high. It can vary considerably and conditions can favor deterioration in a chip pile. Heating in chip piles can cause fire hazards, and fungal spores can cause health hazards to mill employees. A continuing debate also concerns the use of silos for chip storage at pulp mills. The main deterioration processes in chip piles are the following[35]:

- *Respiration* reactions in live parenchyma and cambium cells consume nutrients and release carbon dioxide, water, and heat. When wood is in small particles, the diffusion of oxygen into it and carbon dioxide outside it increases. This accelerates the reactions. Respiration starts at 0°C and increases up to 40°C. At higher temperatures, cells start to die. They typically cease functioning at 60°C.

- *Microbial reactions* are primarily the result of four basic types of microorganisms. Stain and mold *fungi* (*Ascomycetes* and *Fungi imperfect*) cause discoloration problems, soft rot, and allergic reactions in humans through the dissemination of spores. Wood decay fungi (*Basidiomycetes*) are mainly responsible for the decomposition of wood chips. They are white rot fungi that attack carbohydrates and lignin or brown rot fungi. Soft rot fungi (*A scomycetes* and *Fungi imperfecti*) attach themselves to the surface layer of wood chips making them soft. Only rarely have soft rot fungi been isolated from stored wood chips. Their attack proceeds slowly, but it can be extensive because they can grow at low and high moisture levels that can be inhibiting to other fungi. The fourth category contains yeast and bacteria, which appear to have less importance in chip storage.

- Most *decay fungi* have optimum temperatures of 20°C–30°C and are inhibited when the temperature rises above 40°C. In large chip piles, the conditions in the center of the pile are usually too hot for these fungi. Fast growth of decay fungi occur at 25%–60% moisture content. These fungi also need oxygen that is usually available and bound nitrogen. In wood free of bark and foliage with its small amounts of nitrogen, this can be a limiting factor for decay fungi.

- Various *organic chemical reactions* occur in piles of wood chips. Unsaturated extractives in dead cells readily react with oxygen in exothermic reactions. These direct oxidation reactions gain some importance at approximately 40°C and primary importance at approximately 50°C. If the temperature increases to the point of ignition that can occur in pockets formed by fines, bark, or foliage, the excessive heat is probably due to chemical oxidation reactions. Acid hydrolysis of cellulose components is also a thermogenetic reaction.

Wood chips are hygroscopic. This means they absorb and lose moisture. Rainfall and the shape and size therefore influence *moisture content* of uncovered piles. A conical pile absorbs less moisture from rain than a flat one. When water vapor passes through the surface layer of the pile, it cools and condenses. The top 1–2 m are therefore more moist. Because of the increase in temperature and evaporation, microbial reactions start in the interior of a pile due to the chimney effect. Air from the outside comes in around the base of the pile and becomes heated. The upward current of warm moist air contains abundant air-borne microorganisms that accelerate their activity in the pile.

The *temperature* in a pile of wood chips depends on the ambient air temperature and precipitation, the size and compaction of the pile, and the content and distribution of bark, foliage, and fines in the pile. The temperature in the center of the pile normally increases 1°C–2°C/day during the first month of storage in the climatic conditions of the Nordic countries. With further storage, a maximum temperature of 60°C–70°C is normal, and it can remain constant. Some experiments have shown that the maximum temperature occurs even in 2–3 weeks. Birch chips compact more easily than pine or spruce chips. This promotes the rise in temperature. In tropical conditions, the maximum temperature occurs soon after the accumulation of the chip pile. In the Nordic winter, large parts of the pile may be frozen. Excessive freezing is avoided by introducing warm air or steam at the bottom of chip piles. The rise in temperature causes fibers to break, promotes the discoloration process, and inceases the fungal activity somewhat.

Following prolonged storage, the temperature in chip piles can rise to 100°C or higher, and the risk of charring or spontaneous *combustion* increases. Because this risk is high in piles of *whole-tree chips* destined for energy generation, Thörnqvist[36] recommends the following preventive measures:

- storing assortments of wood fuel in different piles (most fires originate in the border area between assortments or between compacted and noncompacted chips due to different permeability)
- attempting to minimize the dispersal of the moisture content in the pile
- avoiding metallic objects in the pile (assumed to function as a catalyst)
- storing chips in oblong piles with the width of the base equal to twice the height
- avoiding pits or humps along the side of the stack
- arranging piles in the direction of the prevailing wind
- limiting pile height to less than 6–7 m for good drying and minimization of energy losses. While avoiding self-combustion, height of noncompacted hardwood whole-tree chip piles should be less than 12 m and that of compacted piles less than 9 m. The height of piles of softwood chip piles should be less than 7–10 m (7 m for compacted piles), less than 4–7 m for bark piles, and less than 4–6 m for saw dust piles.

CHAPTER 7

Discoloration of chips can occur soon. It is often most extensive in the hotter areas of the pile with low pH. During summer storage in cool and temperate climates, the greatest *wood (dry matter) losses* occur along the sides of piles in temperature ranges of 20°C– 50°C. In winter, the highest wood losses occur in the center of a pile. Wood loss for piles averages 0.5%–1.0% per month in cool and temperate climates during the first 2–3 months, but this can accelerate if the storage period is longer. After 6 months, the chips may have suffered bad damage. Prolonged storage in warm and moist climates leads to wood losses approximately twice as high as in a temperate climates. Although usually no or only minor *relative* changes occur in the cellulose, hemicellulose, or lignin contents of softwood chips during normal chip storage, storage of 24 months showed that arabinogalactan, xylan, and glucomannan mostly degraded in pine and spruce chips[37]. The lignin content increases in birch chips while extracts and xylan decrease.

The effects of chip storing on pulp are the following[38]:

The *sulfate* process:

- Yield of kraft pulp (based on oven-dry wood weight) remains unchanged after 2–3 months of outside chip storage (OCS). One must add the wood loss to the total storage loss. In cool and temperate climates, wood loss has been 0.5%–1.0% per month at the beginning of the storage period.
- Prolonged OCS of coniferous chips up to 24 months reduced the pulp yield by 2%–5%. Wide variation occurred in the chip pile.
- OCS has an unfavorable effect on the yield of by-products such as pine oil and turpentine. The decrease in pine oil yield results of loss of resin in the wood and chemical changes in resin oxidation. Resin becomes more water-soluble.
- The strength of kraft pulp decreases with increasing storage time.
- Acidification of chips due to fungal activities increases the consumption of NaOH needed to neutralize chips.

The *sulfite* process:

- In addition to the above, the brightness of unbleached pulp dramatically decreases with increasing time of chip storage. No problems in bleaching the pulp occurred. Considerable savings resulted in the consumption of bleaching chemicals.
- Besides the quantitative reduction of resin, a qualitative effect occurs. The oxidation of fatty acids decreases the chlorination of the resin. This is an important factor for the quality of viscose pulp and paper production. In viscose pulp, chlorinated resin releases harmful particles. In paper mills, the sticky chlorine-rich resin is largely responsible for pitch problems.

Refiner mechanical pulp milling:

- Ratio of big particles increases.
- Strength properties of pulp decreases rapidly. Even a storing period of one week may markedly influence the strength.
- Brightness of unbleached pulp decreases, and consumption of bleaching chemicals increases.
- Amount of extracts decreases.

Because these are significant factors, storing in a silo is preferable for chips used for producing mechanical pulp.

The *resin content* of chips decreases significantly during storage. This decrease is greater in pine than in spruce or birch chips and may be more than 50% of the total amount during one month. The greatest losses usually occur in the hotter, central parts of the pile. To a large extent, rapid resin seasoning is a result of temperature-dependent chemical reactions. A partial or hypothetical explanation is also due to living cell respiration and fungal and biochemical reactions.

Because pulp mills need homogeneous material for pulping and different kinds of mixtures can be applied, chips from different sources may be stored separately. Wind causes wind-screening of chips especially when using blowers to convey chips or when dropping chips from a height. Some modern chip storage systems therefore use conveyers and stackers for forming the pile from the top. Chips of different qualities will then form layers of uniform thickness. Homogenization happens when chips are drawn down with layers intact in place with the out feed. The reclaimer cuts a section through all chip layers obtaining a uniform quality, size, and moisture content. The pile made by a stacker is not a compacted one and contains considerable air and oxygen. Biological and biochemical processes therefore happen in these modern storage systems more easily than in more compacted, old-fashioned storage systems. Short turnover time is of crucial importance.

If new layers of chips are placed above old ones, fungi rapidly infect the new layers. The best practice is therefore to use *plus and minus piles*. The minus pile is the one used for feeding the pulping process, and no new layers of chips are placed on it. The plus heap is then used for storing new chips and so on.

Deterioration of chips can be reduced in chip piles by the following (34):

- fast and even rotation of chips in storage without mixing older layers with newer ones
- good housekeeping
- size of pile
- control of fines
- silos for storing
- encasing of chips (Oxygen must be below 1% – an expensive implementation)
- storage in water that is expensive or irrigation of piles
- water spraying that is inefficient because spraying decreases temperature but has no other positive effects because of the micloflora typical for chip piles
- irradiation of chips
- chemical treatment (Various chemicals have been proposed and tested in OCS situations including pulp mill chemicals) (Chemicals should not harm the pulping process, cause atmospheric or soil emissions, or human health hazards).

In summary, storage periods at mills are normally so short that modern storage systems, good housekeeping, control of fines, modern conveyers and stackers, and effective rotations are the techniques in normal situations. Other means are either too expensive or ineffective. Chemical treatment of chip piles requires an impermeable base.

5.8 Storing of bark

Seasonal *storing of bark* obtained from water-stored or sprinkled logs without artificial drying is difficult. In trials, the average post-storage moisture content was nearly the same as before. High dry-matter losses and reduction of energy content have also been measured. In addition, the presence of high spore numbers on and in the material is a health hazard[39]. Storing of dried bark is easier. The moisture content of the chips can decrease especially with ventilation of the pile. Dry-matter losses have been low, and the growth of fungi fairly limited[40]. One possibility is to mix bark with some dry material such as shavings before storage[41].

6 Environmental impacts of timber harvesting

6.1 Common conceptions

Logging often suffers the blame for destruction of forest areas and *desertification*. Although logging may be part of the process, it is usually very complicated. For example only some tree species in tropical rainforests are commercially desirable. Forest cover therefore remains when logging is complete. The cutting may decrease biodiversity unless regeneration of removed tree species occurs. Opening forest areas by constructing roads makes it easier for people to enter the forest area unless such access is controlled. Shifting cultivation (slash-and-burn), forest fires, and cattle grazing cause destruction of forests. The naturally thin humus layer soon depletes, and erosion follows conversion to desert.

Water erosion can cause serious problems especially in mountainous regions. The erosive force of water depends strongly on the *rate of flow of the water*. The exponential impact of the flow speed underscores the importance of controlling the speed of water flow for erosion control. Most measures taken to minimize the erosion-promoting impacts of logging therefore incorporate control of the flow rate. *Other important aspects* are as follows:

- protection of riverine forests and banks of rivers and streams (buffer zones, no skidding trails across stream beds)
- appropriate planning of the shape, direction, and size of clear-cuttings and openings to prevent windthrow, etc.
- appropriate logging season (avoiding logging during rainy season)
- appropriate planning of road networks, construction techniques, and maintenance of roads
- proper planning of work sites and logging operations
- appropriate technology, working techniques, and supervision of operations.

The following *environmental impacts* are also relevant:
- emissions from and consumption of nonrenewable energy resources
- removal of nutrients with timber from the forest. This is especially true in whole-tree logging, because the foliage contains a large proportion of the tree's total nutrients. Even partial delimbing, topping the trees, or seasoning of trees with foliage intact before forest hauling are therefore beneficial. Nutrient-poor forest sites are especially sensitive to the removal of nutrients. In conditions sensitive to forest fires, one should collect and remove logging residues or slash to reduce the risk of destructive forest fires.
- erosion and leaching of nutrients following clear-cuttings (buffer zones around lakes, rivers, and other watercourses; avoiding intensive soil disturbances and soil preparation methods in sensitive areas; etc.)
- pesticide and wood-preservation chemical residuals. Avoiding harmful impacts here includes using chemicals that have been tested and recognized to have minimum harm, avoiding use of chemicals by good planning and timing of operations, and using spraying methods confining the chemical to the intended object. Work safety during these actions is also important. Most chemicals absorb in the human body via respiratory organs, alimentary canal, and the skin.
- impacts on landscape, multiple-use, biodiversity, and state of health of forests stands (damage to retention trees and soil disturbances).

CHAPTER 7

6.2 Energy consumption and emissions

Mechanization of wood harvesting has meant a ten-fold increase in *energy consumption*. The total energy consumption for wood harvesting and transportation in Finland is probably only a fraction of 1% of the total primary energy consumption in the country and approximately 5% of the total consumption of diesel fuel. Without the technical developments of the past 15 years, the consumption figures would be one-third higher. Diesel fuel consumption accounts for approximately 30% of the NO_2 emissions and 5% of the CO_2 emissions in Finland. Wood harvesting and transportation therefore account for only 1.5% and 0.25%, respectively, of these emissions[20]. These figures are low, but mechanization of wood harvesting is illustrative of the vast developments that have occurred in other fields of industrialized societies. The energy balance of forestry is very positive. The growth of forests in Finland in 1990 was 103 Tg as CO_2. Emissions from machinery used in silvicultural works, wood harvesting, and timber transportation were 424 Gg[42].

The *energy balance* of the forestry and forest-industries sectors is also positive as Fig. 22 shows. Twice as much solar energy is bound in forest-industry products as the external energy used by the forestry and forest-industry systems. External energy means fuels and electricity used in wood harvesting and saw, pulp, and paper milling processes not produced by woody materials. The possibility for reducing energy consumption in mechanized wood harvesting and transportation does exist. These possibilities include reallocation of long-distance transportation among water, rail, and road transportation and technical development in machine design. A totally sustainable system could result by replacing fossil diesel fuel with plant-based fuels (rape, pine oil, etc.), but today these are more expensive than fossil fuels.

The *oils* used in machines are a source of pollution. Some forest-industry companies have solved this problem by discarding mineral oils for plant-based oils or other biodegradable oils. Although it is more expensive, rape oil has better technical properties such as viscosity for chain and hydraulic oil than conventional mineral oils. Rape oil as a hydraulic fluid does soil the outside of machines and loses its good properties on overheating. Biodegradable mineral oils are also available.

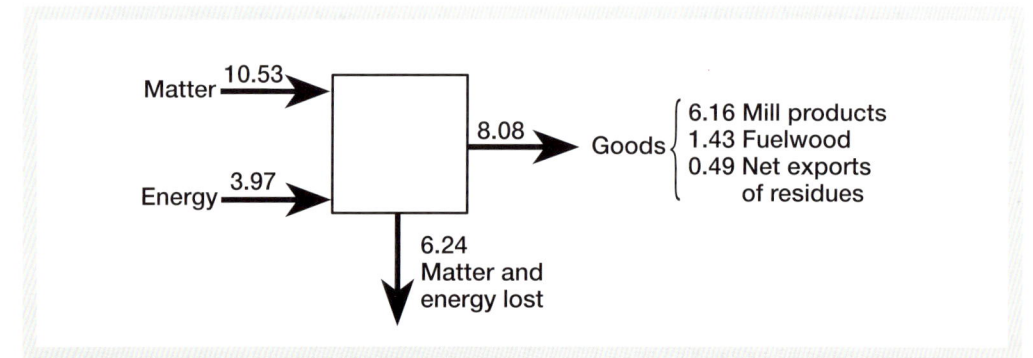

Figure 22. The flow of energy and matter in forestry and forest industries within the ECE region during the 1980s expressed in units of energy (10^6 TJ)[43].

6.3 Impacts on forest site

As a rule, harmful *impacts* fall into three categories especially for thinnings: room occupied by strip roads, damage to retention trees, and soil disturbances. In addition, loss of nutrients, impact of strip roads on selection of trees, opening the way for wind and snow damage, and insect and fungal diseases require attention. The latter are especially difficult to study in the short term.

The commonly accepted maximum level of damage in thinnings in the Nordic countries is 5% of the number of retention trees. Acceptable strip road width is approximately 4 m. The acceptable interval for strip roads is 20–30 m. Ruts deeper than 10 cm are damaging, and the proportion of such ruts in the total length of ruts should be less than 10%. In summary, Table 8 shows the total indirect costs (*loss of stand yield* in the future) caused by logging in first thinning in a spruce stand in Finland[44].

Table 8. Loss of yield in future in a spruce stand following logging in thinning conditions in Finland.

Impact	Discount percentage applied	
	0	5
	Loss of yield, FIM ha^{-1}	
Strip roads (4 m, 20 m)	1 294	488
Damage to trees	633	239
Rut formation and soil disturbances	260	87
Total	2 187	824

The *harvesting gains* for the forest owner are as follows:
- stumpage value of first thinning operation: FIM 6 900 ha^{-1}
- increase in present value of timber in future cuttings can be several times greater than above stumpage value.

Logging always causes some impacts on the growing stock, but the gains far outnumber the losses. *Gains* in thinnings include actual stumpage value of timber to be cut, earlier revenues than would be possible without thinnings, and an increased *value growth* (share of saw logs will be higher in future cuttings). All these harmful impacts can decrease with care and skillfulness by the machine operator and in the planning of logging and the network of strip roads.

Rut formation can damage tree roots. Two technical reasons for rut formation are ground pressure and wheel spin. High, "aggressive" ribs on the tires or tracks are an additional factor. Another important impact on the soil is *soil compaction* measured as water content, bulk density, or macropore space. Measurement can use a penetrometer. Soil compaction is primarily due to the ground pressure of logging machinery. This is especially significant in conditions where ground frost does not reduce compaction.

Although the economic consequences of logging damage are difficult to estimate and depend on several factors, some preliminary attempts compare the *total costs* (direct and indirect costs) of certain logging methods and machines when used in thinnings. In Table 9, the author presents a comparison of a mini-forwarder with tracks (tare 4 Mg), small wheeled forwarder (tare 7 Mg), and a medium-sized wheeled forwarder (tare 10 Mg) in Finnish conditions.

Table 9. A comparison of the total logging costs in thinnings in Finland when using motor-manual cutting and different kinds of forwarders.

Machine type	Direct costs	Indirect costs	TOTAL
	Relative number		
Forwarder (tare 10–11 Mg)	100	66	166
Light forwarder	125	56	181
Light crawler	135	49	184

These results support the idea that medium-sized (10–11 Mg) forestry machines – the most popular in Nordic countries – are the most economical alternatives.

6.4 Impacts on biodiversity, multiple-use, and landscape

In final cuttings, most aspects above are not as important as in thinnings. Erosion and leaching of nutrients, impacts on *biodiversity*, and *multiple-use* of forests and *landscape* are even more important. Many enterprises have implemented or are implementing *environmental policies, directives,* and guidebooks containing instruction for practical harvesting operations. Research concerning the impacts of environmental considerations on *logging costs* are also in progress.

Some calculations have shown that retaining groups of trees for biodiversity reasons, landscape reasons, or both in clear-cutting areas may decrease direct logging costs slightly. The same advantage can result by leaving trees as decaying material and setting aside small key biotypes. This reduction is because decayed trees often remain anyway, and key biotypes are no easy targets for logging and consist of unmarketable timber. The stumpage value of the retained timber can be 4%–6% of the total value of the timber cut[45, 46]. In addition, these measures may have a harmful impact on future yields. Neglecting final cuttings and practicing only selective cutting will give significant increases in logging costs and decreases in stand yields in the conditions prevailing in the Nordic countries. This is due to the subsequently poor regeneration of the sites, low stocking levels, or both. These impacts depend greatly on tree species and conditions.

Other means of influencing these factors include limitations to the size of clear-cut areas and practicing of selective cuttings or natural regeneration whenever feasible considering growth and yield. This generally means increased logging costs and sometimes decreased future yields of timber. Preliminary research results indicate that all the new silvicultural practices represent a potential decrease of 15% in Finland's allowable cut.

7 Timber trade

7.1 Purchasing timber on domestic markets

If a forest-industry company or other timber user lacks adequate forest resources of their own to satisfy their need for wood raw material, they must buy their timber on the domestic or international markets. Several ways to organize the *purchasing of timber* are the following: (i) establishment of the company's own wood department, (ii) starting a separate company for the purpose of wood procurement, or (iii) resorting to the services of an agent to deliver the required wood. The three primary *ways to buy* wood are as follows:

- purchasing of standing trees (marked stand); buyer attends to logging
- delivery sales; forest owner or go-between attends to logging, bringing wood to the roadside or to the mill
- forest concession; a company leases a forest area for a long period of time and attends to forestry activities.

A company can also exchange timber assortments unsuitable for its own use or timber located far from its own mills for timber belonging to other companies. Legislation and *contracts* between buyer, seller, and mediators form the basis for the timber trade. Knowing when the title to a batch of timber changes from seller to buyer is important. Other key considerations are who is responsible for the damage caused by logging (contracts with contractors and subcontractors); what are the rights concerning timber storage, who is responsible for the damages to a third party; what is the length of the period allocated for logging and storing of the timber; when, where, and by what methods timber measurement is to occur; who is responsible for regeneration; etc. The foremost paragraphs of trade contracts concern prices and quality specifications pertaining to standing trees, stems or timber assortments, and the scheduling of payments. Another important document in the timber trade is the *measurement certificate*. This includes the volumes of timber by grade according to quality specifications.

Measurement of timber can occur on the stump (as standing trees) before or after the trade transaction, during or after cutting in the woods, at the roadside, or in the mill yard. Because measurement costs are minimum when done during cutting (harvester measurement) or afterwards in the mill yard, these methods predominate in many countries. The trade contracts therefore often use visual estimation of the volumes involved, the information provided by forest management plans, or preliminary measurement. Final payment uses a measurement certificate. The above preliminary information is also necessary for planning harvesting and milling operations. Computerized systems are available for *setting the price* of marked stands according to different kinds of timber assortment specifications.

CHAPTER 7

In Finland, *timber purchasing* has traditionally used direct contacts between the forest owner and company representatives. The local forest management associations spread throughout the country provide information on the marked stands in their districts. Sometimes they also serve as agents. Because forest ownership is increasingly changing from farmer-resident ownership to urban-nonresident ownership and forest owners also have uses other than timber production for their forest holdings, new means of purchasing timber are necessary. Modern marketing theory is one starting point in the effort to increase the efficacy of purchasing. The means available include spatial systems to find suitable forestry resources, forest owner registers, segmentation of forest owners, direct mail advertising, other marketing efforts (meetings, information, training events, etc.), seller-satisfaction studies (customer satisfaction), and monitoring systems as components of the quality system of a company. Even interactive decision-support systems are under development.

The results of questionnaires have shown that forest owners *appreciate* the level of timber prices, the reliability of the company, personal contacts, good quality of logging, reliable measurement of timber, excellent crosscutting of timber into assortments, and environmental aspects. Young people especially appreciate the latter aspect[47, 48]. Because the revenues from forestry are often used for investments on the farm or in a private household, young people are more *willing to sell* timber than older people. An increase in *stumpage prices* increases the willingness to sell in the short term but not in the long term[49]. High stumpage prices therefore do not guarantee a steady supply of timber in the long run, although good prices may encourage forest owners to invest in good silviculture.

7.2 International timber markets

Prices paid for timber assortments vary considerably from one country to the next. As an example, the export price in Chile in 1992 for radiata pine pulpwood was the equivalent of FIM 217 m^{-3}, and the domestic wholesale price for coniferous pulpwood in Japan was FIM 358 m^{-3} according to FAO statistics. Exported Eucalyptus pulpwood in 1987 sold for FIM 147 m^{-3} in Brazil, FIM 228 m^{-3} in Chile, and FIM 255 m^{-3} in Portugal. The domestic price at the mill gate in Finland for birch pulpwood at the same time was FIM 239 m^{-3} [50]. Such differences in prices can cover transportation costs between countries. Other more important reasons exist for the international trade in wood.

Although the *transportation cost* of the raw material destined to be made into pulp is higher than that of pulp itself, the international trade in roundwood and chips will continue. The pulp industry requires *heavy investments,* and the depreciation time for these investments is long. In addition, the *large scale* of operation of pulp mills is advantageous. Considerable time is necessary to create new capacity in countries having no current pulping capacity or having limited raw material or investment resources. Investments in sawmills or other mechanical wood industries are easier to accomplish than investments in the pulp industry because of the lower investments and possibilities for small-scale activities in a profitable manner. Importing timber can also keep *domestic timber prices* at a reasonable level by buying *marginal amounts* from abroad. For exporting countries, the roundwood trade is an important source of *capital* for industrial investments.

For the longer term, some experts believe the division in global production system of woody raw material will be more clear than today. Tropical and subtropical zones are a natural place to produce fast-growing hardwood plantations, and the boreal zone is a place to produce profitably long fibers in coniferous forests. According to this vision, pulp mills in these zones will concentrate on milling of these natural raw materials.

A problem in the international roundwood trade is preventing the *immigration of pests* and especially insects from one country to the other. In many countries, strict legislation has this aim. The risk is greater when timber is transported from one continent to another than from within the same continent and climatic zone. Contaminating elements such as heavy metals or traces of radioactive fallout may also be problems. Importers must know these risks and any pertinent legislation in the various countries.

The biggest exporters of roundwood, chips, and wood residues are Malaysia, Russia, USA, Canada, Chile, Indonesia, and some countries in Western and Eastern Europe. The biggest importers are Japan, the Republic of Korea, China, Canada, Finland, Sweden, Italy, and some other countries in Western Europe. The total foreign trade in these raw materials in 1991 according to FAO was 131 million m^3 under bark[51].

CHAPTER 7

References

1. Harstela, P., Silva Carelica 25, Joensuun yliopisto, Joensuu, 1994, p. 113.

2. Ahonen, O. -P. and Lemmetty, J., Metsäteho Review 5:1(1995).

3. Airaksinen, P., in Tapion taskukirja (M. Häyrynen, Ed.), 22nd edn., Kustannusosakeyhtiö Metsälehti, Helsinki, 1994, pp. 451–453.

4. Heiskanen, V. and Siimes, F. E., Paperi ja Puu 41(8):359(1959).

5. Uusvaara, O., in Tapion taskukirja (M. Häyrynen, Ed.), 22nd edn., Kustannusosakeyhtiö Metsälehti, Helsinki, 1994, pp. 518–525.

6. Usenius, A., Halinen, M., and Hemmilä, P., Sommardahl, K.O., VTT tutkimuksia 491:1(1987).

7. Bjurulf, A. and Spångberg, K., Skogforsk Resultat 19:1(1994).

8. Kahala, M. and Rantapuu, K., Metsäteho Report 276:6(1968).

9. Kahala, M., Metsäteho Report 355:7(1979).

10. Sisu logging Oy.

11. Nokka-Tume Oy.

12. Ponsse Oy.

13. Timberjack Group.

14. Dietz, P., Knigge, W., and Löffler, H., Walderschliessung, Verlag Paul Parey, Hamburg und Berlin, 1984, pp. 22–30, and 208–237.

15. Samset, I., Winch and Gable Systems, Martinus Nijhoff/Dr W. Junk Publisher, Dortrecht, Boston, Lancaster, 1988, pp. 146–159.

16. Anon., FAO Forest Harvesting Bulletin 5/93, FAO, Rome, 1993, pp. 6–7.

17. Heinämäki, J., The optimum use of log stacking trucks at wood terminals, University of Helsinki, Dep. Logg. Util. For. Prod., Research Notes 54, 1991, pp. 100–109.

18. Vanhanen, H. and Heikinheimo, L., Productivity in forestry and socio-economic change 1950–1981, TIM/EFC/WP.2/R58, Wien, 1983, p. 10.

19. Hakkila, P., The Finnish Forest Research Institute, Research Papers 557:46(1995).

20. Harstela, P., Silva Carelica 31, Joensuun yliopisto, Joensuu, 1996, p. 138.

21. Heinrich, R., FAO Forestry Paper 33:37(1982).

22. Seifert, K., Holz als Roh-u.Werkstoff 24:81(1966).

23. Seifert, K., Holz als Roh-u.Werkstoff 24:185(1966).

24. Wilcox, W. W., Wood and Fiber 9:252(1978).

25. Söderström, O., The influence of saw log storage on some processes etc in the sawmill industry (Summary), The Swedish University of Agricultural Sciences, Dept. of Forest Products, Report 181, 1986, pp. 1–45.

26. Pennanen, O., Laamanen, J., Lucander, M., Terävä, J., Varhimo, A. , Metsäteho Review 1:4(1993).

27. Haikkala, P., Liimatainen, H. and Tuominen, R., TAPPI 1987 Pulping Conference Proceedings, TAPPI PRESS, Atlanta, p.363.

28. Tienvieri, T., Mekaanisen massan vaatimukset puuraaka-aineelle ja hakkeelle, AEL/METSKO Seminaari 12.-13.11.1996, Metsäteollisuuden Koulutuskeskus, Tikkurila, 1996, pp. 2–6.

29. Lönnberg, B. and Ala-Porkkunen, A., KCL Seloste 1374:1(1979).

30. Hainari-Maula, J. Erään sulfaattisellutehtaan puuraaka-aineen vaihtelun vaikutus sellun saantoon ja laatuun, Diplomityö, Teknillinen korkeakoulu, Puunjalostusosaston selluloosatekniikan laboratorio, Espoo, 1985, pp.1–106.

31. Kahila, S., KCL Seloste 618:1(1963).

32. Assarsson, A., Pulp Pap. Mag. Can. 70(18):74(1969).

33. Bergman, Ö., Deterioration and protection of wood chips in outside chip storage, FAO/TWC/72/26, Rome, 1972, pp. 1–33.

34. Bergman, Ö., Sver. Lantbruksuniv. Inst. Virkesl. Rap. 170:1(1986).

35. Hakkila, P., Utilization of residual forestry biomass, Springer-Verlag, New York, 1989, pp. 376–385.

36. Thörnqvist, T., Sver. Lantb. Inst. Virk. Rapp. Upps. 163:2(1987).

37. Hutton, J. V. and Hunt, K., Tappi 55:95(1972).

38. Anon., The production, handling and transport of wood chips, TF-INT 55 (NOR), FAO, Rome, 1973, pp. 60–79.

39. Fredholm, R. and Jirjis, R., Swed. Univ. Agr. Sc. Dep. For. Prod. Rap. 200:2(1988).

40. Jirjis, R. and Lehtikangas, P., Swed. Univ. Agr. Sc. Dep. For. Prod. Rap. 243:2(1994).

41. Jirjis, R. and Lehtikangas, P., Swed. Univ. Agr. Sc. Dep. For. Prod. Rap. 230:2(1992).

42. Karjalainen, T. and Asikainen, A. Forestry 69(3):216 (1996).

43. Sundberg, U. and Silversides, C.R., Operational efficiency in forestry, Volume 1: Analysis, Kluwer Akademic Publ., 1988, p. 125.

CHAPTER 7

44. Kokko, P. and Siren, M., *The Finnish Forest Research Institute Research Papers* 592:10(1996).

45. Hallenberg, T., Luonnon monimuotoisuuden huomioonottamisen vaikutus korjuukustannuksiin uudistamisaloilla, Pro Gradu, Joensuun yliopisto, Metsätieteellinen tiedekunta, Joensuu, 1996, pp. 39–60.

46. Aldentun, Y. and Sondell, J., Studie av naturvårdshänsyn i praktisk skogsbruk. Summary: A study on nature conservation in practical forestry, Skogarbeten, Redogörelse 1, 1991, pp. 11–12.

47. Sikanen, L., Puunmyyjien käyttäytyminen kauppatilanteessa, Joensuun yliopisto, Metsätieteellinen tiedekunta, Joensuu, Tiedonantoja 27:33(1996).

48. Kärhä, K. and Oinas, S., *Metsälehti* 21:14(1996).

49. Kuuluvainen, J. and Ovaskainen, V., Standing stock, amenities and nonindustrial private forest owners' harvesting decisions: Panel data evidence from Finland, Manuscript, Finnish Forest Research Institute, Helsinki, 1993, pp. 1–33.

50. Anon., FAO Forestry Paper 125, FAO, Rome, 1995, pp. 37–41.

51. Anon., FAO Yearbook, Forest Products, FAO, Rome, 1991, pp.1–334.

Timber procurement

CHAPTER 8

Timber measurement

1	**Principles and concepts in timber measurement**	**365**
2	**Measurement methods**	**367**
2.1	Introduction	367
2.2	Measurement of standing trees	367
2.3	Timber measurement during logging	370
	2.3.1 Mechanized logging	370
	2.3.2 Motor-manual logging	375
2.4	Timber measurement at roadside	378
	2.4.1 Saw logs	378
	2.4.2 Measurement of stacked timber	380
	2.4.3 Weighing timber using grapple-mounted scales on timber trucks	383
2.5	Timber measurement at mill	383
	2.5.1 Introduction	383
	2.5.2 Weighing of timber	384
	2.5.3 Volumetric timber measurement	385
3	**Costs involved in timber measurement**	**388**
3.1	Introduction	388
3.2	Measurement of standing trees	389
3.3	Harvester-based measuring devices	389
3.4	Cutter method	390
3.5	Measurement at roadside	390
3.6	Measurement at the mill	391
4	**Further development of timber measuring methods**	**391**
	References	**392**

CHAPTER 8

Antti Asikainen

Timber measurement

1 Principles and concepts in timber measurement

> Timber measurement is determining the quantity and quality of timber required in a contract for purchase prices and payment of wages. Timber measurement can use standing trees[1] or determination of the number of pieces, size, quality, volume, or weight of timber.

Log length is the shortest distance between the end midpoints of a log. *Diameter* is the distance between two parallel tangents on the plane perpendicular to the longitudinal axis of the log[2]. One or several diameters are measured depending on the accuracy requirement and measurement method. Using as an example a measurement based on length and top diameter, only one diameter is necessary. In mechanized cutting, dozens of diameters are detected over the length of a single log. *Weight measurement or weighing*, i.e., determining the weight, can also be used.

The volume of timber can use stereometric measurement, xylometric measurement, or hydrostatic weighing[2]. Using *stereometric measurement* requires determination of many coordinates on the surface of the piece and then basing the volume computations on these coordinates. In *xylometric measurement*, the piece is submerged in a container of water with determination of the increase in the level of the liquid. Since the dimensions of the container are known, the volume of the piece submerged can be determined. In *hydrostatic weighing*, the piece is weighed suspended in air and in water, and the volume is calculated by subtracting the latter value from the former. In practice, stereometric measurement is most common because of its simplicity[2].

Solid volume is the volume of timber material consisting of pieces without the air space between the pieces. *Real solid volume* is the cylinder's real solid volume using the length of the piece and more than one diameter of the piece. *Technical solid measure* is a cylinder's solid volume using the length and top diameter of the piece. Real solid volume over bark (o.b.) is the common measure of timber volume. Several countries use measurement of saw logs under bark (u.b.).

Solid volume by top diameter measurement is solid volume using the length and the top diameter of the log, and *solid volume by mid-point diameter measurement* is solid volume using the length and the mid-point diameter of the log. Finally, *solid volume by butt and top diameter measurement* is solid volume using the length and the butt and top diameter of the log.

CHAPTER 8

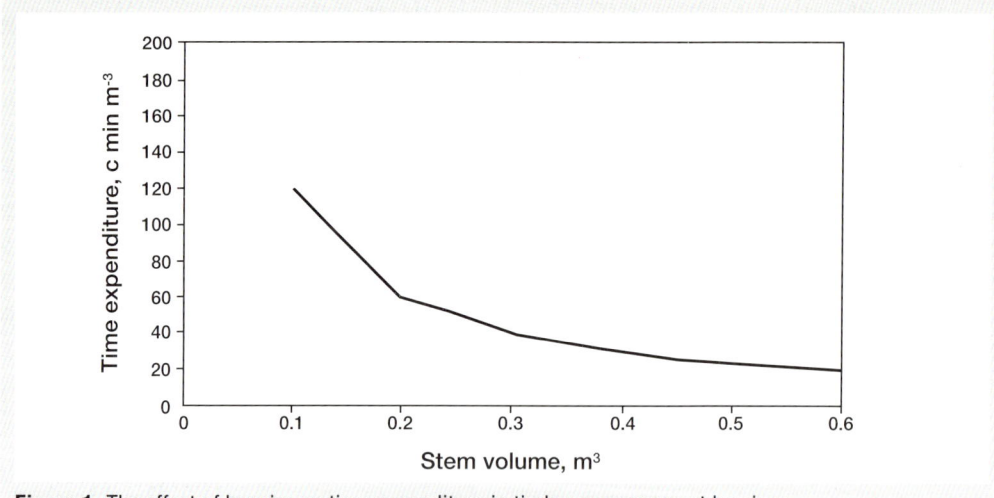

Figure 1. The effect of log size on time expenditure in timber measurement by piece.

Measurement costs per cubic meter can be very high when measuring small logs individually. Several logs or pieces of timber are often measured in stacks, truck loads, or bundles. This is done because in stacks time consumption per log is independent of log volume and because time consumption per cubic meter is very high for small logs when measuring them individually as Fig. 1 shows.

Piled or stacked volume is the frame volume of a pile including the space between the pieces of timber, and *loose volume* is the total volume of a mass of wood, load, etc., including the space between the pieces of wood. Piled volume uses the length and height of the pile and the length of the logs. Piled volume was previously used in the measurement of pulpwood destined for consumption by the pulp and paper industries. Today this measure is converted into real solid volume. Loose volume of chips and sawdust is determined in a container. Fuel wood and fuel chips are still measured in loose volume units especially when delivered to small power plants and to the consumers of firewood.

Measurement accuracy and *measurement errors* cause uncertainty in measurement results. Accuracy is the ability to measure a particular trait correctly, and precision is the ability to measure the trait consistently. Measurement errors cause deviation from the real, exact value. *Random error* means that the errors are randomly positive or negative. When the number of measurements increases, the mean of random errors approaches zero. The variance of random errors depicts the *precision of measurement*[3]. *Systematic error* or *bias* means that the measurement error is usually negative or positive. When this happens, the mean of the measurement errors does not approach zero even when doing many measurements[3]. *Measurement accuracy* is depicted by the term *mean square error (MSE)* that includes bias and random error:

$$MSE(x) = VAR(x) + BIAS^2 \qquad (1)$$

In practice, the concept of measurement accuracy is used somewhat misleadingly to describe the difference between basic measurement by a party of a wood trade and control measurement by an authorized controller. Control measurement gives an estimate that is not the absolutely correct measurement result but merely one random result drawn from a control-measurement distribution. The practice in Finland is to set the highest acceptable difference between basic and control measurements at ±4%.

2 Measurement methods

2.1 Introduction

Measurement of timber has several purposes during the chain of events constituting timber procurement. Before felling the timber, the buyer usually measures and evaluates the timber quality to define the *unit price* offered to the forest owner. The result is used in *planning of harvesting* operations. Measurement of timber determines the *purchase price* and *wages* of workers, logging entrepreneurs, or both. In truck transportation of the timber, *control of payloads* may be necessary. At the mill, timber is measured again to *pay for transportation* and to *control the use of roundwood* at the mill. *Control and calibration measurements* ensure that measuring devices work properly and that the methods are applied appropriately. As a result, a parcel of timber might undergo measurement several times. This can lead to unnecessarily high measurement costs. Today a parcel of timber is usually measured once or twice.

2.2 Measurement of standing trees

Measurement of standing trees is determination of the solid volume of standing commercial timber with bark in stands marked for cutting. The quantity of commercial timber is measured to calculate the purchase price and transportation costs. Measurement by timber assortments such as saw log stems or pulpwood stems allows calculation of logging costs and wages.

Measurement of standing trees has use in all stands except those that are defective due to storm or other damage or those that require logging in some unusual manner due to an abnormality. The measurement of standing trees consists of two stages: (i) planning and tallying including sample tree measurements and (ii) volume calculation.

Planning for the measurement of standing trees is usually part of the harvesting plan. This means determination of work-difficulty factors, strip road networks, and storage places or landings simultaneously. Measurement of standing trees was the first timber measurement method integrated with the planning of harvesting operations. The results can evaluate logging productivity and schedule the harvesting operation. The measurement team in this alternative shown in Fig. 2 can have 1–4 workers. Measurement uses complete *tallying* with measurement of the diameter at breast height (dbh) of each tree to be harvested. Before tallying, the stem assortments should be accurately determined by tree species and timber assortment – pine saw log stems, pine pulpwood stems, spruce saw log stems, etc. Particularly important items to determine are the minimum quality requirements and diameters of saw log stems. Measurement of diameters uses electronic callipers connected to a field computer.

CHAPTER 8

Figure 2. Measurement of standing trees.

Measurement of *sample trees* determines the factors needed for computation of unit volumes and the proportion of timber assortments by sample-tree area, tree species, stem assortment, and diameter class. The diameters and lengths needed in volume computation are measured at the field work stage. Selection of the sample trees uses a pre-selected sampling method. Sample-tree sections are formed if the stem form of the trees varies markedly in different parts of the area being measured. The forming of sample-tree sections uses visual estimation so that each section is homogenous in its range of stem assortments. The number of sample trees is 50–450 according to the total volume – below 100 m^3 and over 4 000 m^3 – of the sample tree section. Stem quality also influences the number of sample trees because defects of large variation of stem curve require additional measurements. The sampling methods used are the line relascope method or sampling at fixed intervals[4].

The *parameters to be measured* from the *sample trees* are diameter at breast high (dbh), upper diameter (d6), height of the tree (h), and the length of the saw timber section of the stem. The dbh and length of saw timber section are always measured. The d6 is measured only when the stem shape is unusual such as for trees in island forests[4]. Sample trees are combined into sample-tree groups for control measurement. This can include 20–60 sample trees. Sample-tree groups should be readily distinguishable from one another. For this reason, the sample trees of a certain sample area are always marked.

The *calculation of results* uses the stem size distribution and sample tree information measured by stem assortment. Use of sample trees determines stem volume, volume of top wastewood, and distribution of commercial timber into assortments. The results obtained for sample trees then apply to all tallied stems. The share of commercial timber in the outturn is determined by applying a taper curve to the minimum top diameter requirement. Distribution among timber assortments depends on the minimum diameter requirements of saw logs determined according to a taper curve[4].

Today the measurement of standing trees has little use in Finland because of its high costs. It was the predecessor of the current "cutter method."

Experience has shown that the measurement results for individual stands obtained in measurement of standing trees differ by 2%–4% from actual solid volumes[5]. The volumes determined by harvester-based measuring devices corrected by control measurement data indicate the actual solid volume of a stand. A method using two parameters (dbh and length) and another method using three parameters (dbh, length, and d6) have also been tested. Table 1 shows the results compared with the real solid volume determined by harvester-based devices.

Table 1. Solid volume of basic measurements and control measurements of standing trees compared with the measurements of a harvester-based device[5].

Measurement method	Solid volume, m³	Solid volume, %	Range in relative solid volume by stand, %
Harvester-based device	22 400	100.0	
Basic measurement of standing trees (2 parameters)	23 074	103.0	96.6–115.3
Basic measurement of standing trees (3 parameters)	22 820	101.9	93.2–109.5
Control measurement of standing trees (2 parameters)	23 145	103.3	96.9–108.8
Control measurement of standing trees (3 parameters)	23 293	104.0	97.9–109.4

One must realize that the measurement of standing trees gives the volume of timber assortments according to given crosscutting rules. If the cutter applies different crosscutting rules, considerable differences can result between the estimated and real value of the timber.

CHAPTER 8

2.3 Timber measurement during logging

2.3.1 Mechanized logging

Method description

The amount of mechanization for all logging work in Finland by the Finnish Forest and Park Service and forest industries was 89% in 1997. Almost all the timber cut using single-grip harvesters is measured during cutting. Figure 3 shows the components of the automated, on-board measurement systems used in harvesters.

A harvester-mounted measuring device measures the length and diameter of harvested timber, computes the volume and number of logs, and then registers the results by species and timber assortment. The device allows correction of the most recently registered data entered by the operator. In addition, a cab-mounted display shows timber lengths and diameters. The device displays and prints results when requested and registers and prints the calibration values for control purposes. The mesuring device must operate in all conditions, and the technical measurement parameters of the device should be controllable and adjustable[6]. To control volume measurement accuracy, the measuring device produces a separate printout with a precision of 0.01 m^3 for the parcel of timber selected for control.

Figure 3. Harvester-based measurement system.

Timber measurement

Figure 4. Determining timber volume using a harvester's on-board measuring device.

Measurement of timber volumes usually uses sections as Fig. 4 shows. Computing the volume of each section uses the equation for cylinder volume or the equation for the volume of a truncated cone with diameters measured at intervals of 30 cm or less and the lengths of these intervals. Diameters are measured over bark in millimeters, and lengths are measured in centimeters[6]. If the device can measure the volume using the length of the sections, this is the method of choice unless the parties involved select another[6]. The volume of a measurement unit of timber can also use piece-by-piece measurement with top- or mid-diameter solid volume measurement.

The operating principle in all harvester-mounted measuring devices is the same. Figure 5 is a photograph of the unit. Diameter is measured according to the position of the stem skimming delimbing knives or feed rolls. At least one delimbing knife has an electronic sensor. Length measurement usually uses a rotating roll forced against the side of the stem being processed. As the stem is fed, the roll rotates along the surface of the stem with length determined by electronic impulses produced by a pulse sensor installed on the roll[7, 8].

CHAPTER 8

Calibration of measuring device

Regular calibration maintains the accuracy and reliability of harvester-based measuring devices. Length calibration can be done by measuring the lengths of a number of logs and then computing the mean difference between these lengths and the actual measured lengths. After entering the control results into the computerized measuring device, the unit automatically calibrates the length measurements to correspond to the actual length. Basic diameter calibration uses the same technique with metal pipes designed for this purpose. Basic calibration occurs again only if the multi-function device is damaged or significantly modified. Diameter calibration usually uses control measurements[9].

Length, diameter, and volume calibrations should occur at the beginning of each stand or block marked for cutting and thereafter at intervals of 500–1 500 m^3 or when necessary. Calibration should focus on tree species in the ratios of their measurement. Definition of the number of logs to be checked uses the standard deviation in diameter and length measurement errors. Standard deviations are the means calculated on the basis of the most recent five calibration results. The diameters of the logs are measured every 1 m beginning from the butt end point of 0.5 m and stored in the memory of the device. The lengths and the solid volumes of the logs are also registered. The same logs are then measured using electronic callipers, and the measurement results are transferred to the measuring device of the harvester. Using these results, the operator determines the need to alter the diameter and length control values of the device and decides whether to change the control values[7].

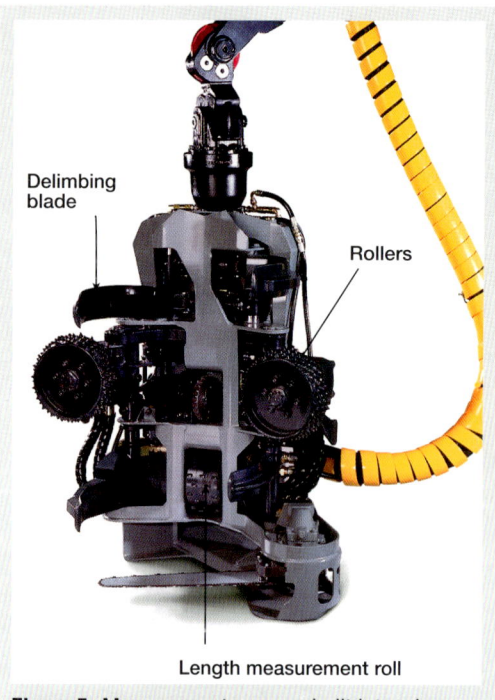

Figure 5. Measurement sensors built into a harvester-based measuring device.

Control over timber measurement

Control over timber volume measurement and the computational accuracy of harvester-mounted measuring devices are integral parts of the timber measurement system. All harvesters measuring timber for purchase price purposes are within the control system. Random tests control approximately 5% of all harvested stands. All stands containing more than 3 000 m^3 of harvestable timber are subject to control measurement. Otherwise, stands for control are chosen randomly. If a stand becomes part of the random sample, the parties involved but not the harvester operator will receive advance

Timber measurement

notification of control measurement. Control measurement usually occurs during logging by the company's representative, i.e., the company for whom the harvester contractor works, or a person nominated for the task.

The control measurement method imitates the basic measurement method as much as possible. If the basic measurement uses the basis of length sections, control measurement should also use the basis of the length sections or mid-diameter solid volume measurement. Figure 6 shows a photograph. In control measurement, the section length is 1 m. The remainder has an accuracy of 0.10 m. The diameter of each section is measured over bark midway along the section in millimeters or centimeters. The solid volume of a log is the sum of the section volumes determined as cylinders. The volume obtained in control measurement depends on the accuracy requirement. The volume of a control measurement parcel of timber is currently approximately 10 m^3 or approximately 100 pieces of timber if the stems are small.

Figure 6. Control measurement.

The measurement result is acceptable if the difference between basic and control measurements for the parcel of timber being controlled does not exceed ±4% for both saw log and pulpwood assortments. If the share of either assortment is less than 10%, a greater difference is acceptable provided the difference of both assortments together is less than ±4%[6].

If the control result is unacceptable, then an extra measurement will be necessary. It targets the timber assortment whose measurement is unacceptable. If the total difference between the basic and control measurements is more than ±4%, then the basic measurement result is corrected. The volumes of each assortment are corrected separately. The correction targets the timber parcel harvested since the most recent calibration and adjustment of the measuring device or since the most recent control measurement. All parties involved receive notification of the control measurement leading to correction of the basic measuring result[6].

For a total result of forest-industry companies' logging operations, 90% of the controlled saw log assortment groups and 82% of the pulpwood assortment groups were measured to an accuracy of ±4%[7]. The average volume of the controlled timber parcels was 8.1 m^3. No statistically significant differences existed in measurement accuracy among different manufacturers' measuring devices[7].

Measurement of pulpwood is more demanding than that of saw logs because of the smaller diameters and more pronounced branching of pulpwood. In addition, a similar absolute error in diameter measurement gives larger relative errors in the volume of pulpwood because pulpwood logs are smaller than saw logs. The accuracy of volume measurement is poorer with birch saw logs and peeler logs, birch pulpwood, and pine pulpwood than with other timber assortments. The mean volume difference of various timber assortments is –0.4%–+0.6% compared with control measurements. The season of the year or the age of the measuring or logging device do not influence the accuracy of the measuring devices. The primary factors influencing measurement accuracy are the harvesting machine and the machine operator[9].

According to results obtained by Ala-Ilomäki[8], the accuracy of a one-grip harvester-based measuring device per piece of timber depends on branches, crookedness, sweep, forks, volume, tree species, and the combination of harvester and measuring device. No single variable has a decisive role, and the error in measuring is not thoroughly understood. More branches attached to a stem give a greater measured volume compared with the real volume[10].

Optimization of crosscutting

Harvester-based measurement systems can be integrated with the wood-procurement information system. An on-board computer helps the operator decide optimum crosscutting points for each processed stem. When *crosscutting to maximize value*, the value of individual stems is maximized. In *cross-cutting to meet demand*, the harvester operator strives to meet the target log distribution given by the sawmill[11]. Both crosscutting methods use a prediction of stem form. Once the harvester has processed the first 1.5–3.5 m of the stem from the felling cut, the first stem curve estimates are produced from the current measurement point to a height of 20 m. This estimation uses regression models, mixed models, and stem form theory. Use of existing data on previously processed stems is also possible. After prediction of the stem curve, the computer compares various crosscutting alternatives and selects the one yielding the highest stem value. Optimization uses a price matrix for different log dimensions. Valuable dimensions receive higher values in the matrix than less valuable ones. When applying crosscutting to maximize value, the matrix values remain constant. For crosscutting to meet demand, the matrix values are automatically adjusted up or down if certain dimensions are absent or there is apparent overproduction of some dimension. In practice crosscutting optimization applies best to tree species with small quality variation along the stem. If the quality of stem varies markedly, cutting points must be given manually.

2.3.2 Motor-manual logging

Cutter-based timber measurement

Cutter-based timber measurement uses the measurement of standing trees described earlier. Figure 7 shows that it uses tallying (measurement of dbh) of trees to be felled and measurement of sample trees. In the initial version of this method, sample trees were measured before logging. Today tallying and sample tree measurements are done during logging. Separate measurement of sample trees is time consuming and requires a two-man team. Solid volumes for purchase prices, payment of wages, and transportation remuneration are calculated according to unit volume values determined by measuring sample trees and a corresponding number of other trees. Cutter measurement was first applied in stands where the stems were small in diameter and primarily suitable for pulpwood. Now saw log stems are also measured using this method[12, 13].

Figure 7. Cutter-base timber measurement method.

If trees in different parts of a stand clearly differ from one another, such a stand is divided into sample-tree areas. Estimation of the number of trees to be cut in the stand defines the sampling density. Various sampling methods are possible. Stratified sampling by stem assortment (pulpwood or saw timber stem) is usually used. Table 2 shows the number of sample trees. The minimum top diameters for the various assortments and marking done on the site are agreed in advance. Electronic callipers measure diameter at breast height to an accuracy of 1 mm. And the length of the assortments is measured to an accuracy of 0.10 m for all trees to be cut[14]. Figure 7 shows an electronic calliper. The measurement device produces a signal to indicate when a sample tree requires measurement. The length of marketable timber and top diameter are measured. If the dbh is greater than 14.5 cm, diameter is also measured at the height of 6.0 m.

CHAPTER 8

Table 2. Number of sample trees when applying the cutter-based timber measurement method[14].

Number of stems to be cut	Number of sample trees	
	Pulpwood stems	Saw timber stems
Up to 999	20	40
1 000–4 999	30	50
5 000–9 999	40	60
10 000 and above	50	70

The measurement results are transferred to a memory module and then to a personal computer for processing according to sample tree area. Sample tree data are then used for computing height curves by tree species and the lengths of marketable timber. The stem curves are next used to compute the volumes of each timber assortment. By summarizing the volumes of each stem by timber assortment and logging unit, the total volume of timber is finally obtained.

Approximately 5% of the stands measured by cutters are controlled. All measurement phases can be controlled, or control can be targeted separately on sampling and measurement of sample trees, tallying and measurement of diameter classes, or volume computation. The control action can be performed at the initiative of the cutter or any other party involved in the measuring[12, 13]. The mean difference of an individual stage in cutter measurement varies from +2.7% (Masser 25 measuring instrument) to +4.8% (Masser 35 measuring instrument) compared with the control measurement result[15]. This volume overestimation was primarily due to computation factors. With an improved computation routine, the volume differences compared with the control decreased to +1.1% (Masser 25) and +0.9% (Masser 35)[15]. The conclusion from this is that the measurement of diameters and lengths causes small errors in the results, but weaknesses in the computation routine can cause larger errors in the volume estimates[15, 16].

Other methods involving manual volume measurement during logging

A method using measurement of the circumference of *grapple bunches* developed and tested in the late 1970s is suitable for measuring logs cut to lengths of 1.6–3.5 m (both set length and approximate length) and 3.0–5.5 m (set length) presented in grapple bunches[17, 18].

The method uses measurement of the circumference along the midpoint of a bunch and the mean length of the logs forming the bunch. The number of stems is also tallied using marked butt logs in the bunches. For single logs, the method uses the mid-diameter and length of the log. Computation of the solid volume of each bunch uses the following equation[18]:

$$V = \sum (I^2) \times L \times k \qquad (2)$$

where V is the solid volume of a bunch [m^3], l is the circumference of a grapple bunch [m], L is the mean length of the logs forming the bunch [m], and k is the density coefficient (according to base length of logs determined at intervals of 0.5 m).

The square of the midpoint circumference of a bunch gives the cross-sectional area of the logs. This corresponds to the solid volume of the bunch when multiplied by the density coefficient. Determination of the solid volume of a bunch results from multiplying the cross-sectional area of the logs by the length of the logs. The length is the length of set-length logs or in the case of approximate length the length estimated using sampling. The mean accuracy of this method of measuring volume varies -3.3%–+0.8% depending on the timber assortment. The deviation among different measurement parcels can be very high, i.e., -20.6%–+16.4%[17, 18].

Log method had wide use in the 1970s and 1980s for volume determination. The method used estimation or measurement of the top or midpoint diameter of the log. Class intervals of 5 cm were common. Volume calculation in this method used the stem height class by tree species and the length and diameter class of the log. Tables gave the solid measure of the log in cubic meters. This method was limited to standard length logs, time-consuming, and expensive since the tops of the logs had to point in the same direction for tallying[17, 19, 20].

Neither of these two methods find use today. The methods were laborious and expensive, and their accuracy was insufficient especially for small timber parcels.

Weighing timber using forwarder's grapple-mounted scales

In motor-manual logging, timber can be measured by weighing it during the forest hauling process. Scales for this purpose mount on the loader grapple. This has an acceleration sensor and the ability for absolute error correction. During loading, each grapple load is weighed, and the total weight of the load is determined by summing the grapple loads. The resultant weight can be converted into volume according to green density values by tree species. Nation-wide density values determined for long-distance timber transportation or values based on sample loads or separate sample measurements can be used in this conversion stage. Timber can be weighed per cutter or per team of cutters. All timber loads can be unloaded to the same storage pile at the landing. Forwarding becomes more effective then because timber delivered to roadside landings can be forwarded as full loads[21]. The accuracy of the devices has been tested by weighing a parcel of timber six times using grapple-mounted scales. The results varied from 8 109 kilos to 8 487 kilos. The mean weight differed only 0.7% from the control measurement value[21].

To determine the solid volume of timber, one must determine the green densities of the various timber assortments. This is possible by measuring the weight and volume of sample grapple bunches. The number of sample bunches depends on the allowed error in green density and the total volume of the cutting block. Allowing an error of 3% in green density in clear cuttings where volume to be harvested is 10 m^3, the number of required sample bunches is 3–6 depending on the timber assortment. It is three for birch pulpwood and six for pine saw logs, pine pulpwood, and spruce saw logs. In clear

CHAPTER 8

cuttings with out turns of 100 m³ and more, the number of sample bundles is 4–9. The measuring method therefore becomes more economical as the volume to be harvested increases. With an acceptable error in green density of 2%, the number of sample bunches doubles, and the method becomes more expensive to use. An allowable error of 4% in green density means that the measurement method will probably fail to meet the accuracy requirement of ±4% of the volume of a parcel of timber[22].

2.4 Timber measurement at roadside

2.4.1 Saw logs

Softwood

Softwood saw logs are measured at the roadside individually as Fig. 8 shows. Determining solid volume uses top diameter and log length[23]. *Diameter* is measured level with the top cut or within 3 cm of the top and no further than 6.1 m from the butt end over bark (o.b.) in the horizontal direction. If a knot or some other swelling exists at the measuring point, diameter is measured at the point where the influence of the swelling ceases toward the butt end of the log. If the bark has eroded, its thickness is estimated. Diameter is measured by callipers or a measurement stick. Electronic callipers are preferred today because they allow direct registration of the measurement result in a computer. Logs showing considerable ovalness require two diameter

Figure 8. Measurement of top diameter of saw logs.

measurements at right angles to each other. The average of the two measurements is then the log diameter.

The length of a log is measured as the shortest distance between the ends of the log. Logs of set or fixed length are cut to agreed lengths. Their length must not deviate more than ±3 cm unless otherwise agreed from these lengths. If length deviation exceeds the agreed value, log length is measured in centimeters and rounded to the nearest class midpoint. Lengths of logs cut to approximate lengths are measured using length classes of 0.3 m[24].

Determination of the solid volume of softwood saw logs uses unit volume values ($m^3 m^{-1}$) that give the solid volume over bark per length for different diameter classes. The values are determined regionally and separately for pine and spruce. Because the average length of logs influences the unit volume values, the total solid volume of saw logs requires correction by a coefficient according to the average length of the logs[24].

Hardwood and special timber assortments

The log measurement method is used when measuring hardwood saw logs with a maximum length of 7.0 m. Top diameter and length are measured the same as with softwood saw logs. The solid volume of these logs is determined according to unit volume values for hardwood saw logs. The total result is then corrected using a correction percentage based on the average length of the logs[24].

When measuring hardwood saw logs applying top diameter, the solid volume of "jump cuts" in each diameter class is calculated by multiplying the total length of the jump cuts with the unit volume value ($m^3 m^{-1}$) of the diameter class in question. *A jump cut* is the short piece of stemwood that fails to meet the quality requirements set for saw logs. The solid volume of acceptable timber is calculated in each diameter class by subtracting the length of jump cuts from the total length of saw logs and then multiplying that length by the unit volume value of the diameter class in question[24].

The solid volume of *pine poles* is determined according to midpoint diameter and the length of the pole. Diameter is measured over bark, and the length is measured as the shortest distance between the cut ends. The solid volume is calculated by using the unit volume values ($m^3 m^{-1}$) of pine poles without any correction and the mean length of the poles[24]. The solid volume of *small logs* uses top or midpoint diameter and log length. Measurement is the same as with normal saw logs. When using top-diameter measurement, the solid volume is determined using unit volume values. When using the midpoint-diameter method, the solid volume uses the equation for computing the volume of a cylinder. With the midpoint-diameter method, the result is corrected using a percentage that depends on the average length of the logs[24].

CHAPTER 8

2.4.2 Measurement of stacked timber

Introduction

Stack measurement measures the solid volume with bark of stacked pulpwood. As mentioned earlier, the unit cost of measuring individual logs increases as their volume decreases. Less valuable, small-diameter timber assortments are therefore measured using the stack method shown in Fig. 9. The method is applied to pulpwood cut to set or fixed lengths or approximate lengths[25].

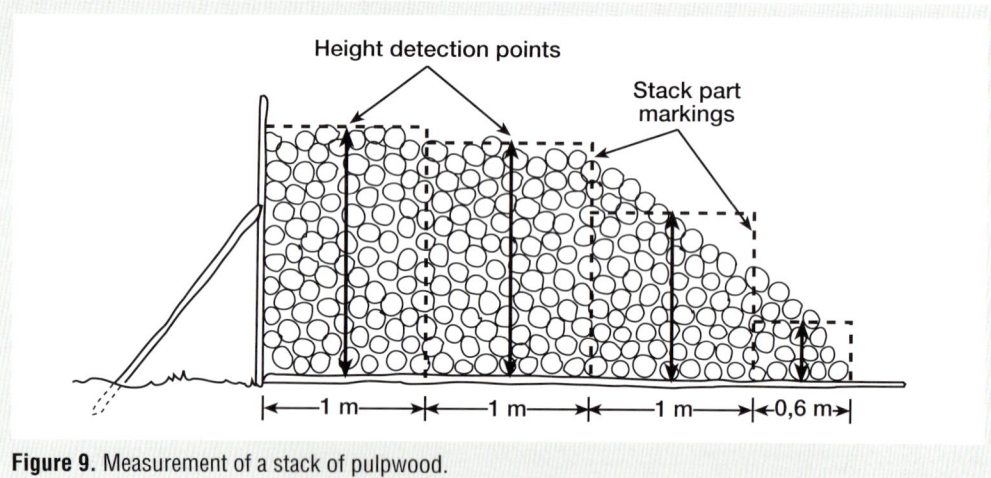

Figure 9. Measurement of a stack of pulpwood.

Determining the solid volume of a stack of pulpwood has the following steps:

- Measurement of length, height, and width of the stack or bundles forming a truck load. The frame volume of the stack or load is calculated using these measurements.
- Determining of the stack-density factors including average diameter, branching, and degree of delimbing, crookedness, and stacking. These factors determine the density of the stack or load.
- Computing the solid volume of a stack or load by multiplying the frame volume with the density.

Measurement of frame volume

The length of the stack is measured at both sides of the pile as the distance between the outermost edges of the outermost rows of logs. If the stack has no end poles or if the end poles are upright, the length is the distance between the outermost edges of the outermost rows of logs at ground level. If the end pole is not upright, the average position of the outermost row of logs is estimated. When measuring a truck load, the width of an individual bundle equals the length of a stack. The width of a bundle is the inside distance between the bolsters at bunk level and the level of the top edge of the bundle[25].

Timber measurement

Height measurement requires dividing the stack into parts 1 m in length. Stacks over 10 m long can be divided into segments 2 m long. The length of the last part is measured using 0.1 m classification. The heights of the individual parts are measured at both sides of the stack or bundle midway along the part or bundle. The heights are measured perpendicular to the length of the stack. Height is the distance between the levelled bottom edge of the lowest row of logs and the levelled top edge of the uppermost row of logs[25]. Stack width or bundle length is the length of fixed-length or set-length logs. If the logs have been cut to approximate lengths, stack width is the estimated average length of the logs. Estimation can be done by measuring a certain number of logs and then computing their average length. With grapple and truck loads, the length of the logs is determined by leveling both ends and then measuring the distance between the ends[25]. The frame volume of a stack or load is the sum of the frame volumes of stack parts or bundles forming the full load.

Stack density

Stack density is the ratio between the solid volume of logs forming the stack and the equivalent frame volume. The density of each stack is determined using the average solid volume percentage and stack-density factors. Stack-density factors are estimated independently. If they clearly differ in different parts of the stack, the stack must be divided into parts complying with these differences[25].

Stack-density factors are as follows: average diameter, delimbing and branching, crookedness, and stacking. *Average diameter* is the arithmetic mean value of the diameter of the log ends. Diameter class is determined without consideration whether the ends are butt ends or top ends. Diameter is determined by measuring the diameters of the end of the logs and then computing the average value. The average log can also be selected visually with this diameter representing the average diameter of all logs[25].

Delimbing and branching are determined from the stack by visual estimation using four classes:

Class 1	No branch stubs or basal swellings
Class 2	Some short branch stubs and minor basal swellings
Class 3	Branch stubs and basal swellings occasionally
Class 4	Considerable branch stubs and basal swellings.

Crookedness is determined using five classes. The last two such classes (classes IV and V) apply only to hardwood from northern Finland. Stacking is similarly determined using four classes with the first class representing good stacking and the fourth class indicating the worst stacking[25].

The solid volume percentage of a stack is determined by summing the effect of stack density factors and adding the result to the average solid volume percentage of the timber assortment in question. Table 3 shows the average solid volume percentages.

CHAPTER 8

Table 3. Average solid volume percentages for softwood and hardwood.

Length of timber, m	Softwood, %	Hardwood, %
2.00–2.50	66	57
2.51–3.50	63	54
3.51–4.50	61	52
4.51–5.50	60	50
5.51–6.00	59	49

Table 4 shows the influence of stack density factors on average solid volume percentage.

Table 4. The influence of stack density factors on average solid volume percentage as percentage points[25].

Stack density factor	Softwood	Hardwood
Average diameter Class, cm		
9	-3	-3
11	0	0
13	+2	+2
15	+3	+4
17	+4	+6
19	+4	+7
21	+5	+8
23	+5	+8
25 and over	+6	+9
Delimbing and branching Class		
I	+2	+1
II	0	0
III	-2	-1
IV	-4	-2
Crookedness Class		
I	+1	+2
II	0	0
III	-1	-2
IV	-2	-4
V		-6
Stacking Class		
I	+2	+1
II	0	0
III	-2	-1
IV	-4	-3

2.4.3 Weighing timber using grapple-mounted scales on timber trucks

Timber-truck drivers control their payloads to avoid overloading their vehicles and have balanced loads per axle. The measurement principle is the same as when using grapple-mounted scales on forwarders. The weights are part of the vehicle loading process. Each grapple load is registered automatically as the load crosses onto the vehicle's bunk or manually by the driver. Weighing optimizes use of load space and also avoids overloading the vehicle. The measurement method has been tested using three different makes of scales mounted on nine timber truck cranes. A total of 20 000 m^3 of timber was weighed during the study, and the results obtained were compared with weight measurement at the mill. The latter represent the actual weight measurement result. The average accuracy of weight measurement per total parcel of timber 0.1%–0.5% depending on make of scales used. For an individual timber-truck load, the weighing accuracy was less than ±4% in 82%–96% of the measurements and depended on the make of scales used. The scales that required no separate entry for every grapple load were the most practical to use. When using such scales, the extra time needed when loading a truck load was 2 minutes. This included preparations, setting the scales to zero, etc. Other scales required nearly 5 minutes extra time per truck load due to the need to enter each grapple load. Even if entering grapple loads did not require this additional time, it could increase the probability of errors[26].

2.5 Timber measurement at mill

2.5.1 Introduction

The share of timber measurement at the mill has increased rapidly in the 1990s. An increased degree of mechanized harvesting, decreased seasonal changes, and faster wood flow from the stump to the mill have made it possible to measure timber at the mill to determine remuneration and purchase prices. In Finland in 1994, 23% or 1.7 million m^3 of the timber sold on delivery basis and 10% or 2.6 million m^3 of standing-sales timber was measured at the mill to determine the purchase prices for parcels of timber. These figures apply to large-scale timber procurement by Finnish forest-industry companies[26, 27]. Note that although most timber is measured at the mill, only part of this total is measured there for determining the purchase price.

Timber measurement at the mill is a modern, reliable, and convenient way of measuring timber. Measurement to determine purchase prices is often cheapest to perform at the mill because measurement takes place at centralized receiving facilities using state-of-the-art, rational measuring methods. Centralized measurement enables larger investments in measuring devices because large amounts of timber are involved per device. Timber is also measured at the mill to control quality, determine the volumes of timber parcels exchanged between companies, and determine truck hauling rates. Saw log measurement can use optical log measuring devices or sampling-based frame measurement. Pulpwood is measured applying stack measurement on the truck, frame-scanning measurement, or sampling-based weight, frame, and bundle measurement.

2.5.2 Weighing of timber

Practically all timber arriving at larger sawmills and pulpmills is weighed using vehicle scales or unloading devices. As a truck arrives at the mill, it is identified and the origin of the timber recorded. The driver moves the vehicle onto the scales where the gross weights of the loaded tractor unit and trailer are recorded. After unloading, the truck is weighed to obtain the tare weight or weight of the empty truck. The payload is the difference between the weights. This result is the basis for transportation payment. Because volumetric units are used when paying the purchase price of the timber, the weight reading obtained is often converted into volume using various sampling procedures. In addition, sawmills control their consumption of raw material in volume units.

In Finland approximately 42% of the timber measured at the mill uses sampling-based weight measurement. This is the most common measuring method used at the mill and applies primarily to the measurement of pulpwood. The method has three stages: weighing of the entire parcel of timber, selection, and measurement of sample bundles or loads.

Timber is first weighed using vehicle scales or scales mounted on unloading devices such as a fork-lift truck[28]. Sample bundles or loads are selected using random sampling. The sampling method can be nonstratified or stratified. Nonstratified sampling is the most common sampling method and accounts for 79% of the timber measured by sampling methods at the mill. In *stratified sampling*, every load or truck load has a particular weight class or stratum according to the estimated green density of the timber forming the load. Sample bundles are selected separately for each stratum, since this increases the efficacy and accuracy of sampling. The green density of the timber in the sample bundles is determined by measuring the weight and the solid volume of the sample bundles as Fig. 10 shows. The solid volume is measured piece-by-piece or by hydrostatic measurement. The mean green density of the timber is computed using these measurements by weight class when using stratified sampling, measurement parcels, or dryness degree depending on the sampling method used[28, 29, 30].

Sampling-based weight measurement meets the accuracy requirement of ±4% per measurement parcel if the sample bundles are chosen per measurement parcel. If the solid volume of a measurement parcel is less than 100 m^3, it is usually not practical to measure the parcel using sampling per measurement parcel. When using stratified sampling, 3–4 strata are sufficient for saw logs and 4–6 for pulpwood. The range of green density of the timber within each stratum is then 70–100 kg m^{-3}. When the received measurement parcels are large, the seasonal variation in the green density of timber is considered by computing the green density coefficient as a mean value using the values obtained for the most recent 5–10 sample bundles[28, 30].

The accuracy of the load's frame measurement using estimated average stack density (stacking of the bundles) primarily influences the accuracy of weight measurement using stratified sampling. These factors influence the basic measurement when allocating the load to the correct stratum. Stratification itself has far less influences on

Timber measurement

Figure 10. Measurement of sample bundles in an automated measurement station (Photo by Antti Asikainen).

measurement accuracy. When using nonstratified sampling, the measurement errors are considerably larger[30].

2.5.3 Volumetric timber measurement

Stack measurement with timber on the truck

Stack measurement with the timber on the truck is similar to the measurement of stacks at the roadside. Timber is measured on trucks or in railway cars by bundles. The bundle frame is measured using a measurement stick or automatically with a video camera or image-processing technique as Fig. 11 shows. The stack density determination uses visual estimation. The method is suitable for small measurement parcels due to its low investment costs and higher labor requirement when compared with the other measuring methods. The method also meets the accuracy requirement of ±4% per measurement parcel[28, 29]. The stack densities of truck bundles are higher than those of stacks of timber measured at the roadside. The same stack density values cannot be applied in stack measurement at the roadside as for stack measurement of truck bundles.

Frame-scanning measurement

Measurement of timber at the mills to determine the purchase price calls for methods that allow identification of the volume of timber parcels originating from different sources. A truck load can contain timber from several sources. These require separation in the measurement process.

CHAPTER 8

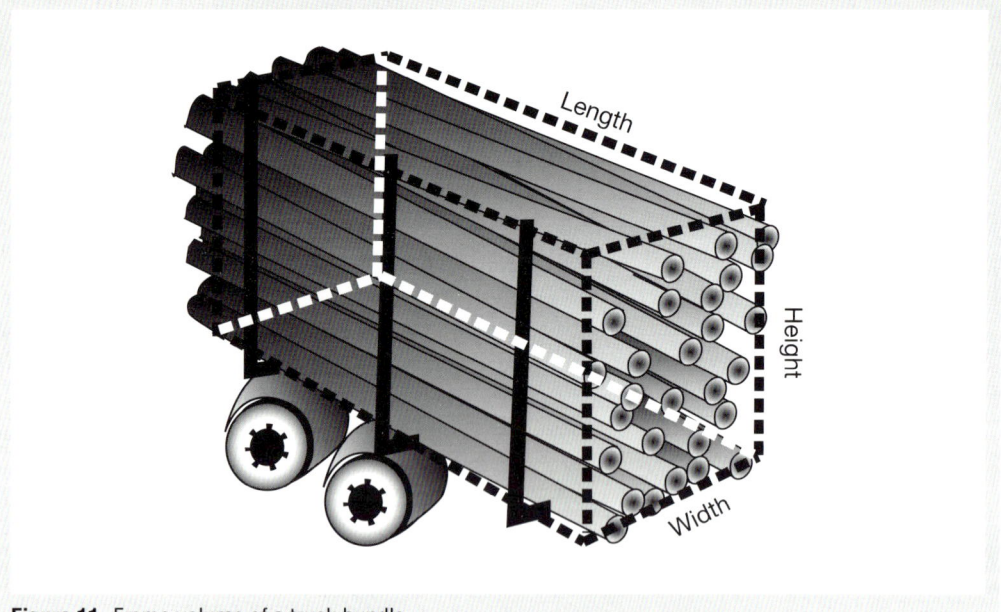

Figure 11. Frame volume of a truck bundle.

Frame-scanning measurement is the most recent method developed for the measurement of timber at the mill. It enables automatic determination of the purchase price of each truck bundle. Earlier methods did not determine the volume of each bundle with sufficient accuracy.

The method uses laser and image-processing technology. The outlines of the outermost logs forming the surface layer of each bundle are determined and recorded on both sides and on the top of each bundle as Fig. 12 shows. The frame volume and solid volume of each bundle are computed using these measurements that include determining the diameters, lengths, and locations of the logs situated on the bundle surface. These measurement occur as the truck passes through the measurement station at 2–5 km h^{-1}. The measurement process takes approximately 30 s per load[31]. Load weight is also recorded.

Monitoring stack density and the green density of the timber continuously controls frame-scanning measurement. Stack density is computed for each bundle measured using the frame-scanning method. The measurement result is acceptable if the stack density value is within limits set for the specific timber assortment – 58%–78% for pine pulpwood. If this is unacceptable, the bundle is measured using the stack measurement method. Green-density control works in the same manner. The green density of the timber forming a truck load should fall between lower and upper limits determined for specific timber assortments. If it does not, the load is measured using the stack measurement method. The accuracy of the frame-scanning measuring method is controlled by annually taking 50–60 truck loads for control measurement. Control measurement consists of measuring the frame volume and solid volume of the bundles forming a

Timber measurement

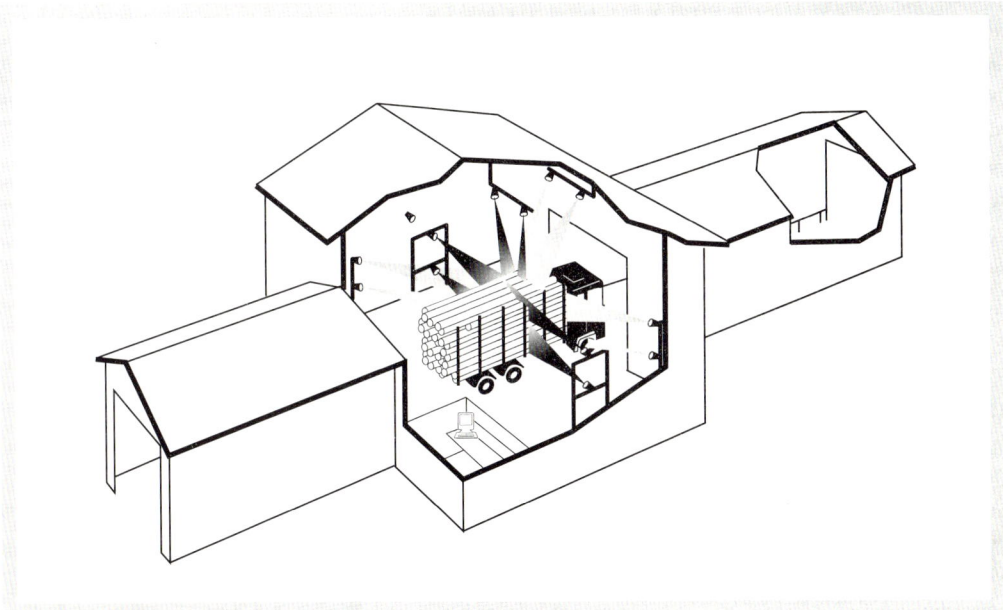

Figure 12. The principle of frame scanning.

load. The frame volume is measured using a measuring stick and solid volume using hydrostatic measurement. The control values of the frame-scanning measuring system are checked against these results[31].

Frame scanning is suitable for measuring the solid volume of small parcels of timber for payment of wages and purchase prices. Due to the high investment costs involved (FIM 3 000 000), the total volumes to be measured annually at one such measurement station should exceed 1 million m^3 [28].

Sampling-based frame and bundle measurement

The sampling-based frame measuring method is another important sampling-based measuring method for measuring timber. The frame volume of all truck loads of timber is measured by truck load and by bundle. The measurements are made per timber parcel. The average load density is calculated using its frame and solid volumes determined using the measurement of sample bundles. The solid volume of the parcel of timber results by multiplying the frame volume by the load density computed using sample bundle data. With this exception, the method is similar to weight measurement. Sampling-based frame measurement is not as reliable as the weight-measuring method because the load densities for a parcel of timber vary more than their green densities[28].

Sampling-based bundle measurement measures floated timber. Floated timber is measured by bundle raft. The number of bundles in the raft are counted, and the mean solid volume of the bundles is determined using the measurement of sample bundles[28].

Log measurement using optical measuring devices

Saw logs and veneer logs can be measured at the mill during grading by using optical measuring devices. Mounted on the log conveyor, these devices measure log diameter by referring to its silhouette at intervals of 0.02 m. Diameter measurement is the result of measurements made from two or three directions applying an accuracy of 1 mm as Fig. 13 shows. Log length is measured applying an accuracy of 1 cm. Logs are measured, and their solid volumes computed using length sections in the same manner as with harvester-mounted measuring devices. Log quality is determined visually during measurement. The results can support quality grading. For instance, indexes for unevenness of log surface can be computed. The accuracy of the current optical measurement devices in measuring log diameter, length, and volume is good. Snow and ice on log surfaces influence the measurement. If 15%–20% of the log surface has snow and ice cover or lacks bark, a volume difference of ±1% is obtained. For reliable results in control measurement, at least 50 logs should be measured[32].

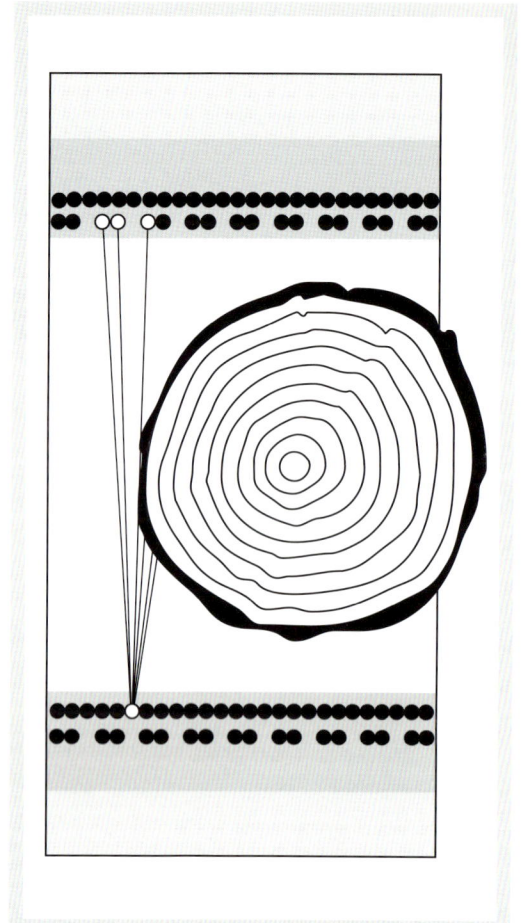

Figure 13. The principle applied in log measurement using optical measurement devices (Rema).

3 Costs involved in timber measurement

3.1 Introduction

The costs involved in timber measurement constitute an important criterion when choosing the appropriate method for a specific situation. This section offers a literature review of the estimated costs of selected timber measuring methods. The costs are presented in terms of functions if these are available. Note that timber measurement costs and harvesting costs depend heavily on prevailing conditions such as stem size and total volume to be measured at a time. In addition, different methods react differently to different conditions. In the measurement of standing trees, stem size is an important factor but in stack measurement, stem size does not influence measurement costs.

Timber measurement

3.2 Measurement of standing trees

Calculation of time consumption in measuring standing trees by a team of two persons is as follows[33]:

$$t_{st} = 0.9 \left(e^{6.9 + \frac{76.4}{V_s} - 0.161 \ln(V_{tot})} \right) \quad (3)$$

where t_{st} is the time expenditure in measurement of standing trees (including travel time and planning of harvesting) [cmin m^{-3}] (workplace time), V_s is the mean stem volume [dm^3], and V_{tot} is the stand out turn [m^3]. The value of 0.9 is the factor for reducing time expenditure because of foreseeable developments in methods and tools.

Using this model and average labor costs including costs of travel to the work site of FIM 140 h^{-1}, the costs of timber measurement in thinnings with mean stem volume at 0.07 m^3 and total outturn within the range of 100–1 000 m^3 varies from FIM 30 to FIM 20 m^{-3}. With mean stem volume at 0.5 m^3, the corresponding measurement cost variation is from FIM 12 to FIM 8 m^{-3}. Adding the costs of training and supervision increases the costs by FIM 1 and FIM 2.

3.3 Harvester-based measuring devices

The average measurement costs involved in present-day harvester-based measurement are determined applying the following default values[34]:

- purchase price of prime mover (harvester), FIM — 2 445 000
- purchase price of measuring device, FIM — 45 000
- machine service life, years — 6
- measuring device service life, years — 3
- annual productivity, m^3 yr.$^{-1}$ — 40 000
- annual depreciation, % of purchase price — 25
- hourly costs, FIM gross per effective hour — 390
- annual maintenance costs of measuring device, FIM. — 10 000

Using these default values, the average cost of measuring timber was FIM 0.47 m^{-3}. The average measurement costs when using a harvester-based measuring device depend on the investment and maintenance costs of the device. If the purchase price of a measuring device is FIM 55 000, measurement costs only increase to FIM 0.51 m^{-3} (annual maintenance costs FIM 10 000). With annual maintenance costs amounting to FIM 20 000 and purchase prices alternatively FIM 45 000 and FIM 55 000, the measurement costs would be FIM 0.72 m^{-3} and FIM 0.76 m^{-3}, respectively.

CHAPTER 8

3.4 Cutter method

The costs involved when using the cutter method depend considerably on the total volume of the stand marked for cutting. If the total volume of the stand exceeds 400 m^3, the total stand volume does not strongly influence the measurement costs. With mean stem size at 0.04 m^3 and the total stand volume marked for cutting below 400 m^3, the cutter method's costs vary from FIM 5 to FIM 10 m^{-3}. In stands with total volumes in excess of 400 m^3, the measurement costs are between FIM 4 and FIM 5 FIM m^{-3} [35].

3.5 Measurement at roadside

The following Equation provides the time expenditure when measuring logs piece-by-piece with a person using electronic callipers[33]:

$$t_l = 1.15(t_{eff} V_l + t_0 V_{tot}) \tag{4}$$

where t_l is the time expenditure of piece-by-piece measurement of logs [cmin m^{-3}] (workplace time), t_{eff} is the time expenditure of scaling one log [cmin] (effective time), V_l is the mean volume of the logs [m^3], t_o is the time consumed in moving, preparations, and computing [cmin/pile], and V_{tot} is the total volume of the logs to be measured [m^3].

If V_{tot} is 50 m^3, t_{eff} is 12 cmin/log, and t_o is 650 cmin, then t_l (time expenditure) is 74 cmin m^{-3}. Assuming labor costs of FIM 140 h^{-1}, the direct measurement costs are FIM 1.7 m^{-3}. The costs of training and supervision increase the costs between FIM 1 and FIM 2.

A model for calculating the time expenditure of measuring stacked pulpwood in bunches by a team of two persons using electronic callipers is the following[33]:

$$t_b = 1.15\left(34.7 + \frac{1095 + t_c + t_0}{V_b}\right) \tag{5}$$

where t_b is the time expenditure of measuring stacked pulpwood in bunches [cmin m^{-3}] (workplace time), V_b is the mean bunch volume [m^3], t_c is the time expenditure in computing, [cmin/bunch] (effective time), and t_o is the time consumed by moving and preparations per bunch [cmin]. If V_b is 50 m^3, t_c is 1 000 cmin, t_o is 600 cmin, and labor costs amount to FIM 140 h^{-1}, then the measurement costs are FIM 2.5 m^{-3}. The costs of training and supervision increase the costs between FIM 1 and FIM 2.

The costs of scales mounted on the grapple of the timber truck's crane consist of the purchase price and operating costs (including interest and extra time consumption in weighing). In average conditions, the weighing costs of using scales mounted on a timber truck's crane varied between FIM 0.43 and FIM 0.62 per Mg (FIM 0.35 and FIM 0.50 m^{-3})

3.6 Measurement at the mill

In 1992–1993, the average costs of measuring timber at the mill were FIM 1.22 m^{-3} [27, 28, 29]. The investment cost involved in a frame-scanning measuring system is approximately FIM 3 000 000, and the annual measuring costs at a mill consuming 1 000 000 m^3 of roundwood varied between FIM 1.0 and FIM 1.5 m^{-3}.

4 Further development of timber measuring methods

Technological development, reduced measuring costs, and changes in timber procurement and legislative requirements are the driving forces behind the development of measuring methods. Rapid development in sensor and data-processing technology has enabled commercialization of new measuring devices and methods. The aim in reducing measurement costs without impairing measuring accuracy favors systems that reduce the number of times a particular parcel of timber is measured and methods that can be automated to reduce labor costs. The faster pace of timber procurement calls for up-to-date data on timber quantities and their location in the timber procurement system. Development of data processing enables complex computations and data transfers. Legislation imposes accuracy requirements and controllability on measuring devices and results.

Timber measurement by harvester-based devices is the most common measuring method used in Finland in determining timber purchase prices in standing sales. In 1997, 89% or 23.1 million m^3 of the timber procured by forest-industry companies was measured by harvester-based devices. The share of harvester-based measurement should remain at least on this level although measurement at the mill will probably increase. This is because timber measurement is also necessary for purposes other than determining purchase prices. Harvester-based devices, timber measurement, and crosscutting of felled trees are important when developing product-oriented timber procurement. This means that stems will be crosscut in compliance with the requirements of the mills and timber assortment distribution will use mill demands.

Measurement at the roadside and at the mill were approximately 10% of the standing-sales timber harvested by forest-industry companies. This share is decreasing slowly because of the high measuring costs and development of other methods.

For timber purchased by forest-industry companies on a delivery basis in 1994, measurement at the roadside was the principal measuring method. Approximately 74% or 5.7 million m^3 of timber was measured at the roadside and approximately 23% at the mill. The current trend indicates that the measurement at the mill will increase for standing-sales and delivery-sales timber. This is due to lower measuring costs and higher accuracy of mill measurement. Monitoring of quality is also easier at the mill, and new measuring methods such as the frame-scanning method and optical log measurement can reliably measure even small parcels of timber.

CHAPTER 8

References

1. Granvik, B. -A., Puu- ja metsäteknologian peruskäsitteitä ja termejä (Basic concepts and terms of wood and forest technology) Part 2: Metsäteknologia, University of Helsinki, Department of Forest Resource Management, Publications 1, Helsinki, 1993, pp. 98–171.

2. Kärkkäinen, M., Puutavaran mittauksen perusteet (Principles of timber measurement), Helsingin yliopisto, Helsinki, 1984, 252 p.

3. Kangas, A. and Päivinen, R., Silva Carelica 27, Joensuun yliopisto, Joensuu, 1994, pp 5–6.

4. Anon., Pystymittaus 1 (Measurement of standing trees – Method 1), Ministry of Agriculture and Forestry 4.5.1992, Helsinki, 1992, 10 p.

5. Ihalainen, A., The Finnish Forest Research Institute, Research Papers 558:3(1995).

6. Anon., Hakkuukonemittauksen ohje (Instructions for timber measurement when using harvester-based measurement devices), Mittausneuvosto, Helsinki, 1991, 9 pp.

7. Ahonen, O. -P. and Marjomaa, J., Metsäteho Review 10:1(1994).

8. Ala-Ilomäki, J., The Finnish Forest Research Institute, Research Papers, 450:1(1993).

9. Rieppo, K., Metsäteho Review 13:1(1991).

10. Rieppo, K., Metsäteho Review 3:1(1989).

11. Ahonen, O. -P. and Lemmetty, J., Metsäteho Review 5:1(1995).

12. Anon., Metsurimittaus (Cutter measurement), Mittausneuvoston hyväksymä ohje 17.2.1989, Helsinki, 1989, 8 p.

13. Anon., Pystymittaus 2 (Measurement of standing trees – Method 2; Cutter measurement), Ministry of Agriculture and Forestry 5.10.1992, Helsinki, 1992, 5 p.

14. Anon., Yksivaiheinen metsurimittaus, Mittaus- ja laskentaohje, Metsäteho Oy ja Metsäntutkimuslaitos, Helsinki, 1996, 24 p.

15. Räsänen, R., Marjomaa, J., and Ihalainen, A., Metsäteho Review 7:1(1995).

16. Nyman, T., Aikio, E., and Mustonen, A., Metsähallituksen kehittämisyksikkö, Tiedote 6:1(1995).

17. Kärkkäinen, M., Folia Forestalia 410:1(1979).

Timber measurement

18. Anon., *Kehämittaus (Grapple heap measurement)*, Ministry of Agriculture and Forestry 5.10.1992, Helsinki, 1992, 6 p.

19. Kärkkäinen, M., Folia Forestalia 542:1(1983).

20. Heiskanen, V., Folia Forestalia 227:1(1975).

21. Naasko, M. and Vuollet, E., Forestry Development Unit, Finnish Forest and Park Service, Bulletin 10:1(1995).

22. Sikanen, L., *Metsätraktorin kuormainvaakaan perustuvan mittausmenetelmän soveltuvuus puutavaran tilavuusmittaukseen (The applicability of forwarder-based weight measurement system for timber volume measurement)*, University of Joensuu, Faculty of Forestry, 1993, 29 p.

23. Anon., *Järeiden puutavarapölkkyjen kiintomittaus (Solid volume measurement of large-sized logs)*, Mittausneuvoston hyväksymä ohje 1.3.1990, Helsinki, 1990, 16 p.

24. Anon., *Puutavarapölkkyjen mittaus (Measurement of timber by logs)*, Ministry of Agriculture and Forestry 30.3.1992, Helsinki, 1992, 8 p.

25. Anon., *Pinomittaus (Measurement of stacked pulpwood timber)*, Ministry of Agriculture and Forestry 30.3.1992, Helsinki, 1992, 7 p.

26. Vuollet, E. and Tiuraniemi, K., Metsähallituksen kehittämisyksikkö, Tiedote 7:1(1993).

27. Anon., *Puutavaran tehdasmittaus (Timber measurement at the mill)*, Metsätehon opas, Helsinki, 1995, 32 p.

28. Marjomaa, J., Työtehoseuran metsätiedote 3(556):1(1996).

29. Oijala, T. and Terävä, J., Metsäteho Review 1:1(1994).

30. Marjomaa, J., Metsäteho Review 8:1(1990).

31. Marjomaa, J. and Sairanen, P., Metsäteho Review 3:1(1996).

32. Marjomaa, J., Metsäteho Review 1:1(1996).

33. Rummukainen, A., Alanne, H., and Mikkonen, E., *Puunhankinta muutospaineessa. Voimavaratarpeiden arviointimalli vuoteen 2010*, University of Helsinki, Department of Forest Resource Management, Publications 2, 1993, pp. 45–59.

34. Rummukainen, A., Alanne, H., and Mikkonen, E., Acta For. Fenn. 248:96(1995).

35. Halinen, M., Metsäteho Review 20:1(1987).

CHAPTER 9

The price of timber

1	Approaches	395
2	The mill price	396
3	Forest management costs	397
	References	402

CHAPTER 9

Pertti Harstela

The price of timber

1 Approaches

In a market economy, the price of timber can be a balance of supply and demand. Markets seldom are perfect. This chapter presents no macro or micro economic theories of price setting. Price is a result of negotiations between the seller and the buyer. Some calculations are necessary to form the basis for such negotiations.

A common practice is to treat the price of standing trees, *stumpage price*, as a residual value. A particular country or enterprise only has limited opportunities to influence world-market prices for industrial products. The highest price a country or enterprise can pay for timber is the market price of products minus all marketing, processing, and timber procurement costs including profit. This concept receives support from the fact that the price of timber is difficult to estimate using production costs. For an area of old forest, the stumpage price is paid after cutting the trees. A minor portion of this goes to establishing regeneration and subsequent tending. The costs of silvicultural and forest improvement work may be 10%–25% of the gross stumpage earnings. Forestry is therefore a special kind of business activity with its very high gross margin but long production time.

Another common way of approaching the subject is to consider the price of an open area and its afforestation costs as an *investment*. The subsequent tending and administration costs are *operating costs*. This is especially relevant in the afforestation of a degraded or former agricultural area. In such a case, the biggest problem is the long production period. The selected *interest rate* has a major impact on the profits realized. A common practice is to compare different silvicultural alternatives by computing their *internal rates of return*.

CHAPTER 9

2 The mill price

The *mill price of wood* varies considerably in different parts of the world because of the variation in climatic and logging conditions, infrastructure, transportation distances, tree species and timber assortments, forest ownership, and domestic forest industries. Table 1 gives an idea of the structure of the mill price.

Table 1. The average mill price of all timber in Finland in 1994[1] with added overhead.

Cost factor	Price, FIM m^{-3}
Cutting	30.8
Forest hauling	16.4
Harvesting, total	47.2
Long-distance transportation	32.1
Procurement, total without overhead	79.3
Stumpage price	147.3
Mill price, total without overhead	226.3
Overhead, 7%	15.8
Mill price	242.1

Procurement costs are high for nonindustrial, private forest ownership and small-sized forest holdings. This means higher harvesting costs due to small operations and the distribution of activities over a wide geographical area. Stumpage prices are also high in Finland and require high *cost-effectiveness* by forest industries. Table 2 illustrates the variation in stumpage prices by timber assortment. The value of the raw material in the mill process, harvesting conditions and harvesting costs, and additionally even historical development of prices influence the stumpage prices.

Table 2. Average stumpage prices paid in Finland in 1995[2].

Timber assortment	Southern Finland	Northern Finland
	Stumpage price, FIM m^{-3}	
Pine saw logs	250	223
Spruce saw logs	198	181
Birch peeler logs	240	-
Pine pulpwood	98	97
Spruce pulpwood	112	98
Birch pulpwood	102	101
Aspen pulpwood	15	-

Stem volume, branchiness of trees and stem taper curve, yield (volume harvested per ha), forest hauling distance, volume harvested per operation, and terrain conditions influence harvesting costs. Especially steep terrain can cause a dramatic increase in the costs. Stem volume is the primary factor as Table 3 for logging costs in 1994 shows[3].

Table 3. Logging costs in mechanized timber harvesting operations in Finland in 1994[3].

Operation	Logging costs, FIM m^{-3}
First commercial thinnings	75–80
Intermediate thinnings	52–57
Final (regeneration) cuttings	30–34

Although yield of timber per hectare and some other harvesting-condition factors are more favorable in final cuttings than in thinnings, stem volume is the main factor influencing logging costs. Remember also that pulpwood obtained from first thinnings is not as valuable a raw material as the timber obtained from final cuttings. Basic wood density and tracheid length in pulpwood from thinnings are smaller and the proportion of bark is greater. Thinnings are clearly beneficial from the silvicultural consideration, and they increase the yield of harvested timber and its value during the rotation.

Labor expenditure in logging operations varies widely according to the conditions and the technology used. In plantation forestry or in the natural forests of the Nordic countries, logging 10 000 m^3 may require the labor of more than 20 man-years when using handsaws, axes, and animal skidding. In motor-manual logging using farm tractors in forest hauling, the corresponding figure could be 5 man-years or only 0.5 man-years in fully mechanized logging using harvesters and forwarders. Manuals, estimated functions, and computer programs allow computation of labor requirement, productivity, and costs of timber harvesting under different conditions[4,5].

In highly mechanized logging operations, *capital inputs* and other machine costs are the predominant cost factors rather than labor. In harvester and forwarder operations, capital expenditures are almost 40% and variable machine costs are approximately 20% of the total costs. Labor costs are only approximately 40%[3,6].

3 Forest management costs

The biggest costs in the management of forests are silvicultural costs. Table 4 shows the *unit costs* of the main *silvicultural works* and road construction in Finland in 1996. Most work occurs only once per rotation in a particular forest stand. The *annual costs* are therefore very different. In addition to the above activities, forest protection, conservation, soil and water management, and administration may require consideration under certain conditions. Conservation seldom causes direct costs but losses of revenues as explained in the section dealing with environmental impacts.

CHAPTER 9

Table 4. Unit costs of silvicultural works, forest planning, and construction of permanent forest truck roads[2].

Operation	Unit costs, FIM ha^{-1} or [2] FIM meter^{-1}
Clearing of regeneration areas	343
Site preparation:	
Harrowing and screefing	661
Plowing and mounding	927
Prescribed burning	1 690
Sowing	963
Planting	3 225
Tending of seedling (young) stands	1 043
Pruning	1 985
Forest fertilization	868
Forest management plan	70
Road construction	41 [2]

Silvicultural works are *labor-intensive* even in industrialized countries. Only soil preparation has been mechanized to a high degree as has planting in favorable, stone-free conditions. In 1982 in Finland, when the mechanization of timber harvesting was in its early stages, the average monthly number of employees in logging was approximately 20 000. Silviculture occupied 5 000 employees. In 1991 after intensive mechanization in timber harvesting, the corresponding figures were 7 000 and 5 500. The figures in Table 5 indicate the *need for manual labor* in the Nordic countries[2]:

Table 5. Need for labor input in the Nordic countries[2].

Measure	Need for labor, man hours ha^{-1}
Prescribed burning	8.0–24.0
Mechanized site preparation	1.0–5.0
Manual planting of bare-rooted stock on prepared site	20.0–30.0
Manual planting of containerized seedlings using planting tube	15.0–20.0
Mechanized planting, former farm land	1.5–2.5
Manual weeding (mechanically)	16.0–30.0
Clearing of young stand using clearing-saw	4.0–24.0

The price of timber

Highly mechanized systems are being developed for planting on normal forest soils and clearing and early thinning of young stands. Their economic profitability is still questionable, and they have little use in the typically stony and uneven terrain of the Nordic countries.

To characterize the nature of forestry as a field of human endeavor and business, Table 6 presents the average *income statements* for jointly-owned large forest holdings located in northeastern Finland, a region representing poor conditions for practicing forestry[7]. Forestry is typified by its *gross margin* and *net income* being high compared with turnover. Net income in forestry can vary considerably annually. *Return on investments* and on capital depend on the prices paid for forest land and several conditional factors[7, 8]. In addition, the computation of investments and on (own) capital is problematic in forestry. As a rule, capital is computed using the value of the land (fixed) and the growing stock. Several computation methods are available. A key factor in estimating the value of growing stock is to estimate the cutting budget for the target growing stock.

Table 6. Income statements for jointly-owned forest holdings in northeastern Finland[7].

Income statements	FIM ha^{-1}
= Turnover from timber sales	162.90
+ Other sales	2.69
= Turnover of forest holdings	165.59
- Logging costs	-32.55
= Turnover of forest property	133.04
- Variable costs (silvicultural activities, etc.)	-10.07
+ Change in value of growing stock	11.84
= Sales margin	111.13
- Fixed costs	-12.57
= Gross margin	98.56
- Depreciation	-1.77
= Net income	96.79
+ Other incomes and costs	15.88
- Change in reservations	-4.11
= Profit-sharing	108.56
- Interest and direct taxes	-1.26
= Surplus	107.30

CHAPTER 9

When practicing *sustainable forestry* as required in many countries, avoiding regeneration costs after cutting is unavoidable. An investment computation can select the most profitable regeneration method. The situation is different when an open area is afforested. The afforestation decision can then use the profitability of the investment. As an example, Table 7 presents an investment calculation for a pine plantation in a Mediterranean country. The poor site is certainly not the best for forestry because it is in a mountainous area. The value of the land is included in the annual costs as an annual net revenue of the alternative use of the land. In this case, the previous land use was pasture.

Table 7. An example of the internal rate of return in the case of afforestation of an open area.

Activity	Costs, FIM ha^{-1}
Afforestation (including weeding and supplemental planting) during 1st year	4 219
Annual administration costs and other annual costs (including average costs of forest fires)	87
1st pruning in 10th year	277
2nd pruning and logging costs of thinning in 20th year	8 531
Logging costs of final cutting in 30th year	39 145
Activity	**Revenues, FIM ha^{-1}**
Sales of timber from thinning in 20th year	11 468
Sales of timber from final cut in 30th year	54 772
Internal rate of return (IRR), real interest	4.5 %

The internal rate of return is very dependent on the soil and climatic conditions, infrastructure, price level, location of forest, timber markets, etc. This calculation therefore cannot be generalized. When appraising an *investment calculation*, one must remember that the real price of timber is rather constant in the long term. Therefore IRR is close to the real interest rate. Because of the long rotation period, investment calculations are very sensitive to *interest rates* used in calculations. For the same reason, it is much more difficult to predict sales revenues than afforestation costs. In the calculation, sensitivity analysis indicated the relation in Table 8 between IRR and change in timber prices.

Table 8. The influence of the change in timber prices on IRR.

Change in timber prices, %	IRR, %
0	4.5
-10	2.6
+10	5.8

Predicting changes in timber prices for the long term is difficult. The weighted real (deflated) *price index* for all forest products in Fig. 1 has been rather steady. The same situation prevails for softwood saw logs, and the real price index for pulpwood has shown a sligtly decreasing trend to 1990[9]. The demand prognoses of forest industry products are positive and therefore indicate continuous growth in demand as of timber[10]. On the other hand potential cut of forest resources easily meets the demand.

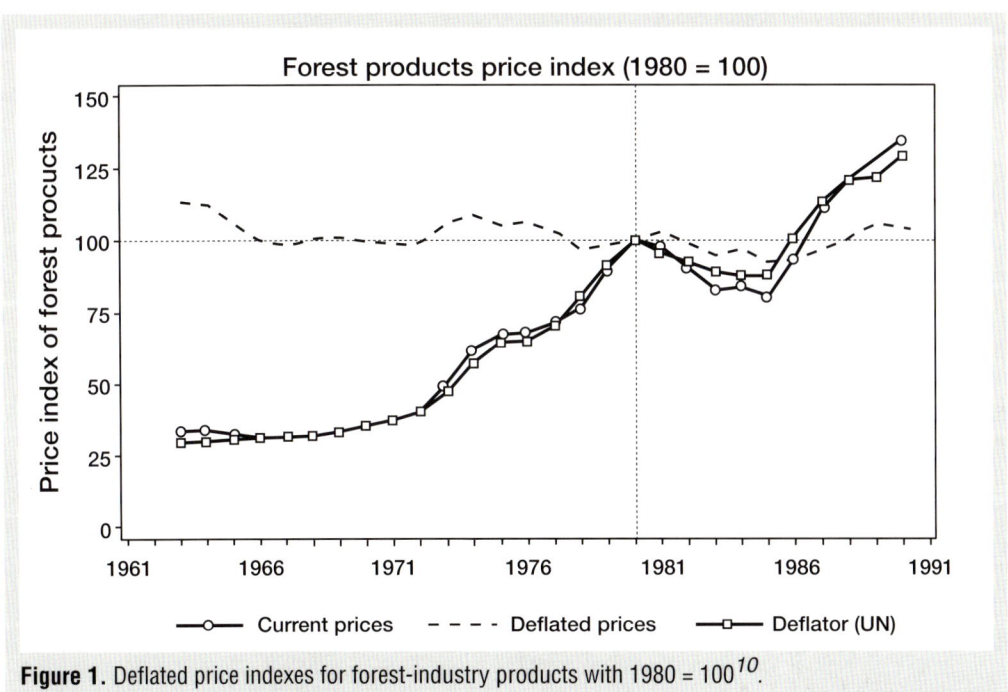

Figure 1. Deflated price indexes for forest-industry products with 1980 = 100 [10].

CHAPTER 9

References

1. Anon., Metsäteollisuuden vuosikirja 1994, Metsäteollisuus ry, Helsinki, 1995, pp. 1–68.

2. Anon., Statistical Yearbook of Forestry (Y. Sevola, Ed.), The Finnish Forest Research Institute, Helsinki, 1998, pp. 125–135

3. Hakkila, P., The Finnish Forest Research Institute, Research Papers 557:49(1995).

4. Anon., FAO Forestry Paper 99, FAO, Rome, 1992, pp. 1–106.

5. Oijala, T. and Rajamäki, J., Metsäteho Review 5:1(1992).

6. Harstela, P., Silva Carelica 25, Joensuun yliopisto, Joensuu, 1994, p. 99.

7. Penttinen, M., Folia Forestalia 799:45(1992).

8. Hyttinen, P., in Yksityismetsätalouden kannattavuusseuranta -Laskentatoimen empiirinen kokeilu (P. Hyttinen, Ed.), Joensuun yliopisto, Metsätieteellinen tiedekunta, Tiedonantoja 35, 1995, pp. 62–63.

9. Anon., FAO Forestry Paper 125, FAO, Rome, 1995, p. 8.

10. Anon., European Timber Trend and Prospects: into the 21st century, United nations Economic Commission for Europe, United Nations, FAO, New York, 1996, pp. 1–103.

The price of timber

CHAPTER 10

The forest sector and national economy

1	Introduction	405
2	Input-output tables	405
3	Production model	407
4	Total output	409
5	Value added	410
6	Employment	411
	Appendix	413

CHAPTER 10

Mikko Toropainen

The forest sector and national economy

1 Introduction

The term *forest sector* covers forestry and the forest industries. The two are closely interdependent. The demand for timber is *derived demand* determined by the demand for forest industry products. The forest industry needs timber and many other items such as machinery, energy, chemicals, and the services of transport and trade. In Finland, the inputs for the forest sector primarily have domestic origin. Changes in the production and input structure of the forest sector therefore have substantial effects on many industries and on the entire economy. These effects are clearly less prominent in industries using large amounts of imported inputs such as the electrical equipment industry.

Input-output analysis is an appropriate method for investigating the interdependencies between industries within the context of a national economy. The following serves as an introduction to the idea behind the analysis and to the basic version of the model. Examples of the economic links of the forest sector are also given. *Input-output tables* reflect the effects and dependencies between industries. In many countries, such tables belong to the system of national accounts. *An input-output model* is an analytical tool using data presented in the tables. For example, the model can investigate how economic changes impact production, incomes, and employment.

2 Input-output tables

An industry x industry input-output table presents the production figures by industries. A row in the industry section shows how the products of that industry are used for intermediate inputs by the industries in the columns (quarter I of Fig. 1), as well as for final products (quarter II). Intermediate inputs are raw materials and other goods and services of short duration needed in production. Final products are goods and services used for private and public consumption, for investments, and for exports. Increases in stocks and statistical discrepancy act as balancing items. A row in the primary input section indicates the use of that primary input for production and for final products (quarters III and IV). Primary inputs include imports of commodities, commodity taxes minus corresponding subsidies, wages and salaries, other indirect net taxes, the consumption of fixed capital and the operating surplus of the firms.

	Industry n + 1 columns	Final products m columns	Σ
Industry n rows	Output for intermediate products (I)	Output for final products (II)	
Primary inputs k rows	Primary inputs for production (III)	Primary inputs for final products (IV)	
Σ			

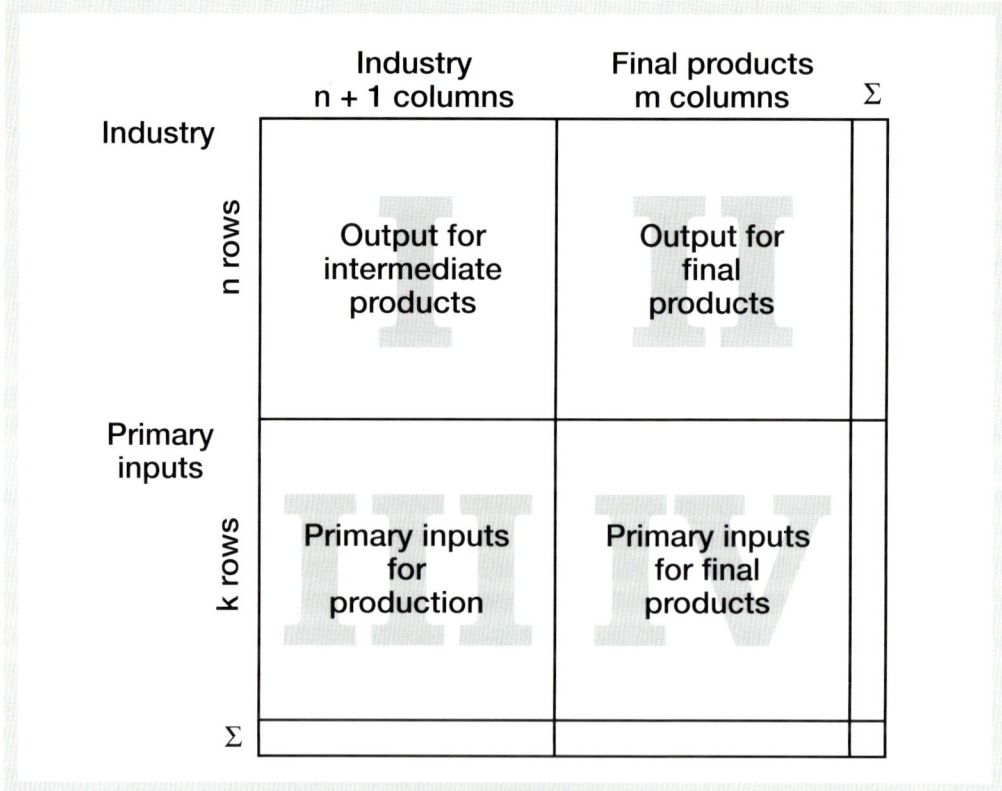

Figure 1. The structure of an industry x industry input-output table.

A column in the industry section indicates the use of intermediate inputs (quarter I) and the use of primary inputs (quarter III) in that industry. The column therefore describes the cost structure of the industry. The column sum equals the value of output. The column sum also equals the row sum. Production is always as high as its use. A column in the final product section shows the commodities and primary inputs of which a certain final product category consists. The value added in basic values of an industry is counted by adding the primary inputs without imports and commodity taxes. If only imports are omitted, the value added of purchaser is obtained.

The input-output table is converted into an *input coefficient table* by dividing the elements of industry section (quarters I and III) by the corresponding column sum. A column in the coefficient table indicates the intermediate and primary inputs needed for one unit of production. Column sum always equals unity. Using this table allows calculation of the direct effects that an increase in final demand has on the use of inputs.

3 Production model

The simplest version of various input-output models is the production model using the intermediate input coefficients a_{ij}. Let us take as an example industry number 1. To produce one unit of output, industry j needs a_{1j} units of the output of industry 1. To be sufficiently high to cover the input requirements of all n industries and final demand, the output level x_1 of industry 1 must satisfy the equation

$$x_1 = a_{11}x_1 + a_{12}x_2 + \ldots + a_{1n}x_{1n} + d_1 \tag{1}$$

or

$$(1 - a_{11})x_1 - a_{12}x_2 - \ldots - a_{1n}x_n = d_1 \tag{2}$$

where x_j is the output of industry j, a_{ij} is the input coefficient, and d_1 is the final demand for the output of industry 1.

By subtracting commodities used for intermediate inputs from the total output, we obtain the commodities used for final consumption. A similar equation can be written for all industries. We then obtain a group of n equations that can be expressed in matrix form

$$(I - A)x = d \tag{3}$$

where I is n x n identity matrix, A is n x n matrix of the input coefficients, x is n x 1 vector of the total output of industries 1 ... n, and d is n x 1 final demand vector. The equation has the solution

$$x = (I - A)^{-1}d \tag{4}$$

The solution shows how high the production of each industry must be to satisfy the demand for it. $(I - A)^{-1}$ is *the Leontief inverse matrix*. The input-output method was developed by Wassily Leontief for which he received the Nobel Prize in Economic Science in 1979.

A column of the inverse matrix shows how many production units are necessary from the industry on the row for the industry represented in the column to produce *one unit of final product*. The column sum tells the total output required. A row shows the effect of one unit change in the demand for final product in the column on the output of industry on the row. The Leontief inverse matrix consists of both direct effects (input coefficients) and indirect effects. These are generated because producing intermediate inputs requires intermediate inputs similarly producing those, etc.

With the inverse matrix, the direct and indirect output effects of changes in final demand can be found. It also is possible to use the matrix for investigating the effects of changes in input structure. The matrix also indicates the need to increase imports when final demand increases should there be any shortages of capacity in industries producing the intermediate inputs.

CHAPTER 10

Using the following equation, one can count the direct and indirect employment requirements of one unit of final product and investigate the employment effects of changes in employment coefficients and in input coefficients. The employment coefficient is the number of employed persons divided by the total output. A change in this coefficient means a change in productivity.

$$L = W^\wedge(I-A)^{-1} \tag{5}$$

where L is the labor input required by one unit of the final product -matrix, and $W^\wedge$ is the diagonal matrix from the vector of employment coefficients.

A column of matrix L shows both the direct and indirect employment in the industry indicated on the row as required by one unit of final product of the industry indicated on the column. The column sum gives the total labor input.

The primary input contents of the final product can also be counted. From the national economic viewpoint, the value of final product consists only of direct and indirect primary inputs. There are no intermediate inputs. This derives from the fact that part of the value of intermediate inputs is always from primary inputs similarly of the value of intermediate inputs needed to produce intermediate inputs, etc. When counting the entire chain of production, only primary inputs remain. Solving the primary input contents uses the following equation

$$B = C(I-A)^{-1} \tag{6}$$

where B is the primary input contents of the final product -matrix, and C is the primary input section of the input coefficient-matrix.

A column in the matrix shows both imports and components of value added directly and indirectly included in one unit of the final product. Using the matrix, one can investigate the effects of changes in final demand on essential economic parameters.

The inverse matrix is also usable for environmental analyses. For example, when sulfur dioxide emissions per unit of output of each industry are known, the employment coefficients of Eq. 5 can be replaced by sulfur coefficients to find the direct and indirect emission effects of changes in production.

The model above does not include derived effects. Those are generated when the incomes of households rise because of growth of production and the consequent increase in consumption causes more production. Estimating these derived effects is uncertain primarily due to a lack of proper data. As a rule of thumb, these effects would raise the total effects 1.5-fold compared with the production model adapted here. In addition, there are no investment linkages in the production model. Counting those would require the use of a dynamic model.

The input-output method has been developed for *the analysis of marginal changes in the short run*. The coefficients seldom change much in the short run. Higher coefficients are more stable than lower ones, and only the changes in a few highest coefficients have any importance. If longer time scales are necessary such as when production capacity changes substantially, the average coefficients of any industry will change to some extent because new plants will be more effective than old ones. The coefficients of the newest plant can be used as average coefficients for estimating future production conditions.

4 Total output

The forest sectors in different countries are highly heterogenous in their input and output structures. One may produce mostly expensive printing papers, and the other may produce timber for exports. One may have access to domestic raw materials, and another may import all the timber used and make furniture for export. The following are some examples of the results of an input-output analysis for the Finnish forest sector. The input-output tables are from 1993.

The forest sector consists of seven industries: forestry, sawmilling, other wood products, furniture, pulp, paper, and paper products. The sector contributes significantly to Finnish exports but does not need much imported inputs directly or indirectly.

For comparison, consider similar figures for the electrical equipment industry. This is another important export branch that uses considerable quantities of imported inputs. The home market industries are represented by personal services.

Let us now suppose that the final demand for each industry's products suddenly increases by FIM 100 million. When final products are produced, there is a need for intermediate inputs made by the industry itself and by other industries. Producing intermediate inputs also requires intermediate inputs, etc. Indirect effects are generated, and the increase in total output is greater than that of final products. The change in total output reflects the enlargement and diffusion of economic activity.

Table 1. Effects on domestic output of an increase of FIM 100 million in final demand.

Industry producing the final product	Increase in domestic output, million FIM		
	Direct	Indirect	Total
Forestry	100	12	112
Sawmill	100	90	190
Other wood products	100	81	181
Furniture	100	81	181
Pulp	100	105	205
Paper, paperboard	100	103	203
Paper and paperboard products	100	91	191
Electrical equipment	100	46	146
Personal services	100	52	152

Forestry uses a small amount of intermediate inputs. Table 1 shows that its indirect effects on output are small. The final products of forestry are, for example, fuel wood for households and exported roundwood. The major share of the production in forestry is used as intermediate inputs by the forest industries.

The indirect effects of the forest industries are significant compared with other manufacturing industries, and the growth effects are spread widely. Forestry, energy production, transport, business services, and the chemical and metal industries stand to benefit most. For example, a direct increase of FIM 100 million in exports of sawmill products indirectly increases the output of the industry itself by FIM 4 million, of forestry by FIM 45 million, and FIM 41 million in other industries.

In the pulp and paper industries, the change of total output per unit of final product is approximately the same. Nevertheless, the degree of processing is of great importance. If timber worth FIM 33 million is processed into pulp for exports, there is an increase of FIM 100 million in the final products and of FIM 205 million in total output. If the same amount of timber is processed domestically into paper, the increase in final products is FIM 275 million and in total output FIM 558 million. The importance of demand must not be forgotten, however. The final demand for forest industry products determines the production and therefore the use of timber – not vice versa.

The comparison with the production of electrical equipment shows that it requires considerable imported inputs. Consequently, its indirect domestic production effects are clearly smaller than those of the forest industries. The situation is similar with personal services that require few intermediate inputs.

5 Value added

The change of total output is a gross figure. The sum that an economy finally earns by an industry is the value added (an industry's share of gross domestic product). The forest industries create considerable value added especially indirectly. For example, an increase of FIM 100 million in final products of sawmilling increases value added directly by FIM 32 million and indirectly by FIM 62 million as Table 2 shows. The indirect effect mainly falls to forestry where the direct share of value added in the value of total output is 91%. Table 2 also shows the indirect/direct ratio. For example, in the paper industry for each directly generated unit of value added there are 1.32 indirect units.

Table 2. Effects on value added of an increase of FIM 100 million in final demand.

Industry producing the final product	Increase in value added, million FIM			
	Direct	Indirect	Total	Indirect/Direct
Forestry	91	7	98	0.08
Sawmill	32	62	94	1.94
Other wood products	44	42	86	0.95
Furniture	43	38	81	0.88
Pulp	21	61	82	2.90
Paper, paperboard	34	45	79	1.32
Paper and paperboard products	40	38	78	0.95
Electrical equipment	40	21	61	0.53
Personal services	63	27	90	0.43

Returning to the example of degree of processing, if timber worth FIM 33 million is processed into pulp and exported, the national economy earns FIM 82 million. If the same amount of timber is processed into paper, the value added is FIM 217 million.

In the example of the production of electrical equipment, the creation of value added is clearly smaller than that of the forest industries. This is because of the use of imported inputs. In personal services, the share of value added is high. The need for intermediate inputs is small, and these are mainly domestic in origin.

6 Employment

Input coefficients are considerably stable. This is not true for employment coefficients. The rise of productivity causes substantial changes even in a relatively short period. The employment coefficients are estimated here according to the level in 1995.

Growth in the forest industries causes an increase in employment especially indirectly. Final products of FIM 100 million in sawmilling require a direct labor input of 101 persons and indirectly five persons in the industry. 96 jobs are created in forestry and 72 jobs in other industries as Table 3 shows. Pulp and paper are capital-intensive industries so their employment effects are mainly indirect. In the pulp industry, the indirect/direct ratio is as high as 3.91.

CHAPTER 10

Table 3. Employment effects of an increase of FIM 100 million in final demand.

Industry producing the final product	Increase in number of employed persons			
	Direct	Indirect	Total	Indirect/Direct
Forestry	221	20	241	0.09
Sawmill	106	168	274	1.58
Other wood products	225	120	345	0.53
Furniture	268	140	408	0.52
Pulp	43	168	211	3.91
Paper, paperboard	73	126	199	1.73
Paper and paperboard products	132	121	253	0.92
Electrical equipment	114	72	186	0.63
Personal services	317	92	409	0.29

If timber worth FIM 33 million is processed into pulp and exported, a labor input of 211 persons is necessary. If it is processed into paper, the corresponding figure is 547. The indirect employment effects of producing electrical equipment are smaller than those of the forest industries. Personal services are labor intensive.

In the long run, the increase in labor productivity reduces the need for labor created by the increase in final demand. Besides changing the production structure whose importance was described above, the only way to avoid this problem is to ensure that the increase in production exceeds the increase in productivity. Paradoxically, one way to solve this problem is raising productivity. If the rise in productivity is higher than that of rivals, competitiveness is improved, and there is the possibility of conquering new market shares and market areas.

CHAPTER 10

Appendix

The column and the row of paper industry in the Finnish input-output table, FIM million.

Column

1	agriculture	12
2	forestry	1 828
3	fishing, hunting	0
4	mining, quarrying	116
5	food, beverages, tobacco	380
6	textiles, clothes, leather	13
7	sawn, planed, preserved wood	504
8	other wood products	145
9	furniture	3
10	pulp	5 632
11	paper, paperboard	1 064
12	paper and paperboard products	426
13	printing, publishing	168
14	chemicals, chemical products	1 138
15	refined petroleum	59
16	rubber and plastic products	81
17	non-metallic mineral products	40
18	basic metals	41
19	metal products, machinery	770
20	electrical equipment	28
21	transport equipment	37
22	other manufacture	2
23	energy and water	3 637
24	building	201
25	other construction	5
26	trade	340
27	restaurants, hotels	14
28	transport	1 359
29	communication	114
30	finance, insurance	14
31	housing	0
32	business and real estate services	845
33	personal services	86
1..33	domestic commodities in basic values	19 103
34	imports	3 940
35	commodity taxes, net	149
1..35	commodities in purchasers' values	23 192
36	wages and salaries	4 312
37	contrib. to social security schemes	1 211
38	other indirect taxes, net	-101
39	consumption of fixed capital	4 393
40	operating surplus	2 110
36..40	value added in basic values	11 926
1..40	total	35 118

Row (transposed)

1	agriculture	8
2	forestry	2
3	fishing, hunting	0
4	mining, quarrying	5
5	food, beverages, tobacco	297
6	textiles, clothes, leather	11
7	sawn, planed, preserved wood	8
8	other wood products	61
9	furniture	17
10	pulp	17
11	paper, paperboard	1 064
12	paper and paperboard products	867
13	printing, publishing	1 242
14	chemicals, chemical products	49
15	refined petroleum	6
16	rubber and plastic products	74
17	non-metallic mineral products	40
18	basic metals	46
19	metal products, machinery	46
20	electrical equipment	37
21	transport equipment	7
22	other manufacture	16
23	energy and water	85
24	building	28
25	other construction	11
26	trade	117
27	restaurants, hotels	7
28	transport	45
29	communication	7
30	finance, insurance	33
31	housing	11
32	business and real estate services	123
33	personal services	21
34	imputed bank service charge	0
1..34	industries total	4 409
35	private consumption	174
36	public consumption	219
37	investments	19
38	exports	31 171
35..38	final products total	31 584
39	increase in stocks	-405
40	statistical discrepancy	-470
1..40	total	35 118

Conversion factors

To convert numerical values found in this book in the RECOMMENDED FORM, divide by the indicated number to obtain the values in CUSTOMARY UNITS. This table is an excerpt from TIS 0800-01 "Units of measurement and conversion factors." The complete document containing additional conversion factors and references to appropriate TAPPI Test Methods is available at no charge from TAPPI, Technology Park/Atlanta, P. O. Box 105113, Atlanta GA 30348-5113 (Telephone: +1 770 209-7303, 1-800-332-8686 in the United States, or 1-800-446-9431 in Canada).

Property	To convert values expressed in RECOMMENDED FORM	Divide by	To obtain values expressed In CUSTOMARY UNITS
Area	square centimeters [cm^2]	6.4516	square inches [in^2]
	square meters [m^2]	0.0929030	square feet [ft^2]
	square meters [m^2]	0.8361274	square yards [yd^2]
	square meters [m^2]	4046.86	acres
	square kilometers [km^2]	0.01	hectares [ha]
	square kilometers [km^2]	2.58999	square miles [mi^2]
Density	kilograms per cubic meter [kg/m^3]	16.01846	pounds per cubic foot [lb/ft^3]
	kilograms per cubic meter [kg/m^3]	1000	grams per cubic centimeter [g/cm^3]
Energy	joules [J]	1.35582	foot pounds-force [ft • lbf]
	joules [J]	9.80665	meter kilogams-force [m • kgf]
	millijoules [mJ]	0.0980665	centimeter grams-force [cm • gf]
	kilojoules [kJ]	1.05506	British thermal units, Int. [Btu]
	megajoules [MJ]	2.68452	horsepower hours [hp • h]
	megajoules [MJ]	3.600	kilowatt hours [kW • h or kWh]
	kilojoules [kJ]	4.1868	kilocalories, Int. Table [kcal]
	joules [J]	1	meter newtons [m • N]
Length	nanometers [nm]	0.1	angstroms [Å]
	micrometers [Fm]	1	microns
	millimeters [mm]	0.0254	mils [mil or 0.001 in]
	millimeters [mm]	25.4	inches [in]
	meters [m]	0.3048	feet [ft]
	kilometers [km]	1.609	miles [mi]
Mass	grams [g]	28.3495	ounces [oz]
	kilograms [kg]	0.453592	pounds [lb]
	metric tons (tonne) [t] (= 1000 kg)	0.907185	tons (= 2000 lb)

Conversion factors

Property	To convert values expressed in RECOMMENDED FORM	Divide by	To obtain values expressed In CUSTOMARY UNITS
Mass per unit area	grams per square meter [g/m^2]	3.7597	pounds per ream, 17 x 22 - 500
	grams per square meter [g/m^2]	1.4801	pounds per ream, 25 x 38 - 500
	grams per square meter [g/m^2]	1.4061	pounds per ream, 25 x 40 - 500
	grams per square meter [g/m^2]	4.8824	pounds per 1000 square feet [lb/1000 ft^2]
	grams per square meter [g/m^2]	1.6275	pounds per 3000 square feet [lb/3000 ft^2]
	grams per square meter [g/m^2]	1.6275	pounds per ream, 24 x 36 - 500
Power	watts [W]	1.35582	foot pounds-force per second [ft C lbf/s]
	watts [W]	745.700	horsepower [hp] = 550 foot pounds-force per second
	kilowatts [kW]	0.74570	horsepower [hp]
	watts {w}	735.499	metric horsepower
Pressure, stress, force per unit area	kilopascals [kPa]	6.89477	pounds-force per square inch [lbf/in^2 or psi]
	Pascals [Pa]	47.8803	pounds-force per square foot [lbf/ft^2]
	kilopascals [kPa]	2.98898	feet of water (39.2°F) [ft H$_2$O]
	kilopascals [kPa]	0.24884	inches of water (60°F) [in H$_2$O]
	kilopascals [kPa]	3.38638	inches of mercury (32°F) [in Hg]
	kilopascals [kPa]	3.37685	inches of mercury (60°F) [in Hg]
	kilopascals [kPa]	0.133322	millimeters of mercury (0°C) [mm Hg]
	megapascals [Mpa]	0.101325	atmospheres [atm]
	Pascals [Pa]	98.0665	grams-force per square centimeter [gf/cm^2]
	Pascals [Pa]	1	newtons per square meter [N/m^2]
	kilopascals [kPa]	100	bars [bar]
Volume, solid	cubic centimeters [cm^3]	16.38706	cubic inches [in^3]
	cubic meters [m^3]	0.0283169	cubic feet [ft^3]
	cubic meters [m^3]	0.764555	cubic yards [yd^3]
	cubic millimeters [mm^3]	1	microliters [µL]
	cubic centimeters [cm^3]	1	microliters [µL]
	cubic decimeters [dm^3]	1	liters [L]
	cubic meters [m^3]	0.001	liters [L]

Index

A
abiotic damage 342
absolute error 374, 377
acceleration sensor 377
accumulation of organic matter ... 60, 67–69, 80, 264
accuracy requirement 365, 373, 378, 384–385
active period 56
adult wood 150
afforestation 232, 395, 400
air pollution 108, 300
allogenic succession 221–222
allometry 51, 82, 253
aminoacids 132
ammonium 267, 269
amorphous areas 138, 163
angiosperm 30–33
anisotropic 122
annual cycle 34, 78
annual increment 97, 112, 190, 268, 276
annual ring 30, 144, 168
apical meristem ..... 21, 47, 151 , 125
arabinan 130–131
arctic zone 89
Artificial Intelligence 340
artificial pruning 261
ash 92, 128, 135–137, 174, 178, 180
aspirated pit 141
assimilation rate 41
ATP 38–39, 45
Austrian formula 203
autogenic succession 219, 222, 247
autumnal moth 285, 290–291, 293
auxin 123
average diameter 380–381

B
backward selection 238
bacteria 68, 239, 299–300, 342, 346, 348
bagasse 178–179
balsams 132
bamboos 179–180
bark 31–32, 117–136, 164–175, 194, 285, 303–304, 342–349, 379–380
bark beetle 294, 303–304
basal area 188, 190, 193, 203, 255–256
basic density ... 133, 150, 160–189, 259, 264
beetle 294–295, 303–304
bias 196, 199, 366
biochemistry 38, 133
bioclimatic regions 89
biodiversity 12, 14, 111, 204, 340, 342, 353, 356
biogeochemical nutrient cycle 272
biological control 298–300
biological fixation 72
biomass accumulation 80
biosphere 59, 79, 89
biotic decay 343
blue-stain fungi 343
bole length 188, 194–195
bordered pit 140–141, 145, 147, 156, 158, 164
boreal zone 34, 55–56, 80–97, 106, 109, 221, 343, 359
branch wood 120–121, 135, 153, 155, 159–161
branch 29–55, 117–120, 125, 157–160, 188–189, 259–261
branchiness 397
breeding 236–241, 285
broadleaved species 22, 50, 85, 227, 273
brown rot 344, 348
bucking ... 313, 315, 318, 321–322, 333, 337
bud 30, 47–49
bulldozer 333
bundle floating 328
bundle measurement 383, 387

C
calcium 62, 135–136, 267, 269–270
calibration 322, 367, 370, 372–373
calorimetric heating value 174–175
cambial activity 32
cambial initials 126–128
cambial zone 123, 128, 150–151

cambium   21, 31, 50, 126–128, 144, 146, 151, 158–159, 168, 188, 316, 348
carbohydrate   .. 36, 37, 39, 45, 128, 130, 132, 150, 295. 173, 344, 348
carbon   .. 14, 16, 32, 36–40, 79, 89, 110, 121, 128, 173–174, 343, 348
carbon dioxide 16, 32, 36–38, 40, 121, 173–174, 343, 348
carbon sink 110, 173
carboxylic acids   132
cavities 130, 154, 165, 343
cell wall   . 128–132, 137–140, 145, 154, 163–165, 168, 172–173, 180, 344
cellulose   . 120–139, 150–180, 343–346, 348, 350
charcoal   101–102
chemical composition   120, 122, 128–129, 137, 151, 154–156, 174
chemical control 282, 301
chemical defense 287, 295
chemical energy 36, 38, 44, 61
chemical pulp   . 122, 129, 131, 137, 162, 164, 172, 315, 344, 347
chip storage 344, 348, 350–351, 361
chipping 152, 161–162, 315
chips 122–134, 150–173, 314–366
chlorophyll 34, 38, 132, 173
circular plot   193
climate   12–16, 61, 78, 91–114, 159, 172, 233, 240, 264
climate change 12, 16, 109–110, 114
climax species   ... 50, 221–222, 231, 233, 235
cluster sampling   200–201
coloring   343
combustion systems   174
commercial thinning 169, 171
compartment inventory   193–195
compound middle lamella   139
compression wood 152–156, 182
cones 21–22, 239
coniferous species 22, 50, 90, 92
coniferous zone   89

conservation of forests   106
constitutive defense   289
control measurement 367, 369–370, 373, 376–377, 387–388
cooking chemicals   134
corrosion rot   344
cost-effectiveness 335, 396
cotton fibers   180
crookedness 252, 314, 374, 380–381
crossing 236, 238–239, 333
crown 21, 30, 33, 117–121
crystallites   138
current annual growth 51, 102, 258
current annual increment   190
cutting budget methods 203, 205
cutting cycle   189
cuttings 119, 121–122, 240–243, 270, 312–322, 355–356, 378, 397
cycling of nutrients 61, 65, 272

D
D-glucose   130
dark reactions 38, 40
daughter cells   127–128
DDT   301
decay 65–69, 71, 95, 178, 342–348
decay rate 66–68, 71
deciduous species   90
decomposition 54–71, 156, 173, 251, 267–270, 280, 344, 348
defense tactics   286
defoliation 119, 291–292
deforestation 16, 97, 110
degree of polymerization   129–130
delayed induced defense   291
delimbing 319, 321–322, 342, 353, 371, 380–381
delivery sales 312, 357
dense wood 163, 168, 172
density coefficient 377, 384
density of wood   164–165, 171–173, 189, 264
derived demand   405

desertification 12, 16, 353
developing countries 92, 95, 97, 101–102, 106,
    108, 111, 178, 319
dextrins 129
diameter at breast height 188, 367, 375
diameter class 368, 377, 379, 381
diameter growth .. 49–51, 253, 259–261, 263
diameter of vessels 124–125, 149
diffuse radiation 62
digital terrain models 331
diprionid sawflies 283, 285, 301
direct radiation 61–62
discoloration 342, 346, 348–350
distribution of growth 51
disturbance dynamics 219
ditching 222, 265, 267–268
diterpene 302
division of cells 125
DNA 238
DNA-based selection 238
dominant height 190, 192, 256, 259, 272
dominant trees 93, 119, 275
dormancy 34, 47
dormant period 34, 56
dry forest 92–93, 175
dry matter content 189, 347
drying 154, 159, 161, 164, 342–344,
    347–349, 352

E

earlywood 123–124, 128–129, 139, 145,
    149–150, 154, 156, 165, 172
ecological balance 14
effective heating value 175–176
efficiency of sampling 197, 199
electronic calliper 375
elongation 32, 47–48, 117, 125–128
emission 114, 408
employment 312, 405, 408, 411–412
endangered species 16
energy 42–45, 59–61, 76–78, 117–118, 120–
    121, 172–176, 347–349, 352–354
energy wood 327
entrepreneurs 312, 334–335, 367
enumeration 194–195
environment 12–18, 29, 59, 78, 109, 132, 219,
    232–246, 288, 335
environmental analyses 408

epithelial cells 30, 146, 299–300
erosion 111, 232, 318, 326, 333, 353, 356
establishment of seedlings 56, 227–228, 231
Eucalyptus 92, 150, 287, 321, 358
European pine sawfly 288, 299, 302
evaporation ... 74–77, 91, 264, 342–343, 349
evapotranspiration 75–76, 93, 250, 273
even-aged forestry 203, 209, 222, 225
even-aged stand 189–190, 209
evergreen 32, 92
excavator method 333
extractives 120, 122, 129–137, 156–183, 348

F

farm tractor 319
fascicle 32
fats 37, 45, 132, 134
fatty acids 132, 134, 350
feller 319
felling 189–212, 268, 270, 312, 319, 321,
    342–343, 367, 374
felling residues 268, 270
fermentation 61, 300
fertility 56, 71–85, 172, 192, 232–235,
    258–279
fertilization 16, 56, 72, 171, 222, 271,
    274–276, 279, 281
fiber properties 121–122, 126, 164, 316
fibers 31, 127, 134, 139, 141, 147–148, 152,
    155–156, 160, 162, 165, 168, 171–
    173, 177–183, 242, 344, 349, 359
fibril 150, 154
final demand 406–410, 412
final product 406–408, 410
fine roots 33, 45–46, 277
Finnish forest planning 209
fir engraver 298
fire 67, 84, 97, 109–110, 268, 342, 348
fixed-radius circular plot 193
flavonoids 132, 289
flowering 22, 56, 238
fodder 178–179, 189
foliage 29–34, 42–78, 117–135, 272–275,
    348–349
forest area 84, 95, 97, 100, 187, 225–226, 331,
    353, 357
forest biomass 92–93, 117, 175, 182, 327
forest compartment 189, 203

Index

forest cover 95, 97, 353
forest ecosystem   12, 14, 59–66, 71–85, 219–222, 264–281
forest fire 67, 342
forest health 108
forest inventory   187, 189–190, 192–193, 197, 200–201
forest management planning ... 187, 202, 204
forest pest 282, 301
forest regulation 211
forest resources 12–18, 89, 95, 109, 202, 357, 401
forest road 331, 333
forest sector 405, 409
forest site type 192
forest soil 67–68, 136, 228
forest stand 189, 192–195, 236, 270, 397
forest succession 219
forest-based production 12, 219
forest-based product 16
forward selection 238
forwarder 319, 325–326, 356, 397
forwarding 317, 319, 377
fossil fuel 16, 110 121, 354
FOWL 95, 97, 111
frame measurement 383–384, 387
frame volume 366, 380–381, 385–387
fresh density 189
fresh mass 118, 175, 189
fuel 110, 174–179, 189, 211, 268, 314, 327, 347, 349, 354, 366, 409
fuelwood 101–102, 110–111, 313
fungal and bacterial activity 65
fungal decay 342–343, 345, 347–348
fungi 33, 35, 97, 289–299, 343–352
fusiform initials 126–128

G

galactan 130, 154, 160
galactoglucomannans 130
gelatinous fiber 155
gene 236, 239–240
genetic improvement 236, 240, 242
genetic variation 236–237, 298
genetical resources 14
genotype 29, 35, 219, 233
geochemical cycle 74
Geographical Information System 337

Geographical Positioning System 337
germination 56, 188, 227, 238
global forest resources 89, 95
global forests 14, 97, 110
global land area 93
global model 187
global warming 16, 110
glucan 130
glucomannan 130, 160, 350
glucose 45, 129–130, 132, 138
goal constraints 207–208
goal programming 205, 207–208
gradients 45, 92–93, 284–285, 326
grading system 315
grapples 323
green density 377–378, 384, 387
gross photosynthesis 45, 95
ground water 72, 74–75, 78, 222
groundwood pulping 315–316
growing stock volume 97
growth hormones 132
growth rate 49–51, 73, 79–81, 151, 158, 169, 189, 237–258, 295
growth respiration 44–45
growth response 272, 274–275
growth rings 21, 122–124, 126, 151–153, 165, 168, 171, 188, 264
gum 132, 150
gymnosperm 30, 32–33, 58
gypsy moth 283, 288, 298, 304

H

half-bordered pit pair 140
hardwood hemicellulose 130
hardwood saw logs 313, 379
hardwood tracheids 149
hardwoods 123–179, 346
harvester 312–335, 342, 357, 370–374, 389, 397
harvester-based devices 369, 391
harvester-based measurement 370, 374, 389, 391–392
harvesting .. 16, 83, 100–120, 158, 178–195, 241–281, 311–340, 352–357, 360–389, 396–398
heartwood .. 46, 55, 125, 128–133, 141–168, 183, 242, 316, 342–343, 346
heating value 173–176

height growth 47–49, 51, 241, 253, 263
height of the tree 119, 369
hemicellulose . 128–130, 132, 137–139, 163,
　　　165, 180, 343
herbivore ... 61, 219, 229, 241, 243, 285–308
heritability 237–238, 241
heuristics 205, 208
holocellulose 128, 130, 344
host plant 283, 286–287
humus 61–80, 235, 251–269, 353
hydraulic conductivity 78
hydraulics 322
hydrogen 128–129, 134, 163, 174–175
hydrological cycle 60, 74, 281
hydrolysis 129–130, 134, 348

I
income 14, 16, 203, 206, 399
induced defense 289, 291, 293, 295
industrial wood 91, 112, 327
infiltration rate 78
inner bark . 31, 123, 126–127, 132, 134, 136,
　　　175, 345
inorganic components 134–135, 165
input coefficient 406–407
input coefficient table 406
input-output analysis 405, 409
insect 35, 97, 110, 283–309, 355
interest rate 395, 400
intermediate fellings 270
intermediate input 407
internal rates of return 395
international timber markets 358
inventory system 187
investment 223, 274, 318, 359, 385–408
irradiance 62

J
jute 180
juvenile growth 152, 241
juvenile wood .. 129, 150–152, 165, 168, 172

K
kenaf 178, 180
Klason lignin content 131
knot 150, 158–161, 243, 379

L
labor input 408, 411–412
lamellae 139
landing 311–312, 318–319, 329, 377
landscape 332, 340, 353, 356
larch 32, 171, 238
latewood 123–175, 316
leaching 65, 72, 264, 353, 356
leaf area index 76
leaf mass 190
length measurement 371–372, 379
length of rotation 258
Leontief inverse 407
lepidopteran 291–292, 301, 304
life cycle 34, 150, 219, 238
life span 34–35, 55, 84, 221, 240, 260
light reaction 38
light-demanding trees 119
lignification 125–126
lignin 31, 45, 68, 120–132, 137–139, 150–179,
　　　242, 343–350
linear programming 205, 207
litter 51, 54, 61, 63, 65–69, 71–73, 80, 84, 95,
　　　235, 251, 264, 267–268, 270, 272,
　　　280
log . 315–316, 342–343, 346, 360–361, 365–
　　　367, 373–377, 390–391
log measurement 365, 379, 383, 388, 391
logging 117, 120, 174, 252, 256, 304, 311–398
logging residues 314, 327, 353
long-term induction 291
longitudinal growth 125
loose volume 366
lower heating value 174
lumber 123, 140, 149, 153–154, 158–159,
　　　161, 164
lumen 122–123, 137, 140, 145, 149,
　　　154–155, 172

M
magnesium 62, 135–136, 267, 269
maintenance respiration 44
management goals 219, 223
management of forest pests 282, 297
management process 219, 234
mannan 130–131
manual volume measurement 376
marker genes 238

market value 14
mass density 165
mass of tree 53
mature wood 73, 129, 144, 150–152, 168
mean annual growth 51, 258
mean annual increment 190
mean diffuse irradiance 62
measurement accuracy 366–367, 370, 373–374, 385
measurement cost ... 357, 366–367, 388–391
measurement method 367
measurement of pulpwood 162, 366, 374, 384
measurement of standing trees 367, 369, 375, 388–389, 392
measurement sensor 372
mechanical pulp 162, 351
mechanical pulping . 134, 151, 158, 162, 172
mechanical site preparation 268
mechanized logging 370, 397
mesophyll 32–33, 38
micelle 31, 138
micro organism 240
microbial activity 267
microbiological activity 134
microclimate 61
microfibril 129, 138–139, 154–156, 163–164, 180
micropropagation 240
middle lamella 128, 139–140
mill 162, 174, 311–396
mineral element 128, 136, 174
mineralization 65, 69
molecular biology 239, 242
molecule 31, 39, 129–130, 138
monomer 129, 132
monosaccharide 130
monoterpene 289
mortality 54–60, 76–81, 219, 233, 243–256, 291–299
motor-manual cutting 319
mountain birch 290–293
mountainous conditions 312, 318, 326, 331, 333
multi-stem processing 328
multifunctional forestry 14
multiple-use 14, 340, 353, 356
mycorrhiza 33

N
NADP 38
NADPH 38–39
national economy 405, 411
natural pruning 237, 241, 261, 263
natural regeneration 222, 226–228, 230–231, 256, 356
nature conservation 106, 202, 362
needles 30–49, 54, 63–66, 76–83, 248, 275, 290–302
net annual increment 97, 112
net growth 16, 51, 80
net income 203, 206, 399
net photosynthesis 40–42, 95
nicotine 290, 301
nitrate 267, 269
nitrogen 14, 41, 45, 63–73, 83, 101, 128, 136, 174, 251, 265–292, 348
nitrogen cycling 265
nuclear polyhedrosis virus 299–300
numerical optimization 204, 209
nutrient availability 61, 110, 251, 264–265, 271
nutrient balance 64, 72, 136
nutrient content 65, 69
nutrient cycle 54, 60, 62, 72, 74, 222–223, 251, 264, 270, 272, 278
nutrient flow 62, 64
nutrient storage 72
nutrient uptake 72, 267, 270
nutrients 14, 30–84, 132, 222, 233, 248, 264–278, 290, 348–356
nutrition management 306

O
oligosaccharides 129–130
one-grip harvester 328, 342
operating costs 390, 395
operational planning 337, 340
optical measuring 388
optimal allocation 199
optimal productivity 223
origin 65, 233, 236, 288, 298, 303, 384, 405, 411
output of industry 407
oxygen 38, 41, 128–129, 174–175, 343–344, 348, 351–352

P

parenchyma cell 140, 149–150
parent soil 62, 267
parent tree 227
pathogen 293
PCR 39
peatland 271, 306
pentose 130
percolation 72, 75
periderm 31–32
permeability of wood 156, 164
pest . 108, 219, 241, 256, 282, 284, 298–304
pesticide 353
petiole 33
pH 134, 184, 299, 350
phelloderm 32
phenols 132, 294
phenotypic variation 237
phenylpropanoids 132
phloem 21, 31–32, 45, 49–50, 123, 126–127, 138, 294, 298
phosphorus 136, 270–271
photosynthesis 29, 36–45, 59–61, 71–79, 95, 129, 132, 173, 249, 272–273
photosynthetic rate 40–41
photosynthetic response 249, 273
physiological processes 29, 34, 36, 42, 56, 60–61, 74
phytophagous insect 286
pile .. 134, 343, 346, 348–352, 366, 377, 380
pioneer species 50, 221–222, 231, 235
pit 30, 140–141, 149, 157, 164, 343
pit cavity 140, 149
pit membrane 140–141
pith 122–126, 144, 151–168, 179, 346
planning 16, 187–213, 311–313, 326–367, 389
plantation 111, 171, 232–234, 306, 319, 397, 400
plantation forestry 232, 234, 319, 397
planting 97, 152, 209–244, 265, 398–399
plus tree 237
pole 89, 110, 164, 314, 344, 379–380
pollination 56
polymer 129–130
polyphagous feeder 287
polysaccharide 130, 134
population dynamics 284

potassium 62, 135–136, 267, 270–271
Power Model 244–245
precipitation 61, 72–78, 89–95, 108, 110, 250, 264, 270, 343, 349
precommercial thinnings 248
predators 282, 299
prescribed burning 109, 222, 268–269
price index 401
price of timber 395, 400
primary cell wall constituents .. 128–129, 137
primary growth 125–126
primary input 405, 408
principal elements 128
Principle of Biodiversity 14
Principle of Ecological Balance 14
Principle of Multiple Use 14
production model 407–408
productivity 78–81, 253–280, 317–335, 411–412
productivity of forest ecosystems ... 232, 270
progeny testing 237–238
prosenchymatous cells 125, 128
protected area 106, 109
protection 12, 31, 111, 189, 232, 353, 361, 397
proteins 37, 39, 45, 132, 289, 300
provenance 29, 232–233, 240
pruning 30, 55, 160, 222, 237, 241, 260–261, 263
pulp and paper industry 242
pulpwood 102, 133, 144, 152, 157–173, 189, 247, 255, 313–401
purchase prices 365, 375, 383, 387, 389, 391
pure cellulose 130, 155, 344

Q

quality requirements 314–315, 336, 367, 379

R

radial growth 21, 29–30, 49–51, 126, 253–254
radial surface 122, 147
radiation 29–44, 61–78, 174, 248–250, 268–273, 286
random sampling 197–199, 384
rapid induced defense 293
raw material 91, 112, 117, 120–122, 127, 133–180, 241, 327, 337–384, 396–397
ray 31, 126–127, 132, 134, 145–147, 150, 156

Index

reaction wood 121, 129, 150, 152, 160–161, 348
real solid volume 365–366, 369
rectangular plot 193
reforestation 16, 232, 236, 240, 242
regeneration   12, 16, 33–41, 56–60, 84, 121–122, 189, 209, 212–268, 295, 353–357, 395, 400
regeneration capacity 12
regeneration cuttings 121–122, 223
regeneration method 234–235, 400
regeneration processes 226, 232
regulated forest 203
relascope 193–195, 368
relative density 165
renewable energy 121, 174
resin 30–32, 124–146, 160, 289–297, 347–351
resin canals 124, 146, 160
resin content 134, 351
resistance 43, 76–77, 129, 172, 178, 240, 282–304, 342–343
resistant clones 282
respiration 36–45, 60–61, 79–80, 95, 134, 343–351
respiration reactions 348
rhytidome 31–32, 123
ring width 168–169, 171, 175, 242
ring-porous hardwoods ... 124, 165, 168, 171
rodent 282, 285, 299
root graft 33
root system 33, 44, 58, 72, 121, 162, 270
root .. 21, 29–63, 78, 82, 117–135, 174, 188, 270–289, 355
rot 315–316, 343–344, 346, 348
rotation length   189–190, 192, 211–212, 258, 277–278
roundwood   101–112, 311, 334–367, 391, 409
roundwood consumption 101
roundwood production 102, 112
Rubisco 39–41
ruminants 282
rut formation 355
rutting 318

S

sample bundles 378, 384–385, 387
sample plots 192, 195, 197, 200–202
sample tree 194–195, 367–369, 375–376
sample tree measurements 367, 375
sampling 194–202, 304, 368, 375–377, 384–385
sampling design 197
sampling ratio 194, 197–198, 201
sampling unit 195–197, 202
sapwood ... 46, 125, 129, 132, 141, 156–158, 162, 298, 342
saw log 160–173, 241, 255, 313–315, 343–388, 401
sawmill 127, 144–173, 313, 316, 361, 410
secondary components 137
secondary metabolites 132, 289–290, 301–302
secondary phloem 31, 126
secondary wall 128, 131, 139–140, 180
secondary xylem 123, 126
seed source 232–233
seeding 222, 227, 232, 235–236, 239
seedling stand 58, 84, 188, 226–229
seedlings 32, 56, 206, 211, 222–233, 265–279, 297
seeds   22, 56, 58, 179, 227–229, 231–232, 234
selection forestry 219, 222, 231, 252
selection of tree species 232–234
selective cutting 230, 252, 340, 356
selective thinning 252
self-pruning 160
sequestration of carbon 16
sesquiterpenes 289
sexual regeneration 56, 227
sexual reproduction 56, 240
shade-intolerant species 50
shade-tolerant species 50
shelterwood method 231
shrinkage 138, 154–155, 164–165, 343
simple pit 140, 145, 147
simple random sampling 197–199
sisal 180
site class 192
site fertility .. 71, 85, 192, 233, 235, 258–275
site indices 192
site preparation 222, 265, 267–268
site productivity 97, 190, 192

site properties	84, 234
site quality	68, 189, 229
site type	192, 233
skidding	312, 317, 319, 326, 342, 353, 397
skyline systems	326
soft rot	344, 346, 348
softwood hemicelluloses	130
softwood	122–179, 343, 346
soil compaction	318, 322, 355
soil moisture	48, 78, 222, 250, 267, 271
soil organic matter	65–75, 92, 95, 267–270, 280
soil preparation	223, 229, 231, 353, 398
soil temperature	69, 89, 267, 269
soil texture	78, 233
soil water	78, 248, 250, 267
solar energy	14, 36, 59, 173, 354
solid volume	177, 365–393
spring wood	32, 49
sprouting	188, 227
spruce pulpwood	171, 314–315
stack density	381–382, 384–385, 387
stack measurement	380, 383, 385, 387–388
stand basal area	190, 193, 203
stand characteristics	190, 192–194, 196, 199, 201–202
stand density	55, 61–62, 81, 119, 159, 189, 243–272
stand volume	190, 193, 196–197, 202, 390
standard deviation	196–197, 199, 372
standard error	196–199
standwise forestry	221–222
starch	129, 132, 150, 156, 345–346
stem mass	52, 81, 118–121, 190
stem volume	152, 160, 188–194, 237–243, 279, 317, 369, 389, 397
stocking	14, 79, 83, 119, 190, 192, 225–231, 247, 256, 356
stomata	32–33
storage places	367
storage time	336, 345, 347, 350
strategic planning	202, 335
stratified sampling	194, 199–202, 375, 384
stratum	194–195, 199, 201–202, 384
straw	178
strip road network	340, 342
strobili	22
stump	117–122, 188, 311–312, 330–357, 383
stumpage price	395
succession	55–60, 81–85, 219–222, 246–247, 279
sucrose	129, 132
sugar	38–39, 130, 132, 156
sulfate pulping	120, 122, 134, 158
summer wood	32, 49
sustainability	16, 136, 212, 214
sustainable annual cut	203
sustainable forestry	14, 16, 400
sustainable management	12, 14, 16, 110, 219, 278
symbiotic fungi	294
systematic error	366
systematic sampling	198–199

T

tactical planning	202, 335
taiga zone	89
tangential surface	122, 127, 147
tangible items	12
tannin	31, 132, 150, 289, 290, 343
taproot	117
taxes	405–406, 413
taxonomic isolation	286–288
temperate	18–49, 68, 84, 90–110, 123–139, 165, 179, 219, 221, 245, 253, 350
temperature	34–49, 58–59, 69, 89–95, 110, 134, 159–174, 192, 229–273, 343–352
temperature sum	48, 58, 229, 233, 267
tension wood	152, 155–156
terpenes	132, 175, 294–295, 301
terpenoids	132, 289–290, 302
terrestrial biomass	89, 93
thinning	73, 119, 169–173, 203–223, 244–281, 306, 318, 325, 355, 399
thinning regimes	252
thinning rule	248
throughfall	75, 250
timber assortment	328, 357–396
timber markets	358, 400
timber measurement	357–393
timber production	12, 16, 18, 29, 92, 203–204, 223, 226, 268, 278, 358
timber quality	236, 259–260, 263, 312, 336, 367
timber resources	12, 16

Index

timber trade 357
timber yield 84–85, 255, 279, 281
time expenditure 366, 389–390
tissues . 30–72, 125–138, 156, 179, 289–307
top-diameter measurement 379
total growth 50–51, 61, 72, 80, 225, 244, 254, 279
total output 41, 407–410
total tree biomass 34, 97, 277
toxic compounds 240, 290
trace elements 136
tracheid 30–31, 122–160, 242, 397
traits 236–238, 240, 295
transpiration 36, 42–44, 61, 76–77, 162
transportation of timber 311
transverse surface 147
tree age ... 119, 160, 169, 171, 188, 263, 316
tree basal area 188, 190
tree biomass 34, 73, 79, 97, 117, 135, 270, 277
tree characteristics 188, 190
tree growth ... 35, 54–80, 222, 225, 243–248, 259–278
tree height 30, 151, 157, 188, 193, 260
tree resistance 294, 297, 304
tree size 193, 203, 225, 237–253, 322
tree structure ... 21, 29, 46–47, 51, 53, 65, 82
tree-length method 317–319
tropical forest 95
tropical zone 22, 89–90, 93, 95
truck 318–337, 366–367, 380–387
two-grip harvester 322
two-phase sampling 201–202
tyloses 149, 156, 158, 164

U
uneven-aged forestry 209, 222, 225, 252
unit cost 380
utility function 208–209

V
vaporization 76, 174
variable-radius circular plot 193
vascular cambium 21, 126–127
vascular tissue 125
vascular tracheids 149
vegetative regeneration 58, 227, 295

vegetative reproduction 56, 58
veneer logs 102, 388
vessels 31–32, 124–125, 147–164, 180
visual compartment inventory 193–194
visual estimation 195, 357, 368, 381, 385

W
wages 365, 367, 375, 387, 405
water availability 74, 110, 250
water balance 74–75, 91
water flow 42–43, 69, 72, 353
water storage 345
watershed management 16
waxes 32–33, 134, 175
weathering 65, 72, 267, 269, 342
weevil damage 298
weighing 365, 377, 383–384, 390
weight measurement   365, 383–384, 387, 393
wet deposition 62, 65, 72
white pine weevil 298
white pocket rot 344
white rot 344, 348
whole-tree harvesting 120, 270, 277, 279
whorl 30, 188, 249–250
wilting point 75
winter moth 288, 290
within-species variation 168
within-tree variation 168
wood density 29, 164, 168–169, 172, 184, 242, 264, 397
wood quality 55, 123, 157, 184, 242, 261
wood-rotting fungi 343, 345
woody biomass 117, 125, 128, 135–136
work productivity 312

X
xylan 130–131, 156, 160, 350
xylem 21, 32, 49–50, 123, 125–126, 294
xylometric measurement 365

Y
yield class 192
yield of pulp 120, 133, 149, 151, 162, 178
Yield-Density Model 244, 254

425